Mass Spectrometry for the Clinical Laboratory

Mass Spectrometry for the Clinical Laboratory

Edited by

Hari Nair, PhD, DABCC, FACB
Boston Heart Diagnostics, Framingham, MA, United States

William Clarke, PhD, MBA, DABCC
Johns Hopkins School of Medicine, Johns Hopkins University, Baltimore, MD, United States

AMSTERDAM • BOSTON • HEIDELBERG • LONDON
NEW YORK • OXFORD • PARIS • SAN DIEGO
SAN FRANCISCO • SINGAPORE • SYDNEY • TOKYO
Academic Press is an imprint of Elsevier

Academic Press is an imprint of Elsevier
125 London Wall, London EC2Y 5AS, United Kingdom
525 B Street, Suite 1800, San Diego, CA 92101-4495, United States
50 Hampshire Street, 5th Floor, Cambridge, MA 02139, United States
The Boulevard, Langford Lane, Kidlington, Oxford OX5 1GB, United Kingdom

Notices
Knowledge and best practice in this field are constantly changing. As new research and experience broaden our
understanding, changes in research methods, professional practices, or medical treatment may become necessary.

Practitioners and researchers must always rely on their own experience and knowledge in evaluating and using any
information, methods, compounds, or experiments described herein. In using such information or methods they
should be mindful of their own safety and the safety of others, including parties for whom they have a professional
responsibility.

To the fullest extent of the law, neither the Publisher nor the authors, contributors, or editors, assume any liability
for any injury and/or damage to persons or property as a matter of products liability, negligence or otherwise, or
from any use or operation of any methods, products, instructions, or ideas contained in the material herein.

Library of Congress Cataloging-in-Publication Data
A catalog record for this book is available from the Library of Congress

British Library Cataloguing-in-Publication Data
A catalogue record for this book is available from the British Library

ISBN: 978-0-12-800871-3

For information on all Academic Press publications
visit our website at https://www.elsevier.com/

  Working together
to grow libraries in
developing countries

www.elsevier.com • www.bookaid.org

Publisher: Mica Haley
Acquisition Editor: Tari Broderick
Editorial Project Manager: Pat Gonzalez
Production Project Manager: Julia Haynes
Designer: Maria Inês Cruz

Typeset by Thomson Digital

Contents

List of Contributors

L.M. Bachmann, PhD, DABCC
Department of Pathology, Virginia Commonwealth University, Richmond, VA, United States

W. Clarke, PhD, MBA, DABCC
Johns Hopkins University School of Medicine, Baltimore, MD, United States

J.C. Cook-Botelho, PhD
Clinical Chemistry Branch, Division of Laboratory Sciences, Centers for Disease Control and Prevention, Atlanta, GA, United States

C.A. Crutchfield, PhD
Johns Hopkins University School of Medicine, Baltimore, MD, United States

D. French, PhD, DABCC, FACB
Department of Laboratory Medicine, University of California San Francisco, San Francisco, CA, United States

U. Garg, PhD
Department of Pathology and Laboratory Medicine, Children's Mercy Hospital, Kansas City, MO, United States

D.A. Herold, MD, PhD
Department of Pathology, University of California San Diego, La Jolla;
VAMC-San Diego, San Diego, CA, United States

P.J. Jannetto, PhD, DABCC, FACB, MT(ASCP)
Mayo Clinic, Department of Laboratory Medicine and Pathology, Toxicology and Drug Monitoring Laboratory, Metals Laboratory, Rochester, MN, United States

H. Ketha, PhD, NRCC
Department of Pathology, University of Michigan Hospital and Health Systems, Ann Arbor, MI, United States

P.B. Kyle, PhD, DABCC
University of Mississippi Medical Center, Jackson, MS, United States

K.L. Lynch, PhD, DABCC, FACB
Department of Laboratory Medicine, University of California, San Francisco, CA, United States

I.W. Martin, MD
Dartmouth-Hitchcock Medical Center, One Medical Center Drive, Lebanon, NH, United States of America

H. Nair, PhD, DABCC, FACB
Boston Heart Diagnostics, Framingham, MA, United States

B. Rappold
Essential Testing, LLC, Collinsville, IL, United States

R.J. Singh, PhD, DABCC
Department of Laboratory Medicine and Pathology, Mayo Clinic, Rochester, MN, United States

D.F. Stickle, PhD
Department of Pathology, Jefferson University Hospital, Philadelphia, PA, United States

J. Stone, MT(ASCP), PhD, DABCC
Center for Advanced Laboratory Medicine, University of California San Diego Health System, San Diego, CA, United States

J.Y. Yang, PhD
Department of Pathology, University of California San Diego, La Jolla, CA, United States

Preface

This book was born out of an idea that a select compilation of the illustrated experiences of a panel of expert practitioners of clinical mass spectrometry might be beneficial to those of us who are considering implementation of the art in our own laboratories perhaps for the first time. I would like to thank my mentors and colleagues from the Department of Lab Medicine at the University of Washington as well as those friends that I get to meet at AACC and other venues for their valuable insights that, in part, helped shape the idea for this book.

Personally, I saw this project as an opportunity to learn. I feel extremely lucky to have Bill Clarke as my coeditor and mentor in this pursuit. I am grateful for his kind and effective mentorship, encouragements, collaboration, and vast technical and professional insights.

First we listed the topics that we felt might be of interest to most clinical labs and then requested some of the best known practitioners in the field to tell their stories on those topics. For their belief in this project, their willingness to contribute, and for their expertise that is so valuable to the clinical chemistry community. I reserve my utmost gratitude to the authors of this book.

This project has been in the works for nearly 3 years. Patience and goodwill gestures from many individuals along the way lit its path. Foremost, I would like to thank my wife Rekha and my boys Shreyas and Shree for their patience and for being my inspiration. I am thankful to Ruthi Breazeale, Sr. VP and Dr. Ernst Scheafer, co-founder and medical director at Boston Heart Diagnostics for their encouragements and accommodation.

What a pleasure it has been to work with the incredibly professional and pleasant Team Elsevier! Thank you!!

My hope is that this book will add at least a drop to the ever growing number of resources that we as a community will need to perfect the art of implementing mass spectrometry in the clinical lab.

Hari Nair, PhD, DABCC, FACB

MASS SPECTROMETRY IN THE CLINICAL LABORATORY: DETERMINING THE NEED AND AVOIDING PITFALLS

1

W. Clarke

Johns Hopkins University School of Medicine, Baltimore, MD, United States

1 CLINICAL MASS SPECTROMETRY

Historically, the complexity of instrumentation and sample preparation has relegated LC-MS based assays to specialized laboratories with extensive technical expertize. Until recently, applications of MS in the clinical laboratory were limited to gas chromatography (GC)-MS for toxicology confirmation testing and testing for inborn errors of metabolism, some GC-MS applications for steroid analysis in specialty laboratories, and inductively coupled plasma (ICP)-MS for elemental analysis. In most cases, this testing has been restricted to specialized laboratories within a hospital, or to large reference laboratories. However, with the simplification of MS instrumentation and introduction of atmospheric spray ion sources along with the emergence of routine liquid-chromatography tandem MS (LC-MS/MS), MS has become a viable option for routine testing in clinical laboratories.

1.1 BASIC MASS SPECTROMETRY CONCEPTS

MS is a powerful analytical technology that can be used to identify unknown organic or inorganic compounds, determine the structure of complex molecules, or quantitate extremely low concentrations of known analytes (down to one part in 10^{12}). For MS-based analysis, molecules must be ionized, or electrically charged, to produce individual ions. Thus, MS analysis requires that the atom or molecule of interest has the ability to be ionized and be present in the gas phase. MS instruments analyze molecules by relating the mass of each molecule to the charge; this identifying characteristic is specific to each molecule and is referred to as the mass-to-charge ratio (m/z). Therefore, if the molecule has a single charge ($z = 1$), the m/z ratio will be equal to the molecular mass.

The analytical power of the mass spectrometer lies in its resolution, or the ability to discern one molecular mass from another. The resolution can be determined by examining the width of an m/z peak or the separation between adjacent peaks; a narrow peak with little overlap indicates greater resolution. For two adjacent peaks of masses m_1 and m_2, the resolving power is defined as $m_1/(m_1 - m_2)$. The expression $(m_1 - m_2)$ may also be referred to as Δm. Higher instrument resolution results in increased mass accuracy and the ability to avoid interference from compounds of similar mass that may also be present

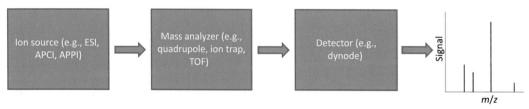

FIGURE 1.1 Schematic Diagram of Mass Spectrometry (MS)

in the sample. Mass accuracy is defined as the mass difference that can be detected by the analyzer divided by the observed, or true mass.

Although there are numerous instrument configurations available, MS system operation can be organized into three main segments: (1) generation of ions; (2) separation of ions based on mass and charge in a mass analyzer; (3) detection of ions and instrument output (Fig. 1.1). Depending upon the type of ionization used, these steps fully or partially occur under vacuum pressure to drive ion movement forward through the instrument.

1.2 COMMON ION SOURCES FOR CLINICAL MASS SPECTROMETRY

There are a variety of ion sources available for mass spectrometers. Some of these ion sources are "direct ionization sources," in which analytes are directly ionized from a surface or from a solution. Other sources, such as atmospheric pressure ionization sources, produce ions from analytes in solution and these are more commonly used in clinical assays due to their compatibility with liquid chromatography. Common atmospheric pressure ion sources include:

- electrospray ionization (ESI)
- atmospheric pressure chemical ionization (APCI)
- atmospheric pressure photoionization (APPI)

A summary of the strengths and weaknesses for these sources can be found in Table 1.1.

1.2.1 Electrospray Ionization (ESI)

ESI is perhaps the most commonly used ionization technique in clinical MS. It is a sensitive ionization technique for analytes that exist as ions in the LC eluent. In ESI, a solvent spray is formed by the application of a high voltage potential held between a stainless steel capillary and the instrument orifice, coupled with an axial flow of a nebulizing gas (typically nitrogen). Solvent droplets from the spray evaporate in the ion source of the mass spectrometer, releasing ions to the gas phase for analysis in the mass spectrometer. In some ESI sources, heat is used to increase the efficiency of desolvation. While ESI is widely used, it is subject to matrix effects, particularly ion suppression, which must be taken into consideration during method development.

1.2.2 Atmospheric Pressure Chemical Ionization (APCI)

APCI uses heat and a nebulization gas to form an aerosol of the eluent from an LC system. In contrast to ESI, ions are not formed in solution or liquid phase. Instead, ions are formed in the gas phase using a corona discharge (high voltage applied to a needle in the source) to ionize solvent molecules and analytes in the aerosol. Ions released to the gas phase are then analyzed by the mass spectrometer. During ionization

Table 1.1 Overview of Three Ionization Techniques Used in Clinical Mass Spectrometry (MS)

Ionization Technique	Advantages	Limitations
ESI	• Sensitive ionization technique for polar analytes or ions generated in solution • Has broad applicability for relevant analytes in clinical MS • May yield multiply charged ions, which allows for analysis of larger molecules (i.e., >1000 Da)	• May be more sensitive to matrix effects compared to APCI
APCI	• Typically less sensitive to matrix effects than ESI • May provide better sensitivity for less polar analytes	• Typically only singly charged ions are formed, limiting the effective mass range, • May be unsuitable for thermally labile analytes • May yield less absolute signal relative to ESI
APPI	• Works well with nonpolar analytes • In some cases will ionize analytes that do not ionize by either ESI or APCI.	• Demonstrates limited applicability in clinical MS to date.

APCI, Atmospheric pressure chemical ionization; APPI, atmospheric pressure photoionization; ESI, electrospray ionization.

in the APCI source, some thermal degradation may occur, which can lead to a greater degree of fragmentation in electrospray ionization. For analysis using APCI, the analytes of interest should be heat stable and volatile for best results. APCI is often less susceptible to matrix effects (including ion suppression) as compared to ESI, and may be considered for a wide range of applications, including measurement of nonpolar analytes.

1.2.3 Atmospheric Pressure Photoionization (APPI)

APPI is an alternative mechanism to ionize analytes eluting from a chromatography system, although it is much less frequently used than ESI or APCI. In APPI, the solvent is first vaporized in the presence of a nebulizing gas (e.g., nitrogen) and then enters the instrument ion source at atmospheric pressure. Once the aerosol is generated, the mixture of solvent and analyte molecules is exposed to a UV light source that emits photons with energy level that is sufficient to ionize the target molecules, but not high enough to ionize unwanted background molecules. Often, an additive to the LC eluent (commonly toluene) is used to increase ionization efficiency in APPI.

1.3 COMMONLY USED MASS ANALYZERS

When coupled to an LC system, the mass spectrometer functions as powerful multiplex detector for chromatography. The analyte of interest is ionized in the source of the mass spectrometer by any of a variety of mechanisms as previously discussed. The ions are then directed to the mass analyzer component of the mass spectrometer, where individual ions are selected according to their m/z. Ions produced from small molecule analytes (<1000 Da) usually possess a single charge; therefore, their m/z is equivalent to the mass of the ion. This is commonly designated as $[M + H]^+$ for protonated (positive) and $[M - H]^-$ for deprotonated (negative) ions. A chromatogram can then be generated in which relative ion abundance of a specific m/z is plotted versus time. Multiple overlapping chromatograms can be extracted from the same analytical run, as the mass analyzers are able to detect multiple ions independently within the time scale of a chromatographic peak.

1.3.1 Quadrupole Mass Analyzers

A quadrupole mass analyzer consists of a set of four conducting rods arranged in parallel, with a space in the middle; the opposing pairs of rods are electrically connected to each other. This type of mass analyzer separates ions based on the stability of their flight trajectories through an oscillating electric field in the quadrupole. The field is generated when a radio frequency (RF) voltage is applied between one pair of opposing rods within the quadrupole. A DC offset voltage is then applied to the other pair of opposing rods. Only ions of a certain m/z will have a stable flight path through the quadrupole in the resulting electric field; all other ions will have unstable trajectories and will not reach the detector. The RF and direct current voltages can be fixed in a way that the quadrupole acts as a mass filter or analyte-specific detector for ions of a particular m/z. Alternatively, the analyst can scan for a range of m/z values by continuously varying the applied voltages.

Single quadrupole mass spectrometers contain a single mass analyzer and can only measure ions formed in the instrument source; these can be intact molecular ions or fragment ions formed by in-source fragmentation. Based on this limitation, single quadrupole mass spectrometers do not provide a large amount of structural information and specificity is limited when compared to tandem quadrupole mass spectrometers. A tandem quadrupole mass spectrometer, often called a triple quadrupole, consists of two quadrupole mass analyzers separated by a collision cell. Precursor ions are selected by the first quadrupole mass analyzer. The selected precursor ion is then fragmented in the collision cell by a process known as collision-induced dissociation (CID). CID results from collisions of the analyte of interest with an inert gas, such as nitrogen or argon. The specific product ions produced by CID are a function of the bond energies inherent in the molecular structure of the precursor ion, as well as the collision gas and energy used. Product ion patterns and relative ion abundance can be highly reproducible if the CID conditions are stable and robust. The product ions are analyzed or selected by the final quadrupole mass analyzer, and then passed to the detector. These pairs of precursor and product ions are called a mass transitions. When the electric fields and collision energy are held constant, only analyte ions having a specified mass transition (precursor/product ion pair) are able to reach the detector, which results the high specificity of tandem quadrupole mass spectrometric methods. This mode of data acquisition is referred to as selected-reaction monitoring (SRM). When multiple transitions are monitored during a chromatographic run, the data acquisition is called multiple-reaction monitoring (MRM).

1.3.2 Time-of-Flight Mass Analyzers

Time-of-flight (TOF) mass analyzers separate ions based on their different flight times over a defined distance or flight path. After generation in the source, ions are accelerated by an electric field into a flight tube, such that ions of like charge have equal kinetic energy. Because kinetic energy is equal to $1/2\ mv^2$, where m is the mass of the ion and v is the ion velocity; the lower the ion's mass, the greater the velocity and shorter its flight time. The travel time from the source of the ion pulse through the flight tube to the detector can be calibrated to the m/z value based on the relationship described earlier.

Unlike a quadrupole mass analyzer, TOF analyzers are not scanning—all ion masses are measured for each ion pulse. Because of this, TOF mass spectrometers offer high sensitivity and a high duty cycle. Historically, TOF mass spectrometers have been used for qualitative experiments in which high resolving power and exact mass measurements are necessary (e.g., metabolite identification or protein sequencing). However, modern TOF mass spectrometers are capable of accurate and precise quantitative measurements as well.

1.3.3 Ion Trap Mass Analyzers

Ion trap mass analyzers use a combination of electric or magnetic fields to capture or "trap" ions inside the mass analyzer. There are multiple configurations of ion traps including 3D ion traps (the Paul ion trap), a linear ion trap (2D trap), and electrostatic trap (Orbitrap), or a magnetic field-based trap (ion cyclotron resonance). The 3D ion trap basically works on the same principle as a quadrupole mass analyzer, using static DC current and RF oscillating electric fields, but the hardware is configured differently, where the parallel rods are replaced with two hyperbolic metal electrodes (end caps) facing each other, and a ring electrode placed halfway between the end cap electrodes; ions are trapped in a circular flight path based on the applied electric field. A linear ion trap uses a set of quadrupole rods coupled with electrodes on each end to facilitate the ion trapping. This configuration gives the linear ion trap a dual functionality—it can be used as a quadrupole mass filter or an ion trap. An orbitrap mass spectrometer consists of an inner spindle-like electrode and an outer barrel-like electrode. The orbitrap stores ions in a stable flight path (orbit around the inner spindle) by balancing their electrostatic attraction by their inertia coming from an RF only trap. The frequency of the axial motion around the inner electrode is related to the m/z of the ion. Last, ion cyclotron resonance (ICR) traps use a strong magnetic field to induce a radial orbit of ions, where the frequency of orbit in the magnetic field is a function m/z for the ion. For both orbitrap and ICR traps, their strength is the ability to trap all ions at once and detect them on the basis of their detected frequencies—a Fourier transform algorithm is required for this signal processing. Mass spectrometers that include an ion trap analyzer are most commonly used for qualitative work (e.g., metabolite identification, protein identification, and screening applications). Although examples of ion trap mass analyzers for quantitative analysis do exist, their use in quantitative clinical MS to date is limited.

1.3.4 Hybrid Tandem Mass Analyzers

There are now many instrument configurations that combine two or more mass analyzers of different design, and are therefore called "hybrid instruments." One such hybrid combines a quadrupole with a linear ion trap; this instrument is referred to as a Q-Trap and has found a niche in toxicology screening for unknown agents, and also can be used for quantitative analyses when the linear trap is operated as a quadrupole. Another type of hybrid combines quadrupole and TOF mass analyzers to form a Q-TOF. In this instrument, the quadrupole mass analyzer performs precursor ion selection, while a TOF mass analyzer performs the product ion analysis. The same strategy has been applied using a quadrupole coupled with the electrostatic orbitrap analyzer. With both Q-TOF and Q-Orbitrap instruments, having a high-resolution instrument as the second mass analyzer quantitative analyses in addition to their use for protein sequencing and proteomic or metabolomics screening. High-resolution mass spectrometry will be covered further in Chapter 12.

The choice of mass spectrometer really depends on its intended use. While tandem quadrupole mass spectrometers are by far the most common type of mass spectrometer used in the clinical laboratory, the use of other types should not necessarily be excluded. Care should be taken when selecting an appropriate LC-MS system for the laboratory. Method sensitivity, throughput, and robustness requirements, as well as system cost, are all factors that must be balanced when evaluating systems for purchase. Numerous mass spectrometer models are available today from a variety of vendors, and they may vary substantially in terms of analytical sensitivity and quality. Acquisition of the highest quality instrumentation (within the laboratory's budget constraints) may reduce method performance limitations and result in fewer instrument problems.

1.4 DETERMINING THE NEED FOR CLINICAL LC-MS

Before incorporating MS applications in the clinical laboratory, careful consideration should be given to the benefits and challenges of implementing this powerful technique. Getting the full benefit of clinical MS will greatly depend on the needs of the patient population served by the clinical laboratory, and the necessary performance characteristics of the clinical application to be performed on the instrument. MS was first routinely used in a clinical setting for newborn screening and genetic disorder programs [1]. One of the early drivers of incorporating LC-MS into the clinical laboratory was the challenge of monitoring an immunosuppressant drug, rapamycin, in solid organ transplantation. At the time, there were no acceptable commercial immunoassays for the drug, and so LC-MS became a default solution for therapeutic drug monitoring. As laboratories that supported transplant centers began to acquire LC-MS/MS instruments, they were looking to utilize the systems fully. These laboratories began to add additional analytes to their immunosuppressant assays (tacrolimus and cyclosporine A). As more laboratories acquired the technology, there were additional assays of interest developed including steroid hormones, antifungals drugs, and drugs of abuse. Currently, LC-MS is utilized for clinical applications, such as drug screening and monitoring, pain management, and the detection of metals and low concentration hormones.

More recently, LC-MS/MS has rapidly replaced ion chromatographic methods for screening and confirmation of genetic disorders and inborn errors of metabolism. This is mainly due to the ability of LC-MS/MS systems to analyze multiple compounds at one time from a single sample. This ability to use one small sample for several dozen types of inherited amino acid and metabolism disorders is particularly attractive when considering the limited specimen volume that is typically obtained from newborns. GC-MS is commonly employed for clinical or forensic drug detection, where it is the gold standard method for toxicology confirmatory testing. The volatile and nonpolar nature of many drugs, coupled with their small molecular weights, make them particularly amenable to GC separation, and the structural information and specificity from MS analysis makes it an ideal detection system for toxicology specimens.

The application of MS to endocrinology for quantification of compounds, such as steroids and other hormones has also seen increased interest, due to the lack of specificity in some commercial immunoassays. In certain patients, treatment is primarily driven by the quantification of hormones, such as progesterone and estradiol, thus increased specificity can improve clinical decisions regarding patient care based on these measurements. A recent publication describes an LC-MS/MS method capable of quantification of nine different steroids simultaneously, while demonstrating the ability to determine lower concentrations than those obtained by immunoassay [2]. When compared to commercially available immunoassays, correlation coefficients for the comparison for these nine steroids ranged from 0.866 to 0.988. As described in the report, the lack of drug interferences and the multiplexed design of this LC-MS/MS assay demonstrates the potential impact of clinical LC-MS/MS assays for quantification of hormones.

While the assays described earlier are already finding use in clinical laboratories, there are many more opportunities for LC-MS to make an impact in patient care. However, there are important considerations for these laboratory developed tests used in a clinical setting that are not present in a research setting. Unlike methods developed for research use only (RUO), methods developed in the United States for clinical use must meet certain regulatory requirements. In 1988, the US Congress passed the Clinical Laboratory Improvement Amendments (CLIA '88) which established standards for all clinical laboratory testing. Adherence to these standards is meant to ensure the accuracy, reliability, and timeliness of patient test results regardless of what laboratory performed the test. Aside from the extensive clinical and analytical validation required for assays used in patient care, one of the most significant differences between a research and clinical laboratory is the extensive documentation requirement. Clinical laboratories are required to maintain detailed and meticulous records, documenting every step in the testing process from method development and validation, to patient sample pickup and processing, all the way to instrument maintenance. Because of these requirements, it is critically important to carefully consider the clinical need and the ability of an LC-MS method to meet that need.

The primary drivers for expansion of clinical LC-MS are the well-known limitations of immunoassays for low molecular weight compounds, reduced reagent cost relative to commercially available clinical assays, and simplified workflows and higher throughput as compared to traditional high-performance liquid chromatography (HPLC) method and GC-MS [3]. Before creating a plan for method development, a few basic issues should be addressed. First, a new clinical LC-MS assay should provide a less expensive and/or more effective means to diagnosis and manage patient care [4]. In some cases, the new assay will be more expensive but the benefits of a new test to patient care management could outweigh the increased cost of an additional test. Second, the laboratory must determine whether the new test is expected to provide increased specificity and increased sensitivity, and/or improve turnaround time. If a new LC-MS/MS method (or any clinical method) does not have the potential to improve at least one of these areas, the impact on patient care is likely to be negligible, and the new assay is not worth the investment of time or money to develop.

Starting an LC-MS laboratory or project can be cost intensive, so prior to beginning, a projected cost analysis should be completed to help determine feasibility of the project and to determine whether the test should be set up in-house or sent to a reference laboratory. To perform this type of analysis, considerations include the estimated test volume, estimated personnel hours required for the assay, and capital equipment cost and the cost of reagents and other supplies. While the capital cost for acquiring LC-MS/MS may be intimidating, the overall savings on reagents (compared to other methods, e.g., immunoassay) over a defined duration of test performance may be favorable. However, it is also important to include line items, such as the cost of instrument maintenance (LC system, MS instrument, etc.), waste disposal, proficiency testing (PT), database management, and the cost for technicians to maintain continuing education requirements. Also of significant importance is the estimated test volume. Development of a method for low volume tests may not prove beneficial in a clinical setting if the need for the result is not acute, or, for example, if there is a significant loss of reagents due to expired materials. Last, it is important to remember that LC-MS/MS methods typically require a higher level of expertize and technical skill, and the availability of personnel with the required skill set should also be evaluated.

1.5 POTENTIAL PITFALLS AND ADDRESSING THE CHALLENGES

Despite the significant advantages that can be gained from implementation of MS into the clinical laboratory, considerable challenges exist. Even using for laboratories using similar LC-MS methods for some assays, significant laboratory-to-laboratory variability for the same analyte has been observed. There is currently limited assay standardization for MS-based methods, and much of the assay variability can be attributed to the lack of commercially available calibrators; that is, each clinical laboratory must formulate its own calibrators. Differences in chromatographic methods from site to site lead to variable matrix effects during analysis. When compared to commercial clinical laboratory analyzers, MS instruments are significantly less automated requiring more direct interaction by the analyst, and assay throughput (in terms of number of assays per hour) is significantly lower. In addition, many laboratories verify their assays using various protocols in accordance with different regulatory or industry standards.

1.5.1 External Quality Assessment and Quality Control

The process of external evaluation of method performance is often referred to as external quality assessment (EQA) or PT. Performance of EQA facilitates an evaluation of the total error of a single measurement using blinded specimens and enables laboratories to verify their results relative to other laboratories using similar methods for specific assay. Ideally, EQA/PT programs are administered by professional organizations, such as the College of American Pathologists (CAP) or the American Proficiency Institute (API), which have delegated authority from CMS to evaluate the regulatory compliance of clinical laboratories. For administration of EQA or PT, providers circulate a set of specimens among a group of laboratories, and each laboratory tests of these specimens in the exact manner that patient specimens are tested; the results from each laboratory are then sent back to the EQA provider for analysis. For a laboratory to be CLIA certified and reporting patient results, it is a regulatory requirement that the laboratory must participate in an EQA program. Evaluation of the results can be based upon comparison of the laboratory result to a peer-group mean, or it can be accuracy-based.

An inherent challenge to peer-group based EQA is that even though the laboratory result may agree with the peer group mean (within the limits of acceptability), the peer-group mean may not be analytically accurate. An inherent challenge to accuracy-based EQA is commutability of the proficiency testing material relative to laboratory testing methods. In other words, due to matrix effects, the assigned value from a reference method may not match with the laboratory-generated value if the two methods do not respond in the same way to the matrix material. External EQA programs, however, may not always be available, particularly if an LC-MS/MS test is new to clinical use or if analyte instability makes it impractical to use an EQA program. In these circumstances, a sample exchange between clinical laboratories can be established. All results from EQA programs, including any necessary corrective action, must be documented in the laboratory's records.

A common assumption for LC-MS/MS methods is that due to the nature of the technology, assay results from different clinical laboratories will be harmonized; however, this assumption does not necessarily hold true. The interlaboratory differences can be attributed to differences in calibration (e.g., matrix-based calibrators vs. solvent-based calibrators), the source of calibration material, sample preparation and recovery, and selection of ions or ion transitions for monitoring. While the issue of assay harmonization is not unique to LC-MS/MS methods (e.g., creatinine analysis) [5], it is something that must be addressed for this technology to find widespread adoption into clinical laboratories.

Standardization of approaches to calibration and the availability of certified reference materials for calibration are one way to move toward harmonization of clinical LC-MS/MS assays.

In addition to the periodic EQA, quality control (QC) specimens must be analyzed at least once each batch or day of analysis to ensure the ongoing performance of the analytical system. QC procedures must be designed to detect immediate errors due to instrument failure, adverse environmental conditions, and/or operator performance. Control and calibration materials indirectly evaluate the accuracy and precision of patient results. As such, QC materials should be tested in the same manner as patient samples. For LC-MS/MS assays, QC materials often consist of pooled residual samples spiked with pure compound to achieve appropriate concentrations of QCs. QCs are typically prepared in advance, aliquoted, and stored at suitable conditions until needed. Thus, it is crucial to determine the stability of each QC and, when new pools of QCs are made, to define the acceptability criteria for the new batch of QC material. The values of each QC should be charted and monitored to ensure that they fall within the established criteria for acceptance (often within 2 SD of the mean) [6]. This visual representation allows for the identification of QC problems that may otherwise go unnoticed, allowing the laboratory to investigate and solve a potential problem before a QC failure occurs. Other criteria evaluated during method development and validation, such as signal-to-noise, can also be used to judge the quality of an analytical run. Each laboratory must establish policies and procedures for remedial actions for QC failures, and ensure that any corrective action is documented.

1.5.2 Matrix Effects and Interferences

As discussed previously, mass determination is an extremely selective analytical parameter; therefore, MS methods are not affected by many of the common interferences seen with typical clinical analyzers. MS analysis is not based on recognition of a part of the molecular structure, but is based on the analysis of the whole molecule, and is therefore not compromised by issues, such as cross-reactivity with structurally-similar but nonisobaric compounds as with common clinical immunoassays. In addition, the specificity protects it from interference from formation of endogenous colored end products or interference from hemolysis, lipemia, or icterus that may affect immunoassays and/or spectrophotometric measurements. However, it is important to note that sample matrix effects are still observed with MS methods and should be considered. Coeluting matrix may suppress or enhance ionization which could result in inaccurate analyte quantitation.

Before consideration and evaluation of matrix effects and interferences, it is most important to determine what is the most appropriate matrix (analytically and clinically) from which to extract the analyte of interest. The most common matrices used in clinical LC-MS/MS are plasma, serum, and urine; the sample matrix chosen should be the one that yields the most clinically relevant information. For example, drugs of abuse are distributed rapidly via blood throughout the body, and typically converted by the liver to water soluble metabolites. The kidneys subsequently remove these metabolites from the blood and they are excreted in urine. Metabolism of these drugs can be a relatively fast process, resulting in a small window of detection if blood is chosen as the sample matrix. Consequently, urine is often the matrix of choice when developing methods to detect illicit drug exposure, as it typically has a longer window of detection as compared to blood. If more than one specimen type will be used for patient care, the LC-MS method must be validated independently for each matrix type. Laboratories must also consider what source of matrix will be used for the preparation of calibrators and controls. Matrix "stripped" by charcoal adsorption to remove the analyte can be generated in-house or purchased commercially. Calibrators and controls prepared in such matrices, however, can behave differently than clinical samples and must therefore be thoroughly evaluated during assay development and validation as well as continuous evaluation postimplementation during any reagent lot changes [6].

One of the most critical step in dealing with matrix effects during development and implementation of a clinical LC-MS assay is determining the appropriate sample preparation procedure. Each of the matrices described earlier is compatible with a variety of sample preparation procedures, which contributes to the removal of materials that may contaminate the column, elimination of compounds, such as phospholipids that are known to cause ion suppression or enhancement, and concentration of the analyte to increase sensitivity. Sample preparation can be accomplished by protein precipitation, solid phase extraction (SPE), or liquid-liquid extraction (LLE) [7,8]. The method selected for sample preparation must balance the need for ease of use and minimization of labor with the need to account for or eliminate matrix effects. Protein precipitation by organic solvent, or even simpler "dilute-and-shoot" sample preparations are among the most popular, however, they are most susceptible to impact from molecules other than the analyte of interest or other matrix components.

When simpler sample preparation methods do not adequately address matrix effects, LLE and SPE can be chosen for the cleanup of samples of low molecular weight analytes, such as drugs and hormones, but these approaches are costlier than protein precipitation and require more time and effort for sample cleanup. LLE allows compounds in a sample to be concentrated and removes more potential contaminants than protein precipitation, however, identifying the proper solvents for LLE can make developing such an extraction method labor intensive. In addition, when considering this approach to sample clean-up for clinical application, solvent cost and disposal should be considered. The lack of automation for LLE also makes this technique highly laborious, especially in a clinical setting. SPE is used to concentrate and purify samples using sorbent cartridges based on their physical and chemical properties. The wide range of formats for this technique has facilitated automated offline and online sample processing, thereby reducing labor requirements as compared to LLE, while still achieving cleaner and more concentrated extracted samples than protein precipitation [9].

Another important way to compensate for matrix effects on LC-MS assays is the use of appropriate internal standards (IS) during analysis. While research studies employing LC-MS/MS will often use absolute analyte response to establish a standard curve for calculation of analyte concentration in a sample, the high degree of variability of patient samples in the clinical laboratory makes this a challenging proposition. Ionization efficiency of the analyte of interest can be affected by the temperature and pressure of the ion source, as well as a variety of other variables that can change from day-to-day—important factors when considering longitudinal operation of a clinical laboratory. To help in correcting these differences, internal standards have become a critical component of quantitative LC-MS/MS assays, as such variables only affect absolute response and not the response ratio between analyte and IS [10]. When considering the appropriate internal standards, it is important to note that while ionization is influenced by matrix components that can either increase or decrease analyte signal, it is also influenced by the chemical structure of the analyte itself; consequently, great care should be taken to choose an IS that is similar in structure to the analyte. An ideal IS is either a structural analog or stable isotopically labeled analog that is at least 3 mass units heavier than the analyte [6]. In addition, when using a stable isotopically labeled analog, the purity of each lot number of IS should be established to ensure that there is no cross talk between IS and analyte. Appropriate IS selection and evaluation helps to minimize the adverse effects of ion suppression on quantitation, helping to ensure accurate measurement of the analyte of interest and thereby decreasing the risk of patient harm by reporting an inaccurate result.

In addition to matrix effects, other analytical interferences can result from either endogenous or exogenous sources. Endogenous interferences are substances that are already present in the sample matrix (e.g., lipids or metabolites). Exogenous interferences result from substances introduced to the sample from an outside source (e.g., drugs or herbal supplements). Potential interferences that should be considered and evaluated during method validation are isobaric molecules, medications that may be prescribed for the targeted physiological condition, commonly prescribed medications and their known metabolites [7] especially within the disease state where the analyte of interest is investigated, as well as matrix interferences, such as hemolysis (lysed red blood cells), lipemia (elevated lipids), and icterus (elevated bilirubin). Commonly consumed substances that may be present in a population tested in a clinical setting, such as over the counter drugs including aspirin, ascorbic acid (vitamin C), ibuprofen, and pseudoephedrine may also be sources of interference [6]. For analysis performed on blood samples, the collection tube used must be considered as a potential source of error or interference. Each collection tube type can have a variety of additives, any one of which could interfere with the LC-MS/MS assay. Collection of a specimen in the wrong collection tube could potentially lead to reduced sample stability or ion suppression/enhancement [11]. Thus, if more than one type of collection tube is to be deemed acceptable, the validity of results obtained from each tube type for the validated method must first be confirmed.

1.5.3 Technical Hurdles

While assay performance and potential interferences are important considerations for LC-MS in the clinical laboratory, there are other technical challenges that stand in the way of widespread uptake and implementation of LC-MS for clinical testing. One of the main issues is that current LC-MS instrumentation is not highly automated when compared to current clinical laboratory instrumentation. This impacts not just the labor needed for operation, but leads to decreased assay throughput—both of which increase the per test cost of LC-MS assays. Another major challenge for clinical MS implementation, is the level of technical proficiency required by laboratory staff in order to achieve minimum assay downtime due to instrument failure. While MS vendors are improving their service support for testing in a clinical environment, there is still a significant gap between the turnaround time for service of a traditional clinical chemistry analyzer and a clinical MS. Over time, laboratory staff will develop the expertize required to effectively repair minor instrument problems and triage major problems to dispatch advanced service. During early implementation, though, when laboratory staff is unexperienced, a clinical MS laboratory, may have trouble maintaining instrument performance.

While almost all LC-MS systems include an autosampler in their system configuration, this feature is only one of many automated pieces that are needed for routine clinical analysis. As more clinical laboratories have implemented LC-MS/MS, manufactures have focused on developing instruments designed to better fit into the pre- and postanalytical workflows of the clinical laboratories. These developments include robotic liquid handlers and LC systems for preanalytic considerations, as well as capability to transmit results from the postanalytical data processing on the instrument to the clinical laboratory information system (LIS). In addition, it is essential that any routine clinical LC-MS system include the ability to read bar codes on sample containers—this is something currently lacking in almost all LC-MS systems. The incorporation of bar code reading into the LC-MS workflow minimizes the potential for transcription errors when tracking specimens through the testing process, as well as decreasing the hands-on interaction with specimens. In addition, it is vital for any clinical LC-MS system to include a robust, fully functional interface to the LIS that will allow for bidirectional communication between the LIS and the LC-MS operating system. This functionality (coupled with the ability to read bar code labels) will allow the system to automatically determine which test has been ordered for a particular specimen, and implement that method for analysis. This LIS interface will also allow direct transfer of the test result to the electronic patient record, which minimizes the opportunity for transcription errors, and increases patient safety. Apart from advances in information transfer, a desirable feature for a clinical LC-MS system would be the ability to aliquot, process, and dilute specimens in an automated fashion. Anything that decreases manual interaction with the sample, increases walk-away time, and standardizes analyses between multiple operators will increase the accessibility of LC-MS systems for use in routine clinical laboratories.

Assay throughput is another significant challenge for the implementation of LC-MS in routine clinical laboratories. Even with rapid or condensed chromatographic methods, the best throughput for a single channel LC-MS system is 20–25 samples per hour. In comparison, a routine clinical analyzer can have a throughput of hundreds of samples per hour for colorimetric assays or immunoassays. This significantly higher throughput comes from the ability to process samples in parallel—the immunoassays and colorimetric assays can be started every few seconds and incubated in parallel, and then an assay endpoint read every few seconds. The nature of routine LC-MS/MS analyses makes it a serial technique that is limited in the throughput it can achieve. However, the use of multichannel chromatography systems [12] as well as alternative sample introduction approaches, such as LDTD [13] or direct SPE-MS [14] have made progress toward increasing the throughput of LC-MS systems as needed for routine clinical testing. Another approach to increase throughput is to minimize the amount of time needed for calibration of the system. There are multiple investigations of alternative calibration approaches [15–17] to reduce the number of calibrators in LC-MS/MS assays (in some cases entirely internal calibration), and increase amount of time spent analyzing patient specimens.

It is important to note that technical problems associated with LC-MS systems are more prone to cause extended downtimes compared to standard clinical chemistry analyzers [8]. For example, failure of one of the vacuum pumps on the MS has the potential to cause a downtime of several days, as the response time of MS service centers is significantly longer than the response time provided for standard analyzers. Ideally, a laboratory would have access to at least two identical LC-MS instruments, providing a backup in the event that one fails, however in most laboratories this is not feasible. Given that instrument downtimes impact the ability of a laboratory to deliver timely results, they also have the potential to adversely affect overall patient care due to the inability of a physician to use test results to make important healthcare decisions. It is important for laboratories considering LC-MS to develop a contingency plan during the earliest stages of method development for how patient samples will be run when extended downtimes occur, especially for critical assays.

1.5.4 Practical Considerations

When setting up a new LC-MS testing system, a CLIA certified laboratory must have written standard operating procedures (SOPs) for each method on the system that describes all aspects of a test—from obtaining the sample to reporting results to the ordering physician. Each SOP must describe proper specimen collection, processing, and storage of patient samples, as well as a step-by-step description of the procedure. A detailed listing of information that should be included in an SOP can be found in Table 1.2. The procedure for proper specimen storage and preservation should also be described to ensure specimen integrity until testing is complete, in addition to describe a contingency plan (system backup) in the event that a test system goes down (is not operable). Where appropriate, critical values

---

Table 1.2 Key Components of an Standard Operating Procedures (SOPs) for Liquid Chromatography-Mass Spectrometry (LC-MS) Methods

- Preparation of mobile phases
- Type and handling of chromatography columns
- Interpretation of results
- Description of calibration and calibration verification
- Reportable range of patient test results
- Control procedures (QC)
- Actions required if calibration or QC does not be established acceptability criteria
- Downtime procedures
- Critical action values (if established)

QC, Quality control.

(results needing immediate clinical action for treatment/resolution) and the protocol for reporting patient results that fall into this range should also be defined.

Another critical issue for successful implementation of LC-MS in the clinical laboratory is instrument performance and maintenance. For an LC-MS system, the performance of each component within that system must be verified individually on a regular basis. Given that many factors can influence the day-to-day performance of LC-MS systems, it is critical for each laboratory to establish protocols to monitor and detect any deviations from normal performance. One simple way to monitor the LC-MS system is to check the absolute response, peak shape, and retention times of the IS. Given the large day-to-day variations in MS response, this is typically done by visual review for an injection (or multiple injections) of a system suitability sample (analyte dissolved in mobile phase/solvent rather than biological matrix) instead of establishing formal QC rules. Problems with individual samples may also be identified by monitoring the response of the internal standard within each batch. Confirming that the LC-MS/MS system operates at its optimal performance potential before each run helps to ensure precise and accurate measurements of patient samples. Oftentimes sensitivity is the first parameter to be affected when there is a failure to confirm that both the LC and MS meet the minimal acceptance criteria for performance.

Instrument maintenance plays a key role in maintaining optimal instrument performance. Following the manufacturer's recommendations for instrument maintenance is generally sufficient, and must be performed as defined by the manufacturer and at the specified frequency. These activities must be documented and may include cleaning the source, cleaning the first quadrupole, checking the oil level of the roughing pumps, etc. Completion of daily, weekly, and monthly maintenance must be monitored; implementing a check-off sheet or other type of standard form with signature and date and maintained in the instrument logbook is the best practice. Maintaining a rigorous and consistent approach to instrument maintenance is also critical to ensure the LC-MS/MS system is in peak operating condition in order to give the highest quality patient results.

It is also important to consider storage and transport conditions for patient specimens relative to each analyte tested by LC-MS; although this is not specific to just LC-MS assays. Often in research settings, samples are collected over an extended period of time (e.g., months to years), stored (typically at −80°C), and then analyzed as a single batch. In contrast, results from clinical tests are used by clinicians to guide patient care in real-time, so samples are typically analyzed in a continuous fashion or in smaller batches within a day rather than in one large batch. For some lower volume tests or tests where resulting turnaround time is not clinically critical in nature, however, samples may be batched and run at specific intervals (e.g., once per week or when a specific number of samples have been received). In either case, if samples are not analyzed immediately, stability studies should be performed to evaluate the stability of the analyte under a variety of conditions, such as temperature (room temperature, 4°C, −20°C, and −80°C) and time.

It is not uncommon for results to differ significantly when comparing analyte measurements from an immunoassay and an LC-MS method for the same sample. In addition, as previously discussed, most LC-MS assays are not standardized or harmonized, so results from one laboratory may not be directly comparable to results from another institution, even if the same platform is used. Given the increased specificity of MS over immunoassays, it is not recommended to apply reference ranges determined for an immunoassay to an MS method [6]—this may even apply to switching from one LC-MS method to another. When a laboratory switches from a less specific method (e.g., immunoassay) to an LC-MS/MS method, clinicians should be notified of the transition and of any changes in reference ranges and/or potential differences in laboratory test result interpretation. It is important to remember that it is the responsibility of the clinical laboratory to effectively communicate the impact of LC-MS on the concepts of sensitivity, selectivity, predictive value, and test interferences, with the goal of ensuring quality patient care.

REFERENCES

[1] Garg U, Dasouki M. Expanded newborn screening of inherited metabolic disorders by tandem mass spectrometry: clinical and laboratory aspects. Clin Biochem 2006;39(4):315–32.

[2] Guo T, Chan M, Soldin SJ. Steroid profiles using liquid chromatography-tandem mass spectrometry with atmospheric pressure photoionization source. Arch Pathol Lab Med 2004;128(4):469–75.

[3] Grebe SK, Singh RJ. LC-MS/MS in the clinical laboratory—where to from here? Clin Biochem Rev 2011;32(1):5–31.

[4] Ferreira-Gonzalez A, Garrett CT. Laboratory developed tests in molecular diagnostics. In: Coleman WB, Tsongalis GJ, editors. Molecular diagnostics for the clinical laboratorian. New York City: Humana Press; 2006. p. 247–57.

[5] Miller WG, Myers GL, Ashwood ER, Killeen AA, Wang E, Thienpont LM, Siekmann L. Creatinine measurement: state of the art in accuracy and interlaboratory harmonization. Arch Pathol Lab Med 2005;129(3): 297–304.

[6] Honour JW. Development and validation of a quantitative assay based on tandem mass spectrometry. Ann Clin Biochem 2011;48(Pt 2):97–111.

[7] Kushnir MM, Rockwood AL, Roberts WL, Yue B, Bergquist J, Meikle AW. Liquid chromatography tandem mass spectrometry for analysis of steroids in clinical laboratories. Clin Biochem 2011;44(1):77–88.

[8] Vogeser M, Kirchhoff F. Progress in automation of LC-MS in laboratory medicine. Clin Biochem 2011;44(1): 4–13.

[9] Vogeser M, Seger C. Pitfalls associated with the use of liquid chromatography-tandem mass spectrometry in the clinical laboratory. Clin Chem 2010;56(8):1234–44.

[10] Stokvis E, Rosing H, Beijnen JH. Stable isotopically labeled internal standards in quantitative bioanalysis using liquid chromatography/mass spectrometry: necessity or not? Rapid Commun Mass Spectrom 2005;19(3): 401–7.

[11] Bowen RA, Hortin GL, Csako G, Otanez OH, Remaley AT. Impact of blood collection devices on clinical chemistry assays. Clin Biochem 2010;43(1–2):4–25.

[12] Shintani Y, Harako K, Motokawa M, Iwano T, Zhou X, Takano Y, Furuno M, Minakuchi H, Ueda M. Development of a miniaturized multichannel high-performance liquid chromatography for high-throughput analysis. J Chromatogr A 2005;1073(1–2):17–23.

[13] Wu J, Hughes CS, Picard P, Letarte S, Gaudreault M, Levesque JF, Nicoll-Griffith DA, Bateman KP. High-throughput cytochrome P450 inhibition assays using laser diode thermal desorption-atmospheric pressure chemical ionization-tandem mass spectrometry. Anal Chem 2007;79(12):4657–65.

[14] Highkin MK, Yates MP, Nemirovskiy OV, Lamarr WA, Munie GE, Rains JW, Masferrer JL, Nagiec MM. High-throughput screening assay for sphingosine kinase inhibitors in whole blood using RapidFire® mass spectrometry. J Biomol Screen 2011;16(2):272–7.

[15] Taylor PJ, Forrest KK, Salm P, Pillans PI. Single-point calibration for sirolimus quantification. Ther Drug Monit 2001;23(6):726–7.

[16] Olson MT, Breaud A, Harlan R, Emezienna N, Schools S, Yergey AL, Clarke W. Alternative calibration strategies for the clinical laboratory: application to nortriptyline therapeutic drug monitoring. Clin Chem 2013;59(6):920–7.

[17] Rule GS, Rockwood AL. Alternative for reducing calibration standard use in mass spectrometry. Clin Chem 2015;61(2):431–3.

APPLICATION SPECIFIC IMPLEMENTATION OF MASS SPECTROMETRY PLATFORM IN CLINICAL LABORATORIES

H. Nair

Boston Heart Diagnostics, Framingham, MA, United States

1 MASS SPECTROMETRY VERSUS AUTOMATED IMMUNOASSAY PLATFORMS

Clinical laboratories have seen significant increase in the role of mass spectrometry in the last two decades with the prospect of expanding that role much further in the years to come. Currently, by many estimates, only a small proportion (about 5%) of the LC-MS instruments sold by the different manufacturers are used in the clinical laboratories and mass spectrometry in clinical labs represent only about 5% of the reported tests [1,2]. Mass spectrometry assays are typically implemented in a clinical laboratory when the analytical needs, such as sensitivity and specificity indicated for the clinical application is not met by the automated analyzer or the cost and throughput considerations do not permit sending tests out to the reference mass spectrometry laboratories [3–5]. Considering the inherent benefits of mass spectrometry in clinical diagnosis [6,7], the role of mass spectrometry in clinical laboratory is expected to keep expanding in the years to come.

Mass spectrometry is operationally very different from the automated immunoassay analyzers (and biochemistry assays using optical detection) that are the current vanguards of clinical testing in the majority of clinical laboratories, a status that has been gained mainly due to its operational efficiency and the need for minimal human intervention and training needed for routine operation. This operational efficiency is owed to the capability of biochemistry and immunoassays to measure analyte signatures practically directly from the complex biological matrix of the patient sample with minimal sample processing requirement prior to analysis [8]. Immunoassays rely on the innate ability of the reagent antibodies (or antigens) to seek out its target directly in the complex biological matrix of the patient sample. Biochemistry assays can directly detect optical signals from the patient sample. Because automatable assays do not require considerable preparation of samples prior to detection, immunoassays and biochemistry assays are much more amenable to automation of analytical and quality assurance processes and allows for "blackbox" type operation and standardization of the many tests across laboratories. In a mass spectrometry assay, on the other hand, the detection occurs very distant from the biological matrix from which the analyte must be separated and purified prior to analysis. Purification of the analyte for detection by mass spectrometry requires significant upfront sample preparation, such as lengthy and multistep analyte extraction and the separation of the coextractants by a subsequent chromatographic step. Therefore, the success of mass spectrometry analysis depends primarily on the degree of purification of analyte from the matrix prior to the analysis. For this reason, the choice of the front end sample preparation components and the effectiveness of the protocol for analyte extraction from biological matrix and separation of analytes from the extract are arguably more critical to the success of the mission than the choice of the mass spectrometry instrumentation itself.

With the advent of newer technologies, such as ambient ionization techniques [8], discussed in Chapter 13 in further detail, some stages of the sample preparation process required for mass spectrometry testing, such as extraction and separation may be simplified or streamlined in future. Ambient ionization techniques have begun to allow mass spectrometry to be used as aids in real time clinical analysis [9] just as it has been used in environmental and forensic applications for some time. You may have noticed the negative ion mass spectrometry instruments at the airports in the United States and other parts of the world employed for screening travelers during security clearance using direct desorption mass spectrometry from paper swabs [10]. It is conceivable that in the near future such devises will find diagnostic applications in the physician's office.

Currently though, for clinical applications, mass spectrometer must be regarded simply as a detector that records the m/z signal(s) of the signature molecules or fragment(s) of the marker presented to it in a purified and in an ionization friendly formulation by the front end sample preparation and separation components. A mass spectrometry based assay is only as successful as the sample preparation protocol that precedes the analysis. An old adage has it that a chromatographer may think of mass spectrometer as just a detector and mass spectrometrist may view chromatography as just an accessory to the mass spectrometer. But in a clinical laboratory the two must marry each other and must each seek out and understand each other in order to successfully develop, validate, launch a clinical assay, and maintain the validated performance throughout its life. The success of the routine clinical performance of the assay, in big part, depends significantly on the extent of control the laboratory has on the protocols employed for the extraction and separation of analytes from the biological matrix. Typically, mass spectrometer is the most capital expensive component of the analytical platform. However, choosing the "right mass spectrometer" is less consequential than the identification of the "right extraction and separation techniques" and developing and managing the in-house resources and expertise necessary to develop, validate, and control the assay in production [11]. Mass spectrometry instrumentation has been evolving over the past decades to require less and less intervention by the technologist performing the routine testing but unlike the automated analyzers, is not a "black box" just yet.

Today, most commonly, mass spectrometry is implemented in clinical laboratories for the analysis of small molecules [12]. It is primarily in this area that mass spectrometry offers superior value to immunoassay based automated platforms, such as RIA and has allowed the technique to establish strong feet hold in the clinical laboratory. For many small molecule analysis, mass spectrometry currently offers a more sensitive and specific alternative to immunoassays. It is in the area of small molecule analysis (endocrinology, pain management and toxicology, and therapeutic drug monitoring) that mass spectrometry has gained significant prominence over immunoassays and offers improved clinical and analytical utility over immunoassays. Mass spectrometry has also established itself as a formidable tool in the analysis of large protein fingerprints in the clinical microbiology laboratories for accurate, rapid, inexpensive identification of bacteria, fungi, and mycobacterium [13].

Targeted analysis of proteins and peptides [14–16], is emerging as an alternate technique with the promise and potential to overcome the many limitations of immunoassays, such as inadequate sensitivity and poor specificity due to cross reactivity from interferents and antireagent antibodies, and platform to platform variations arising from the different isoforms of the molecule targeted by the different immunometric reagents. However, routine clinical use of mass spectrometry for targeted protein analysis is at the moment limited by the lack of adequate automation of sample preparation, operational complexity, and the unavailability of adequate vendor support necessary for the routine operation in the setting of a high throughput clinical laboratory operation. Additionally, the technical expertise and the training needs for routine analysis of targeted proteins by mass spectrometry far exceeds such needs for small molecule mass spectrometry. This field of mass spectrometry of targeted proteins is still evolving to the production realities of the clinical laboratory.

2 EARLY DECISIONS IN THE IMPLEMENTATION OF MASS SPECTROMETRY IN THE LABORATORY

The rigor and skills necessary for implementing a mass spectrometry assay would depend in part on the performance demands of the assay. For instance, the extraction and separation techniques will need to be driven by the performance requirements for the assay, such as sensitivity and selectivity. At every stage in the workflow, many decision points exist. Even among triple quadrupole mass spectrometry, many manufacturer specific features and performance levels exist as reflected in their relative costs. A search of the websites of popular mass spectrometer manufacturers would indicate that each vendor supplies a variety of triple quadrupole mass spectrometers that vary considerably in the performance characteristics. For example, higher end triple quadrupole mass spectrometers, in addition to being more sensitive, can process many more multiple reaction monitoring (MRM) transitions with minimal crosswalk (both discussed in more detail further in this chapter) than the lower end mass spectrometers of the same configuration. Benefit of minimal crosswalk may be critical for an assay that is designed to report a large panel of markers simultaneously such as a steroid panel (described in further detail in Chapter 10). Consideration of the sensitivity of the detector is an early decision that goes hand in hand with the considerations for sample preparation.

The skills that a laboratory needs for developing the most suitable sample extraction and preparation protocols are often learnt and refined over a period of time in the laboratory. So the decision to implement a mass spectrometry assay in a laboratory for the first time could also double as a decision to put the laboratory on track to become a center of excellence in mass spectrometry in the years to come. As discussed in more depth in Chapter 1, plan for implementation of mass spectrometry platform should begin with the consideration to assess and recognize the current level of expertise in the laboratory and what the laboratory would need to acquire in order to successfully develop and validate the assay and perform routine testing.

A key early step in the preparation for mass spectrometry life is to choose or hire interested and capable technologists that have a flair for assay development and a keen interest in analytical instrumentation. Even without prior experience in the platform, the right individuals can develop the skills needed for routine use of mass spectrometer by utilizing a variety of sources, such as training packages from the manufacturers, guidance from experienced individuals in the laboratory, or from consultants and with help from the many published work on the topic [17] including hopefully this book. Involving as many technologists as possible in the initial platform implementation decisions and assay development process can help promote the in-house expertise necessary for successful assay development and the routine operation of the analytical platform.

Unlike mass spectrometry manufacturers who sell only instruments to the laboratories (and not assays), most immunoassay platform manufacturers also market many assays as FDA cleared or approved or as RUO kits. So the technical assistance available to laboratories for implementation and the routine use of immunoassays from the vendors can far exceed what is available for laboratory developed tests run on mass spectrometry platforms. When trouble shooting an immunoassay, typically the kit manufacturer offers significant help to the laboratory but when trouble shooting an LC-MS assay, the laboratory would have to depend, in big part, on its own in-house expertise. Instrument manufacturers of the mass spectrometry platform do not have much control or expertise over the specific sample preparation or components of the assay developed or adopted by the laboratory that may be at the root of the problem at hand. The manufacturer of the mass spectrometry platform typically cannot offer direct help beyond the use of their standard formulations and performance tests to ensure that the instrument is functioning according to their specifications. While this is helpful and often provides critical initial information necessary for further troubleshooting, an in-house capability to troubleshoot sample preparation components is needed to solve the problem and to prevent significant downtimes. It can take a motivated and skillful technologist 3–6 months to accumulate enough experience to independently troubleshoot problems with the assay. Every troubleshooting effort must be fully documented and the information gained from successful and the unsuccessful efforts must be used whenever possible to further define and improve the assay characteristics. For the laboratory developed mass spectrometry assays, the assay characterization process does not stop at validation; it is a continuous improvement process that goes along with patient testing. For high complexity laboratory developed tests, such as a mass spectrometry assay, every routine QA and QC review event (described in more detail in Chapter 1) should be viewed as an opportunity for process improvement. Any changes made to the assay or workflow as a result of process improvement efforts must be carefully validated against the predecessor to ensure that patient care is not impacted by the potential change in the longitudinal reporting of the results.

The choice of the mass spectrometry platform components, such as extraction and separation, would depend primarily on the chemistry of the analyte of interest, the composition of the matrix that hosts the marker and the analytical sensitivity and specificity required for the clinical application. Performing urine drug screen testing for some analytes, for instance, might require just a simple dilute and shoot approach with or without the need for a high sensitive mass spectrometer. But measuring low levels of free testosterone in females or children or a steroid panel may require significant experience in order to develop and perform rigorous protocols that may span several hours to overnight and would require more expensive and ultrasensitive mass spectrometer for the sensitive and multiplexed analysis. Scheme 2.1 outlines the key steps in a mass spectrometry workflow where critical application specific decisions must be made when implementing mass spectrometry platform.

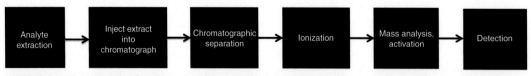

SCHEME 2.1 The Key Steps in a Mass Spectrometry Workflow Where Critical Application Specific Decisions Must be Made When Implementing Mass Spectrometry Platform

3 IMPORTANCE OF INTERNAL STANDARDS IN MASS SPECTROMETRY

An important choice to make at the beginning of the assay development process is the identification of the appropriate internal standard(s) that represent the analytical behavior of each analytes assayed. The absolute response of the mass spectrometry detector is not a quantitative entity; minor changes in the analytical environment in any of the steps in the analytical workflow—sample extraction, separation, ionization, mass analysis, and detection—can potentially significantly alter the detector responses between two injections of the exact same sample. To translate these nonquantitative responses to a quantifiable response ratio, choice of an internal standard that represent the analyte's interactive behavior with all of the components and the accessories employed in the protocol is critical. Such an internal standard is used in an identical fashion, in the calibration standards as well as patient samples to normalize the detector response of the analyte. No mass spectrometry assay is quantitative without the use of an appropriate internal standard to normalize the variations in the detector response between injections. When possible, it is best to use a dedicated internal standard for each of the analytes measured in a multiplexed assay. Fortunately for mass spectrometry detection, isotopes of the analytes offer an excellent choice of internal standards. [It is important to ensure that the response from the isotopes of the internal standard and the response of the isotopes of the analyte do not cointerfere with each other. To minimize this possibility, the internal standard is chosen such that sufficient m/z gap exists between the internal standard(s) and the analyte(s)]. However, quite often, prohibitive cost and/or the lack of availability of an isotopically labeled analog of the molecule may necessitate the use of a structurally similar analog as an internal standard. When employing such an analog of the analyte as internal standard, more rigorous validation studies must be performed to characterize the relationship between the analyte and the internal standard. The structural analog used as internal standard may differ from the analyte in its efficiency of extraction from the biological matrix, separation, and ionization efficiencies and can result in inefficient normalization of the mass spectrometry response of the analyte [18].

4 MATRIX EFFECTS, EXTRACTION EFFICIENCY, AND ION SUPPRESSION/ ENHANCEMENT

As mentioned earlier in this chapter and in Chapter 1, the absolute mass spectrometry detector signal, unlike the signal produced by spectrophotometric techniques, is not quantitative. A myriad of noncritical to critical components from pipette tips to the components of instrumentation, such as ionization source, electrostatic lenses, the analyzer, and the detector that the analyte encounters on its journey from the biological matrix to the mass spectrometry detector can significantly alter the injection to injection consistency of the responses between two analyses of the same sample. The detector response generated from the same lot of the quality control sample, for example, can be significantly different prior to and after cleaning of the ionization source or the electrostatic lenses that direct the beam of ionized molecules to the mass analyzer. The collective term "matrix effect" is evoked to describe the variations in detector response experienced by an analyte under different analytical environments. Except in the case of the most extreme cases, such as visible lipemia, hemolysis etc., variations in chemical and biological environments of the specimen cannot be recognized and hence cannot be controlled easily. The matrix composition and the extraction strategy employed can impact the "extraction efficiency" of the analyte from patient to patient. Extraction efficiency of a technique can be assessed by recovery studies using post extraction addition experiments performed in as many patient samples as possible [12,19–21].

A common approach employed to assess extraction efficiency begins with splitting the patient sample into two aliquots. To one aliquot a known concentration of the analyte standard is spiked and extracted (direct extraction), while the other aliquot is first extracted and then stoichiometrically equivalent analyte standard is spiked to the extract (post extraction). Detector responses (not internal standard normalized responses) of both extracts (direct extraction and the post extraction) are analyzed very carefully in sequential order (to control for variations in experimental conditions) and the results compared to determine variations in extraction efficiencies. Recovery of the analyte can also be impacted by the suboptimal chromatographic conditions that may not sufficiently resolve interferents present in some patient samples. A major contributor to matrix effect, particularly when employing electrospray ionization (ESI) is the ion suppression/enhancement triggered by neutral or ion-pairing agents present in the droplet during the ionization process [10,21–23]. This is particularly a problem for ESI which necessitates the use of ESI friendly solvents and additives in the preceding chromatography step.

Quality of HPLC separation is affected by numerous factors, such as mobile phase composition, temperature, ionic strength, pH, column dimensions, and stationary phase chemistry among others. Optimal control of ionic strength and pH requires the use of nonvolatile buffers and salts and strong ion pairing agents. If nonvolatile buffers, such as phosphate buffer and ion-pairing agents, such as TFA that allow for improved chromatographic separation are used, significant ion suppression will result during ionization by ESI and the nonvolatile components will coat and render the ionization source practically unusable. Presence of high levels of buffers and additives in the shrinking electrospray droplet can distort the normal process of ion desorption and lead to poor ionization efficiency for the analyte. This is a key analytical disadvantage of LC-MS technique [24–26] compared to an LC technique that uses a sprectrophotometric detector. To overcome ion suppression, ESI and other soft ionization techniques can only use solvents with lower buffer capacities and additives that are ionization friendly. Although, other atmospheric ionization techniques, such as atmospheric pressure chemical ionization (APCI) and atmospheric pressure photoionization (APPI)—both techniques discussed in more detail in Chapter 1 and later on in this chapter—can handle stronger buffers better than ESI can, but these techniques are by no means free from ion suppression/enhancement.

A key rationale for the use of patient samples earlier on in the method development process is to determine the extent of matrix effects and address it. When ion suppression/enhancement is observed, the extraction and separation steps as well as ionization parameters may need to be reoptimized using more ionization friendly reagents and/or improved analytical conditions, such as the use of longer chromatographic elution times and more extensive cleaning of the column after elution of the analyte. It is an unfortunate reality that in order to minimize ion suppression and to promote efficient ionization, chromatography in LC-MS is performed with compromised efficiency. In LC-MS/MS, however, the specificity loss from the subpar chromatographic separation in LC can be overcome by the use of tandem mass spectrometry at the back end that can provide improved analytical specificity in spite of the poor front end chromatographic separation.

The extraction, chromatographic separation, ionization efficiency, and mass analysis parameters are all interdependent and the process of identification of the optimal parameters for each stage should follow a reiterative path. Assay characterization experiments, such as recovery studies are best conducted when extraction, separation, and mass spectrometry parameters are all optimized and the performance verified with a large number of patient samples. Use of a well-designed internal standard remains the best tool in the hands of the assay developer to control for the impact of matrix effects during routine analysis of the patient samples.

5 ANALYTE EXTRACTION

When the patient sample reaches the bench for the analytical part of the clinical testing most often the first major step is the extraction of the analyte from the patient specimen. All extraction techniques result in some degree of coextraction of the multiple components present in the matrix—biological and reagent related. Therefore, analyte specificity is driven in the subsequent analytical steps—GCMS or LC-MS/MS. Extraction step is a major source of analytical error in routine analysis and a key contributor to the assay imprecision. Therefore, whenever possible, it is desirable to automate sample extraction by adopting automatable formats, such as the use of 48 or 96 well plate format and the use of auto samplers instead of processing samples individually using Eppendorf tubes. The quality of manual portion of the process must be maintained via constant evaluation and training of the analyst(s).

As mentioned in the earlier chapter, the most common extraction techniques employed are the protein precipitation, solid phase extraction and liquid–liquid extraction techniques (single step or multistep) either performed manually or by automated approaches [18,25,26]. It is sometimes necessary to achieve additional degree of extract purity to improve the performance of the separation process. This may be achieved by evaporation of the extract and reconstitution the analyte in a solvent that is more compatible with the LC method. Use of the solvent system with similar composition as the LC equilibration solvent is a good starting point for optimization of the composition of the reconstitution solvent. The next chapter (Chapter 3) is dedicated to sample preparation for LCMS and discusses extraction techniques in further detail.

For analytes that are present in millimolar or micromolar levels in the patient sample, when coupled to a sensitive mass analyzer, a simple, one step extraction technique may be sufficient. For example, new born screening (NBS) by tandem mass spectrometry performed in the different state laboratories from a dried blood spot specimen use a one-step methanolic extraction of relatively abundant (micromoles) amino acids and organic acids [27]. Often sometimes the sensitivity of the mass spectrometry detector and/or the high levels of analyte present in the specimen allow for a simpler single step extraction protocol. For analytes that are present at much lower levels (nanomoles, femtomoles, or lower) in patients, such as for free testosterone [28], a much more rigorous extraction techniques must be employed and the use of techniques, such as dialysis, liquid-liquid extraction followed by evaporation and reconstitution may be necessary. Often, additional techniques may be necessary to ensure that the extract is optimal for the subsequent sample preparation steps. For example, lipids, predominantly phospholipids are ubiquitous in biological specimen and can be abundantly present in many extracts. Phospholipids usually present in organic and nonpolar extracts can pose significant assay interference and/or matrix effects and significantly impact the performance of the assay. Phospholipid removal plates can be useful to minimize the adverse impact of phospholipids [24]. Many additional sample preparation approaches including the use of immunosorbents and magnetic beads are summarized in Ref. [24] and the references therein.

6 CHOICE OF SEPARATION TECHNIQUE: GC VERSUS LC

Once the analytes are efficiently extracted from the specimen matrix, the analytes must be separated from its coextractants by either gas chromatography or liquid chromatography. Mechanically, a key difference between LC and GC is in the ease with which a liquid extract can be introduced into the chromatograph. While liquid extract can be easily introduced into a liquid flow in LC, prospect of introducing liquid extract into a gas flow in GC can pose significant technical limitations for many analytes (1 mL of water produces 1.3 L of vapor at RT). Injection of small volumes of liquid extract (a few microliters) into GC column is achieved through the use of a heated port at the head of the column. A calibrated microsyringe is used to deliver up to a few microliters of the sample through a rubber septum and into the vaporization chamber which is typically heated 50°C above the lowest boiling point of the sample and subsequently mixed with the carrier gas for transportation into the column.

While volatile analytes can be easily introduced into the gas chromatograph, nonvolatile analytes cannot be introduced into the system unless they can be made volatile by chemically modifying the analytes by derivatization. For those analytes that cannot be easily derivatized to volatile analogs, GCMS remains a prohibitive tool in the clinical laboratory. The type and the amount of sample that can be introduced into the GC column is an analytical limitation of the technique compared to the more versatile LC. In order for the sample extract to be amenable to analysis by GCMS, all of the extract's constituents must be volatile. Because only a small volume of extract can be introduced into the column, the analytes must be present at a relatively high concentration often achieved by the use of preconcentration of the analyte after initial extraction. It is also critical that the sample introduction process must be optimized to ensure that the chromatographic separation is not compromised. Unlike LC-MS, GCMS offer uncompromised chromatographic resolution—the key driver in the continued relevance of GCMS in clinical laboratories and in other applications. Fortunately, many small molecule drugs and metabolites of interest in toxicology, pain management, and endocrinology are either volatile or can be made volatile by simple derivatization and are well separated by GC. This allows GCMS to be an invaluable tool for screening patients suspected of drug overdose, drug mix up, and to direct intervention in a hospital setting [29]. Because of the superior chromatographic separation offered by GC, GCMS is considered the method of choice (compared to LC-MS) in the analysis of isobaric isomers of sterols and fatty acids for applications in the clinical nutritional and cardiovascular risk assessments [30]. Fig. 2.1 (based on https://www.agilent.com/cs/library/selectionguide/Public/5989-6328EN.pdf; https://www.chem.agilent.com/cag/cabu/gccolchoose.htm; http://www.chromatographyonline.com/ hplc-column-selection; http://www.spectroscopyonline.com/mass-spectrometry-metabolomics-addressing-challenges-0) shows the classes of biomolecules that are analyzed by GC/MS and LC/MS.

7 CHOICE OF STATIONARY PHASE

Choice of the type and specifications of chromatography column for both GC and LC would depend on the analyte, the production requirements (injection time, sample volume, etc.), and the clinical performance required of the assay. Manufacturers of the GC columns (and LC columns) usually offer detailed guides for choosing the column that is optimal for the application [30]. Key features to consider when selecting the column include the type of column (packed vs. capillary), column material (inert, fused silica, etc.), stationary phase (film) of the column (alumina, DVB, etc.), film thickness, column length, and the inside diameter.

For LC, there exist many choices for stationary phases and many variations therein. Hydrophobic stationary phases like C18 and C8 are commonly used stationary phases in part because of the compatibility of the traditional reverse phase mobile phases with the ionization techniques, such as ESI. To enable enhanced separation of molecules with polar functional groups (vitamins, steroids) modified alkyl phases are often used. Traditionally, polar molecules are separated by normal phase chromatography

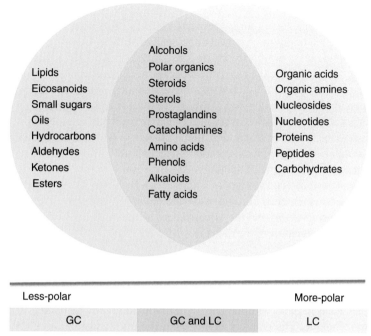

FIGURE 2.1 The Classes of Biomolecules That are Analyzed by GC/MS and LC/MS

that uses nonpolar solvents that are not ionization friendly. To separate highly polar molecules with "ionization friendly" mobile phases, use of Hydrophilic interaction liquid chromatography (HILIC) columns have been demonstrated as a practical alternate to normal phase chromatography [31].

Column length is a key determinant of the peak resolution while the internal diameter of the column is an indicator of the speed and sensitivity of analysis. Particle size of the stationary phase support dictates the overall column efficiency. Use of stationary phase support particles that are sub 2.0 μm (ultra-high pressure liquid chromatography—UPLC) requiring supporting instrumentation that can handle very high back pressures (typically 15,000 psi and above) are now common in the clinical laboratories. UPLC offer very high efficiencies of separation at much higher speeds than the traditional HPLC. Switching a method from HPLC to UPLC is typically a straightforward transition and detailed guidance is offered by the manufacturers of the UPLC column and the instrumentation. However, the type and level of some commonly used additives in traditional HPLC (such as THF) may not be ideally suited for use in the UPLC and needs to be evaluated carefully.

When developing a new laboratory developed test, a good strategy is to first identify a working chromatographic condition using standards of analytes and internal standards and then optimize the extraction, ionization, and mass spectrometry acquisition conditions. For mass spectrometry analysis, it is important and sometimes critical that the internal standard and the analyte should have identical retention time. If there is a retention time difference between the analyte and the internal standard, however small, there exists an opportunity for unfettered matrix affects that may not be normalized adequately by the internal standard. In situations where internal standard is not the labeled isotope of the analyte, additional caution must be exercised during validation. As discussed earlier, the mass spectrometry response for a given analyte/internal standard ([A]/[IS]) can vary with the analytical environment. Differences in the retention times between the analyte and the internal standard can reduce the ability of the internal standard to normalize the response of the analyte effectively. It is possible to use internal standards that may not share the exact chemical and analytical properties of the analyte. For example, many laboratories manage to use cheaper structural analog of immunosuppressant drugs, such as cyclosporine as an internal standard in the TDM LC-MS/MS assays. When using a structural analog of the analyte as internal standard, the recovery of the analyte must be examined closely using as many patient samples as possible. One approach to examine the recovery of the analyte in patient sample is to spike a known amount of analyte (standard addition technique) into the patient sample with known endogenous concentration of the drug. If spiking the analyte is not a viable analytical option, then low and high patient samples whose concentration is known can be mixed in different proportions to create a series of pools with incremental concentrations. Recoveries can be calculated from the expected versus the measured concentrations and the effectiveness of the internal standards in normalizing for the matrix effect can be accessed from the data [12,19–21].

8 CHOICE OF IONIZATION TECHNIQUES

A decision that goes hand in hand with the choice of chromatography technique is the choice of the suitable ionization technique. The types of ionization techniques adaptable to GCMS are limited to those ionization sources that can directly ionize analytes and matrix presented by the chromatograph only in vapor state. Electron impact (EI) ionization technique [32] allows ionization of molecules in vapor phase and therefore, most clinical laboratories that use GCMS employ EI ionization technique. At the EI ionization, source sample and matrix molecules eluted from GC in vapor state are intercepted by a curtain of electrons ejected from a filament typically of 70 eV energy. This results in the formation of molecular ions when the electrons from the highest energy orbital of the analyte molecule are dislodged by the bombarding electrons. The electronic excitation often causes the molecule to disintegrate into structurally rich diagnostic fragments at the ionization source that allow for the structural determination of the molecule by a single stage mass spectrometry. Because the energy requirement for the excitation of electron in a molecule is discrete, the fragmentation spectra generated by EI at a given energy (e.g., 70 eV) are highly reproducible across laboratories and manufacturers. Therefore, GCMS spectra generated using identical energy of electrons is highly reproducible and amenable for the identification of unknown compounds using automated database finger printing. Where applicable, the combination of superior GC separation coupled with the excellent reproducibility of the diagnostically rich EI mass spectrum, offers a powerful tool for many clinical applications. GCMS is routinely used for screening patients in the toxicology and drug overdose settings without the need for sign out by an on-site mass spectrometry trained toxicologist [28].

A less commonly used ionization technique that can also be coupled to GC is the chemical ionization (CI) technique [33] in which ionization is accomplished by ion-molecule reactions between the analyte molecule and a reagent gas introduced into the EI source. The choice of reagent gas depends upon the relative proton affinities and the ease of energy transfer between the analyte and the reagent gas. For example, in order to promote protonation of the analyte a reagent gas with lower proton affinity than the analyte is employed.

Eluents from the liquid chromatography can be directly coupled to mass analyzers, such as quadrupole mass spectrometer via a host of atmospheric pressure ionization (API) techniques, the most predominant of them being ESI [34], APCI (http://www.chromacademy.com/chromatography-understanding-APCI.html), and APPI [35]. Choice of ionization technique for analyzing LC eluents depends upon the analyte chemistry (class), functional groups, volatility and thermal stability of the analyte, separation chemistry, the solvents, and the flow rate used for, in chromatography.

ESI is a highly versatile ionization technique that can ionize a variety of molecule types, small and large. LC elution is directly coupled to a small metal capillary in the ionization source that is held at a high electric potential (kV) at atmospheric pressure. The droplet that emerges at the capillary tip is distorted into a Taylor cone which emits charged droplet into the atmosphere. Naked ions are desorbed from this droplet via a variety of mechanisms. A generally accepted mechanism [34] for the desorption of small molecules from the charged droplet is the ion evaporation model where the droplet shrinks as it traverse in the atmospheric pressure region due to the evaporation of the solvent molecules. When the Raleigh limit is reached, columbic explosion of the droplet occurs which causes the charged molecules to eject as free ions. For larger molecules the predominant mechanism of ion formation is thought to be represented by charge residue model and/or chain ejection model. In charge residue model, the droplet containing a single large molecule is evaporated to dryness transferring the charge(s) on to the molecule. In the chain ejection model, large folded proteins unfold in a shrinking droplet exposing more sites that can be ionized. Unfolding and the resulting change in conformation make it unfavorable for proteins to reside in the droplet thus ejecting from the droplet carrying a distribution of charge states with it. It is likely that all these mechanisms are at play during an ESI event. Unlike EI mechanism discussed earlier, which is a hard ionization technique that can cause extensive fragmentation of the molecule prior to mass analysis, ESI is a soft ionization technique that allow desorption of cold intact molecule for analysis by mass spectrometer.

APCI and APPI are techniques that are complementary to ESI. APCI (http://www.chromacademy.com/chromatography-understanding-APCI.html) is a type of chemical ionization which has an interface design that is similar to ESI but the ionization and ion desorption mechanisms are very different from ESI. LC eluents are coupled to an APCI probe consisting of a heated (350–550°C) ceramic tube into which the solvent and sample molecules are sprayed using a gas, such as N_2 and evaporated. A corona discharge needle with a high voltage applied to it is introduced into the spray. This creates plasma in the ionization source with very high energy ions (such as N_4, O_2, H_3O^+, OH^-) from air, water, and the solvent. Ion-molecule reactions (such as proton transfer and electrophilic addition) occur between these high energy ions and the gas phase sample molecules resulting in the chemical ionization of the analyte molecules. Although, compared to EI, APCI is a softer ionization technique, it is harder than ESI and produces singly charged ions for smaller and more thermally stable molecules. APCI is best suited for the ionization of small, nonpolar lipid soluble molecules (such as steroids and vitamins) that typically do not ionize well in solution. Since APCI ionization process occurs fully in the gas phase, a notable benefit of the technique is that it is much less prone to ion suppression compared to ESI. Also, ESI performs optimally at lower flow rates (below 500 µL/min) while APCI operates optimally at higher flow rates (above 500 µL/min) allowing the technique to be better suited for coupling with conventional LC. Many manufacturers offer instrumentation with the dual ionization source option capable of operating in ESI and APCI modes.

APPI [35] is a relatively newer alternative soft ionization technique that allows for more efficient ionization of some molecules that are not as efficiently ionized by ESI or APCI. In clinical laboratory, APPI finds application in the analysis of steroid hormones; some steroids are demonstrated to be more efficiently ionized by APPI than with APCI [36]. A part of the appeal of APPI is that it produces cleaner chromatograms because of its higher selectivity for the analyte compared to solvent and sample matrix components. Like with APCI, the first step in APPI is to completely vaporize the LC eluents. Then, in the place of the corona discharge used in APCI, a photon beam generated from a discharge lamp (Krypton, Xenon, or Argon) is made to interact with the vapors of the LC eluents. Photons generated from Krypton have energies that are lower than the ionization energies (IE) of the solvent molecules but higher than the analyte molecules. The molecules (M) absorbs a photon ($E = h\nu$) and become an electronically excited molecule: $M + h\nu \rightarrow M^*$. When the IE of molecule is lower than the energy of the photon, photo ionization occurs by release of an electron from the molecule resulting in an odd electron cation ($M^* \rightarrow M^{*+} + e^-$). During the development of multiplexed methods, such as steroid panel analysis by LC-MS/MS, it is beneficial to evaluate multiple ionization approaches.

Another soft desorption ionization technique is matrix assisted laser desorption ionization (MALDI) [37] that is used primarily in the clinical microbiology laboratory to identify microorganisms, such as bacteria and fungi based on the organism specific membrane protein fingerprints. In MALDI, the sample is cocrystallized onto the target plate (typically a modified or unmodified stainless steel, or gold coated plate) with a matrix that is capable of absorbing energy from a laser and transfer some of that energy into the molecule without decomposition and "assist" with its desorption into the mass analyzer.

A variety of mass analyzers can be used to analyze the intact ions generated by these soft ionization techniques and can be analyzed by a single mass analyzer such a quadrupole or a time of flight (TOF) analyzer. MALDI generates large clusters of peaks in the low m/z range (typically below 500 m/z) with a peak at almost every mass rendering the simple mass spectrum in the small m/z range practically useless for qualitative and quantitative analysis. Both biological matrix as well as the reagent chemicals contributes to these interferences. For MALDI, in addition to the contribution from the biological matrix, these low m/z peaks predominantly originate from the synthetic matrix that is used to promote desorption of large molecules for mass analysis. The utility of MALDI is therefore limited to the molecules in the large molecular weight (MW) range where the impact of interfering small m/z matrix peaks is minimal. Large intact ions generated by MALDI are easily pulsed into a TOF analyzer with an unlimited mass range (in principle). MALDI-TOF is therefore ideally suited for the qualitative analysis of fingerprints of large intact proteins from biological specimen. This technique finds excellent utility in the area of molecular biology for the rapid identification of bacterial fingerprints in infected specimen.

In ESI mass spectra, in addition to the presence of a large variety of small biomolecules from the specimen, the chemical reagents (buffers, detergents, additives) used in the extraction and separation generate a large array of small m/z peaks. A dramatic example of the reagent peaks dominating an ESI mass spectrum is the distribution of detergent peaks observed when detergents, such as Tween 20 are used in the sample preparation steps. Chemical and biological interferences make it impractical for highly specific analysis of small molecules by single stage low resolution mass spectrometry. ESI generated intact biomolecules can be analyzed with very high specificity using tandem mass spectrometry (coupling of multiple low resolution mass spectrometers that have the ability to activate the ions further to generate and analyze highly specific fragment ions) or by the use of single stage high-resolution (HR) mass analyzers, which have the ability to resolve very small differences in m/z ratios between the molecule of interest and the interfering peak. With any of the ionization approaches, the proportion of ion transmission from ionization source to the mass analyzer is inefficient (10–15%) and continues to be a subject of considerable research by the mass spectrometry community [38]. Choice of ionization and the mass spectrometry techniques based on the polarity and MW of analytes is outlined in the Fig. 2.2.

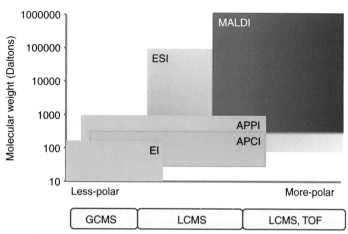

FIGURE 2.2 Choice of Ionization and Hyphenated Mass Spectrometry Techniques on the Basis of Analyte Polarity and MW

9 CHOICE OF MASS ANALYZER

Soft ionization techniques, such as ESI, APCI, and APPI are coupled to the LCMS workflows while EI, a hard ionization technique is used in the GCMS workflows. Soft ionization techniques present cold, intact molecules for mass analysis. When analyzed by a low resolution mass analyzer, such as quadrupole, the intact molecules do not offer diagnostic specificity. High resolution (HR) mass analyzers on the other hand can offer structure specific information from intact molecule based on their superior mass accuracy. In the commonly used low resolution mass spectrometers, such as quadrupole, an additional activation step is required to produce structure specific fragments from the intact molecules generated by the soft ionization technique. In a quadrupole based mass analyzer this is accomplished by the use of two or more mass analyzers arranged in tandem (tandem mass spectrometry).

Many types of mass analyzers [39] are adapted in clinical laboratories. Mass analyzers can be classified in many ways including scanning versus trapping (based on their working principle) or low resolution versus high resolution (based on their performance) etc. Quadrupole mass analyzers [40], the most common type of the scanning mass analyzers applied in the clinical laboratories, use the stability of ion trajectories in oscillating electric fields (DC and RF) to separate the ions according to their m/z ratios. A quadrupole assembly consists of four cylindrical rods set parallel to each other. DC and RF potentials are alternatively applied to the quadrupole rods such that only one ion with a select m/z can pass through the quadrupole in a given scan time frame (typically microseconds per scan). All other ions collide with the rods, and are neutralized and do not register a signal at the detector. Thus a quadrupole analyzer is a mass filter that filters in or filter out ions of specific m/z. Mass spectrum can be acquired in m/z range spanning many hundreds of mass units (full scan mode) to a fraction of a mass unit. To acquire a mass spectrum in a specified mass range, the DC and RF potentials are scanned in constant ratio across the range.

Depending upon the scan range, a full scan cycle can take many seconds to a minute (scan rate) to complete. The wider the scan range the slower the scan rate. Therefore, while scanning a large scan range the analyzer may only sample a typical chromatographic peak a few times. In order to measure a peak reproducibly from injection to injection, it is desirable to have at least 10–15 samplings (points per peak) of a chromatographic peak. This may not be possible in the full scan mode and thus this mode is not practically useful for quantitation. Full scan mode is traditionally used mainly for qualitative, explorative applications. Full scan mode, however, can be a very useful ally in the early phase of assay development—to explore a wider range of the LC elution profile to characterize the analyte and coeluting peaks as well as to qualify reagents used for assay development, especially internal standard lots for the presence of unlabelled material.

For routine clinical analysis, a much higher sensitivity and precision is required of the mass analyzer than what the full scan mode is capable of. In order to increase the number of scans (typically 10–15 scans) of the eluting chromatographic peak during the mass analysis, a narrow (targeted) mass range around the specific m/z of the analyte(s) and internal standard(s) is selected for acquisition in the first set of quadrupoles. The selected ion is directed to a quadrupole based collision chamber (second quadrupole in a triple quadrupole configuration) for activation where the intact molecule is subjected to structure specific fragmentation. Most commonly, in the collision chamber, the ion is made to collide with a neutral gas (usually an inert gas like Nitrogen, Argon, or Helium) that is introduced into the chamber. This ion activation process is referred to as collision induced dissociation (CID) or collision associated dissociation (CAD) [41,42]. The ion-neutral collisions results in the conversion of part of the translational energy of the charged molecule into its internal energy which trigger the dissociation of molecule into fragments (reaction). Diagnostic fragment(s) that is unique to the molecule is then mass selected by the third analyzer (third set of quadrupole in triple quadrupole configuration) for detection. Thus the first quadrupole is programed to select ions of specified narrow m/z ranges around the m/z of the analyte (allowing very high scan rates of the selected m/z range) that emerge from the LC and direct the ion(s) one at a time into the collision chamber where no mass filtering occurs (set at RF only mode) to facilitate fragmentation of the selected ion. The fragmentation spectrum from the third analyzer can be acquired in full scan mode during method development stage to select the most specific and sensitive fragments. After the most intense and specific fragment ion(s) is identified from the full scan fragmentation spectrum, the third mass analyzer can be set to analyze only the selected fragment m/z. This mode of selecting an intact molecule m/z in the first analyzer, activating the molecule in the second analyzer to generate reaction products and analyzing a select fragment (reaction product) using the third analyzer is referred to as single reaction monitoring (SRM) or MRM when the acquisition method contains many channels of SRMs. MRM mode allows for the quantitation of multiple peaks from a single chromatographic run allowing for simultaneous detection and quantitation of many analytes (multiplexing). A chromatogram obtained for a single molecule-fragment pair is referred to as a "transition." Branching ratio (ratio of intensities of two transitions of an analyte) can be used to further monitor the specificity of the assay during routine analysis [26]. The ability of the triple quadrupole mass spectrometers to allow for the use of this highly sensitive, specific, multiplexable, and fast analysis mode together with the ease of coupling the soft ionization techniques with liquid chromatography makes MRM the most commonly used approach for routine high throughput multiplexed clinical testing by tandem mass spectrometry.

In order to preserve the high level of specificity offered by this tandem mass spectrometry approach, it is critical to ensure that there is no "cross talk" (when ions from one scan event are still present in the collision cell when a second SRM transition is taking place) between two transitions [11,12,26]. The prospect for cross talk can increase with the number of transitions in the acquisition method. "Cross talk" criteria can be used to evaluate the different mass spectrometers during the platform selection process. Some instruments are less prone to cross talk than others and some instruments can handle multiple transitions better than other instruments. This must be verified with a carefully designed test during instrument demonstration by the manufacturers. One has to take extreme precaution and diligent reassessments when adding or subtracting a transition to an acquisition method. MS acquisition method parameters for a triple quadrupole mass spectrometry platform must be carefully validated for cross talk and optimal number of MRM transitions.

Other types of mass analyzers used in the clinical laboratories include trapping mass devices, such as quadrupole ion trap (cylindrical and linear), Orbitrap and Fourier transform ion cyclotron resonance (FTICR), and TOF mass analyzers. Orbitrap, FTICR, and TOF are high-resolution mass spectrometers (HRMS), which offer superior specificity because of their superior mass resolving power and mass accuracy [43]. Chapter 12 is dedicated to the topic of HRMS in clinical laboratories.

In a TOF mass analyzer [44], soft ions generated by MALDI or other ionization processes are pulsed into the drift tube at which point all the ions entering the drift tube has the same kinetic energy. In the drift tube, ions separate according to their drift velocities dictated by the m/z ratios. Smaller ions travel faster and the larger ions travel slowly. An array detector is used to record the event. Ions hit the detector at different times dictated by their m/z ratios. The drift times are then translated to the m/z ratios and the spectra is generated for that pulse. Theoretically, the TOF analyzer has unlimited mass range because in principle, any TOF for any ion can be recorded and converted to the m/z. In practice, however, the mass range is determined by the length of the flight tube which is limited by practicality. The flight length of a drift tube is increased by the use of "reflectron" mirrors.

Orbitrap [43,45], the most recent avatar of the HRMS introduced to the clinical laboratory in the last decade resolves ions based on the frequency of ion motion in electrostatic field while in an ion cyclotron resonance analyzer, the ions are separated based on their motion in a magnetic field.

Decision to choose a specific mass analyzer depends upon the application, the desired performance of the assay and the cost. For routine LC/MS quantitation the key analytical considerations are sensitivity of the mass analyzer (application specific—ultrasensitive analyzers may be needed for the quantitation of low levels of analytes while moderately sensitive analyzers may be sufficient for applications, such as drug monitoring), the speed of analysis (to allow sufficient points per peak in the chromatographic time frame), linear dynamic range (to ensure that the detector response is proportional to the concentration of the analyte across a desired target range), mass accuracy, and mass resolving power within the mass scale.

While triple quadrupole tandem mass spectrometers offer adequate to optimal sensitivity (in MRM mode), optimal analysis speed and optimal linear dynamic range (which makes these instruments the most versatile mass analyzers used in clinical laboratory), compared to HR mass analyzers they have limited mass accuracy and resolving power. Because of the lower resolving power and limited mass accuracy of the quadrupole mass spectrometers, specificity of the assay that uses these analyzers must be validated very carefully and rigorously before launching the test. Any extra efforts invested in the investigation of assay interferences during validation will pay off handsomely during routine patient testing. A good example of an application that requires extensive interference study during validation is the analysis steroids panel by LC-MS/MS. Many steroids differ in 2 mass units and are prone to cointerference from the many biological isomers and isobars as well as from isotopes [12]. In this case the chromatography and the mass acquisition parameters need to be optimized carefully and validated with simulated samples that represent as many different clinical scenarios as possible. Relative levels of the different steroids vary significantly from patient to patient.

Table 2.1 The Analytical Performance Characteristics of the Different Mass Analyzers

Mass Analyzer Type	Resolving Power	Mass Range	Mass Accuracy	Sensitivity	Speed	Dynamic Range	Cost
FTICR	xxxxx	xxxx	xxxxx	xxx	x	xxx	$$$$$$
FT-Orbitrap	xxxx	xxx	xxxx	xxx	xx	xxx	$$$$$
TOF	xxx	xxxxx	xxx	xxxxx	xxx	xxxx	$$$
Quadrupole[a]	x	x	x	xxxxx	xxxxx	xxxxx	$
Quadrupole ion trap	x	x	x	xxx	xxx	xxx	$

FTICR, Fourier transform ion cyclotron resonance; SRM, single reaction monitoring; TOF, time of flight.
[a]In triple quadrupole configuration, SRM mode.

Compared to quadrupole mass analyzers, HR mass analyzers offer superior mass accuracy and mass resolving power across a wide mass range but may not offer sufficient linear dynamic range or the speed necessary for precise measurement of analytes that elute from a LC or GC. However, because of the ability of HR mass analyzers to separate ions with very high resolution, many qualitative and semi-quantitative applications in clinical laboratory favor the use of HR mass analyzers [43].

Table 2.1 shows the analytical performance characteristics of the different mass analyzers. The overall analytical performance of an assay depends in large part on the suitability and the level of optimization of the sample preparation methods.

A key consideration in the choice of HRMS for routine clinical application is the high initial capital expense of the platform. Fig. 2.3 shows the cost versus performance of the mass analyzers of interest to the clinical laboratories. Hybrid instruments that combine quadrupole mass analyzers with high resolution and higher mass range mass analyzers, such as Q-TOF, Q-Orbitrap offer some of the benefits of both the instrument types.

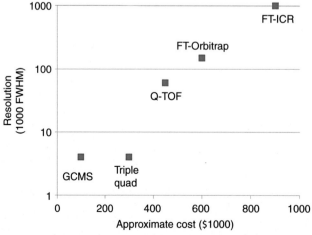

FIGURE 2.3 The Cost Versus Performance of the Mass Analyzers of Interest to the Clinical Laboratories

10 SUMMARY

Because of the significant benefits that a mass spectrometry platform brings to the clinical laboratory, there is very little evidence to the contrary that mass spectrometry will reach most, if not all clinical laboratories in the near future. By offering some technical insights and pointing to some key resources on the various topics, this chapter aims to add to the discussions among laboratorians who are considering implementation of a mass spectrometry platform in the clinical laboratory for the first time. A key goal of this chapter, as many have pointed out in the past and some referenced here, is to emphasize that while the mass spectrometry and the allied instrumentation may be the most capital expensive components of the analytical platform, choosing and adopting the "right sample preparation strategies" is likely to be more consequential to the end goal than choosing the "right mass spectrometer." Decision to implement mass spectrometry for the first time in the laboratory is a decision to significantly expand the human skills and the technical capacity of the laboratory. Each laboratory must determine its own unique technical and logistical realities very diligently when implementing mass spectrometry platform and validating the assay. Soundness of those initial decisions will keep the assay running for years with the performance metrics established during the validation.

REFERENCES

[1] Shushan B. Is automation the key to LC-MS/MS migrating from research and reference labs to hospitals and clinics? Chromatogr Today 2015. On line publication: https://www.chromatographytoday.com/articles/mass-spectrometry-and-spectroscopy/41/dr_bori_shushan/is_automation_the_key_to_lc-msms_migrating_from_research_and_reference_labs_to_hospitals_and_clinics_/1839/

[2] Shushan B. LC-MS offers significant upside for clinical labs. Genet Eng Biotechnol N 2013. On line article: http://www.genengnews.com/keywordsandtools/print/1/32733/

[3] Janetto P. Liquid chromatography tandem mass spectrometry: the Swiss army knife for clinical laboratories. Clin Lab N 2015. AACC online article: https://www.aacc.org/publications/cln/articles/2015/july/liquid-chromatography-tandem-mass-spectrometry

[4] Zhang V. Should we bring this test in house? Clin Lab N 2015. Online article: https://www.aacc.org/publications/cln/articles/2015/june/should-we-bring-this-test-in-house

[5] Nair H, et al. Liquid chromatography–tandem mass spectrometry work flow for parallel quantification of methotrexate and other immunosuppressants. Clin Chem 2012;58:943–5.

[6] Arnaud CH. Mass spec welcome in clinical labs. Chem Eng N 2015;93(20):32–4.

[7] Rockwood A. Future of clinical mass spectrometry. Clin Lab N 2015. AACC Online article: https://www.aacc.org/publications/cln/articles/2015/february/future-of-mass-spec

[8] Domin M, et al. Ambient ionization mass spectrometry. Cambridge: The Royal Society of Chemistry; 2015. P001-P004. http://www.bc.edu/sites/libraries/facpub/domin-ambient/book.pdf.

[9] Balog J, et al. Intraoperative tissue identification using rapid evaporative ionization mass spectrometry. Sci Transl Med 2013;5(194). 194ra93.

[10] Boland M. The science of airport bomb detection: mass spectrometry. 2015. http://scitechconnect.elsevier.com/airport-bomb-detection-mass-spectrometry/. Elsevier online article: http://scitechconnect.elsevier.com/airport-bomb-detection-mass-spectrometry/

[11] Clarke W. Challenges in implementing clinical liquid chromatography–tandem mass spectrometry methods— seeing the light at the end of the tunnel. J Mass Spectrom 2013;48:755–67.

[12] Grebe S, et al. LC-MS/MS in the clinical laboratory—where to from here? Clin Biochem Rev 2011;32(1):5–31.

[13] Murrey PR. What is new in clinical microbiology—microbial identification by MALDI-TOF mass spectrometry. J Mol Diagn 2012;14(5):419–23.

[14] Scherl A. Clinical protein mass spectrometry. Methods 2015;81:3–14.

[15] Grantt RP, et al. From lost in translation to paradise found: enabling protein biomarker method transfer by mass spectrometry. Clin Chem 2014;60(7):941–4.

[16] Henderson CM. Measurement by a novel LC-MS/MS methodology reveals similar serum concentrations of vitamin D–binding protein in blacks and whites. Clin Chem 2016;62:179–87.

[17] Vincente FG, et al. Training and competence in LC-MS/MS laboratories. Clin Lab N 2015. AACC online article: https://www.aacc.org/publications/cln/articles/2015/august/training.aspx

[18] Sargent M. Guide to achieving reliable quantitative LC-MS measurements. 1st ed. London, UK: RSC Analytical Methods Committee; 2013.

[19] Hetu PO, et al. Successful and cost-efficient replacement of immunoassays by tandem mass spectrometry for the quantification of immunosuppressants in the clinical laboratory. J Chromatogr B Analyt Technol Biomed Life Sci 2012;883–884:95–101.

[20] Marney LC, et al. Isopropanol protein precipitation for the analysis of plasma free metanephrines by liquid chromatography–tandem mass spectrometry. Clin Chem 2008;54(10):1729–32.

[21] Strathmann FG. Current and future applications of mass spectrometry to the clinical laboratory. Am J Clin Pathol 2011;136:609–16.

[22] Annesley TM. Ion suppression in mass spectrometry. Clin Chem 2003;49(7):1041–4.

[23] Volmer DA, et al. Ion suppression: a major concern in mass spectrometry. LC GC N Am 2006;24(5):495–510.

[24] Cappiello A. Use of nonvolatile buffers in liquid chromatography/mass spectrometry: advantages of capillary-scale particle beam interfacing. Anal Chem 1997;69(24):5136–41.

[25] Bylda C, et al. Recent advances in sample preparation techniques to overcome difficulties encountered during quantitative analysis of small molecules from biofluids using LC-MS/MS. Analyst 2014;139:2265–76.

[26] Vogeser M. Pitfalls associated with the use of liquid chromatography–tandem mass spectrometry in the clinical laboratory. Clin Chem 2010;56(8):1234–44.

[27] Chase DH, et al. Use of tandem mass spectrometry for multianalyte screening of dried blood specimens from newborns. Clin Chem 2003;49(11):1797–817.

[28] Rhea JM, et al. Direct total and free testosterone measurement by liquid chromatography tandem mass spectrometry across two different platforms. Clin Biochem 2013;46(7–8):656–64.

[29] Nair H, et al. Clinical validation of a highly sensitive GC-MS platform for routine urine drug screening and real-time reporting of up to 212 drugs. J Toxicol 2013;2013:7.

[30] Scheafer EJ, et al. In: De Groot LJ, Chrousos G, Dungan K, editors. The measurement of lipids, lipoproteins, apolipoproteins, fatty acids, and sterols, and next generation sequencing for the diagnosis and treatment of lipid disorders and the prevention of cardiovascular disease. South Dartmouth (MA): http://www.endotext.org/"MDText.com, Inc; 2016. p. 1–69.

[31] Rappold B, et al. HILIC-MS/MS method development for targeted quantitation of metabolites: practical considerations from a clinical diagnostic perspective. J Sep Sci 2011;34(24):3527–37.

[32] McLafferty FW, Turecek F. Interpretation of mass spectra. 4th ed. Sausalito, CA: University Science Books; 1993.

[33] Munson MSB, Field FH. Chemical ionization mass spectrometry. I. General introduction. J Am Chem Soc 1966;88(12):2621–30.

[34] Wilm M. Principles of electrospray ionization. Mol Cell Proteomics 2011;10(7).

[35] Wang C. The ionization technology of LC-MS, advantages of APPI on detection of PPCPs and hormones. Austin Chromatogr 2015;2(2):1032.

[36] Stolz BR. An improved micro-method for the measurement of steroid profiles by APPI-LC-MS/MS and its use in assessing diurnal effects on steroid concentrations and optimizing the diagnosis and treatment of adrenal insufficiency and CAH. J Steroid Biochem Mol Biol 2016;162:110–6.

[37] Murray PR. What Is new in clinical microbiology—microbial identification by MALDI-TOF mass spectrometry. J Mol Diagn 2012;14(5):419–23.

[38] Page JS. Ionization and transmission efficiency in an electrospray ionization–mass spectrometry interface. J Am Soc Mass Spectrom 2007;18(9):1582–90.

[39] Smith GH. In: Barner-Kowollik Christopher, Gruendling Till, Falkenhagen Jana, Weidner Steffen, editors. Mass analysis: mass spectrometry in polymer chemistry,. 1st ed. Weinheim, Germany: Wiley-VCH Verlag GmbH & Co. KGaA; 2012.

[40] Miller PE, et al. The quadrupole mass filter: basic operating concepts. J Chem Educ 1986;63(7):617.

[41] Jhonson AR, et al. Collision-induced dissociation mass spectrometry: a powerful tool for natural product structure elucidation. Anal Chem 2015;87(21):10668–78.

[42] McLuckey SA. Slow heating methods in tandem mass spectrometry. J Mass Spectrom 1997;32:461–74.

[43] Brenton AG. Accurate mass measurement: terminology and treatment of data. J Am Soc Mass Spectrom 2010;21(11):1821–35.

[44] Vestal ML. Modern MALDI time-of-flight mass spectrometry. J Mass Spectrom 2009;44(3):303–17.

[45] Hu Q, et al. The orbitrap: the new mass spectrometer. J. Mass Spectrom 2005;40:440–3.

SAMPLE PREPARATION TECHNIQUES FOR MASS SPECTROMETRY IN THE CLINICAL LABORATORY

3

J. Stone

Center for Advanced Laboratory Medicine, University of California San Diego Health System, San Diego, CA, United States

1 WHY IS SAMPLE PREPARATION NEEDED FOR LC-MS/MS?

LC-MS/MS sample preparation is needed:

1. To simplify complex biological matrices (serum, urine, etc.) for optimal:
   a. Method analytical performance (precision, detection limits, selectivity). The focus is on the visible and well-defined analyte(s).
   b. Method robustness by removing sample matrix such that LC columns have acceptable lifetimes, LC components do not require unscheduled maintenance, and the MS/MS has a predictable maintenance-free interval. The focus is on the often invisible and incompletely defined matrix.
2. To adjust analyte concentration(s) for the detection limit of the LC-MS/MS (concentrate or dilute).
3. To exchange the sample matrix—rich in proteins, lipids, and polar, metabolic waste products—to a simpler solvent/water injection solution compatible with the LC method.

2 SELECTING A SAMPLE PREPARATION PROTOCOL

The amount of effort to spend on developing a sample preparation protocol and the acceptable sample preparation cost per reportable test can be a controversial topic between clinical chemists who use LC-MS/MS. Some advocate that the latest generation of LC-MS/MS instruments can tolerate minimal sample cleanup while maintaining good performance and others are firm believers that routinely investing in more extensive sample cleanup will pay off in methods with greater robustness and more predictable instrument operation. This debate is not easily resolved and is addressed at several points in this chapter to provide the broadest perspective.

New users of LC-MS/MS usually want to keep sample preparation simple—dilution (DIL) for urine and protein precipitation (PPT) for serum/whole blood—as long as method performance goals can be met. With higher-sensitivity LC-MS/MS instruments in particular, simple sample preparation protocols

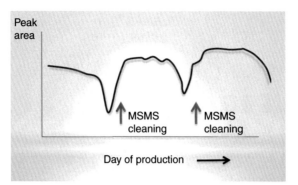

FIGURE 3.1 The Time Course of MSMS Signal Degradation Caused by Injection of Extracted Biological Samples and Recovery After Cleaning the MSMS Interface

Internal standard peak areas from each batch are plotted by date. As additional matrix is deposited on the MSMS interface, signal decreases and eventually falls below an acceptable threshold. The MSMS is vented, interface hardware is cleaned, the MSMS is pumped down, and the expected signal is regained. Then the cycle repeats. Typically 12–24 h of production, at a minimum, are lost with each cleaning.

may work well, for example, testing for drugs of abuse in urine, immunosuppressants in whole blood, or 25-hydroxy vitamin D in serum. As less analyte can be detected with higher-end instruments, less matrix is injected. Whatever the degree of sample cleanup, the natural history of LC columns and MS/MS hardware is a loss of performance over time as additional samples with residual biological matrix are injected. LC peak area, shape, width, resolution, and retention time (Rt) degrade and the MS/MS delivers less signal as matrix components adhere to the column stationary phase and to the MS/MS source, interface region, and eventually, quadrupole rail. When a new column is installed or after the MS/MS interface is cleaned—performance can be restored. Then the cycle repeats, as diagrammed in Fig. 3.1.

Quantifying robustness can be difficult because a sample preparation protocol may work well initially, and only as matrix gradually accumulates on the instrument over days and weeks—will MS/MS sensitivity degrade. This lag phase complicates the attribution of sensitivity loss to the sample preparation protocol. Of course when the lag phase is long enough and the degradation of instrument performance occurs after an acceptable interval—the method becomes robust by definition.

This chapter recommends the fit for purpose approach for developing and validating sample preparation protocols. Minimal sample preparation may be appropriate for a given analyte, instrument, and set of performance and operational expectations. Under other circumstances, extensive sample cleanup may be worthwhile to avoid unscheduled instrument downtime and meet more demanding performance criteria.

Sample preparation protocols are available from LC-MS/MS and extraction media vendors, as well as the scientific literature, but the long term effect of the protocol on instrument performance is rarely stated. Protocols with proven robustness that are recommended by clinical colleagues are particularly valuable. Suggestions for metrics to track as a measure of robustness are given in the section on the evaluation of protocols.

3 DESCRIPTION OF COMMONLY USED CLINICAL SAMPLE PREPARATION PROTOCOLS

Sample preparation protocols can be categorized as simple and low cost (DIL and PPT), more complex but still relatively low cost (liquid–liquid extraction or LLE) and having variable complexity but higher cost because commercially available extraction media are required (solid phase extraction or SPE, supported liquid extraction or SLE, phospholipid removal media or PLR) [1]. A new commercial product, released in 2014, is immobilized coating extraction media (AC plate) from Tecan [2]. This product has potential as an easily automated sample preparation technique for LC-MS/MS but is not discussed further here because too few applications are available at this time.

SPE can be integrated into LC systems and performed online or through integration of a sample preparation robot with the autosampler (Gerstel MultiPurpose Sampler-MPS). TurboFlow and Symbiosis are commercially available online extraction systems that automate most sample preparation steps and can run on multiple LC systems connected to one MS/MS (LC multiplexing) to increase LC-MS/MS throughput [3,4].

A variety of novel and more elaborate sample preparation protocols with increased selectivity, such as immunoaffinity extraction and molecular imprinted polymers (MIPs) have been developed but are not yet widely used in clinical laboratories and are not discussed in this chapter [1].

3.1 SIMPLE—DILUTION SAMPLE PREPARATION PROTOCOL (DIL)

DIL is appropriate for sample types with low protein matrix, such as urine or cerebrospinal fluid. As with all sample prep methods, the first step in a DIL protocol is addition of internal standard (IS). Sample, and diluent (water or mobile phase) are added, and enzyme if glucuronide hydrolysis is being performed. Containers are sealed and mixed, incubated at higher temperature if necessary to facilitate hydrolysis, and centrifuged before injection on the LC-MS/MS. Centrifuging is an important step to prevent over pressure from particulates clogging inline filters, guard columns, and columns.

The advantages of DIL are simplicity, minimal sample loss (few or no transfers), inexpensive reagents, relatively small amount of labor involved, applicability to a broad range of analyte chemistries, and ease of automation. Disadvantages are the inability to concentrate to achieve a more sensitive lower limit of quantitation (LLOQ) and more ion suppression than would occur with SPE, SLE, or LLE [1]. Depending on the dilution factor and whether enzymatic hydrolysis is used, DIL may yield a relatively dirty injection matrix (e.g., 5 times dilution or less), that can shorten column lifetimes and require more frequent cleaning of the MS/MS or, in contrast a relatively trouble-free protocol if higher dilutions are feasible (e.g., 100 times dilution or more), particularly if used with a divert valve program that directs LC flow to waste except when analytes are eluting from the LC column to the MS/MS.

3.2 SIMPLE—PROTEIN PRECIPITATION SAMPLE PREPARATION PROTOCOL (PPT)

PPT is appropriate for high protein matrices—serum/plasma/whole blood. Acids, organic solvents, salts, and metals are used as precipitating agents [5–7].

Precipitation occurs because the change in pH or hydrophobicity alters interactions between the protein and the aqueous environment or through binding of salts or metals to protein functional groups such that intramolecular interactions are disrupted and the proteins denature, aggregate, and fall out of solution [5]. Centrifugation or filtration is used to create a particulate free supernatant or filtrate.

In a typical PPT experiment—serum and IS are mixed and equilibrated. The precipitation reagent is added, the container is sealed, mixed, and centrifuged before introducing the sample to the LC-MS/MS. An automation friendly option is precipitation and mixing in the wells of commercially available 96-well PPT plates. After mixing, vacuum or positive pressure is applied to the plate, precipitates are retained above the frit at the bottom of the wells, while particulate-free filtrate moves through to a collection plate [8].

Including the IS in the precipitating agent to reduce the number of pipetting steps is common [9–12]. One concern with adding IS to the precipitation reagent, if the analyte has extensive or high-avidity protein binding, is the potential for differential or more variable recovery of the analyte versus the IS.

The advantages of PPT are simplicity, minimal loss of sample (few or no transfers), inexpensive reagents, relatively small amount of labor, applicability to a wide range of analyte chemistries, and ease of automation [1].

Disadvantages of PPT are the inability to concentrate analytes, significant ion suppression caused in large part by serum phospholipids, and lack of robustness [1,13–23].

Phospholipids are present in high (mg/mL) concentrations in serum, they are at best minimally removed in PPT protocols, and are notorious for causing ion suppression with electrospray ionization (ESI) [13–22]. The degree to which ion suppression and contamination of the LC-MS/MS with matrix is an acceptable or unacceptable consequence of PPT sample preparation is debated within the clinical laboratory community, as many laboratories use PPT to test dozens to hundreds of serum 25-hydroxy vitamin D and whole blood immunosuppressant samples a day. The difference that details of the PPT protocol, as well as LC conditions, may make to the robustness of a method, as well as variable definitions of robustness are factors in this debate. These issues are discussed in the Section 4.

3.3 MORE COMPLEX—LIQUID–LIQUID EXTRACTION SAMPLE PREPARATION PROTOCOL (LLE)

In LLE, an immiscible organic solvent and an aqueous body fluid are mixed. Nonpolar compounds partition to the organic phase, leaving polar moieties in the aqueous phase. LLE is compatible with serum, urine, and many other sample types.

In the LLE workflow, samples are mixed with IS and a buffer. The immiscible organic solvent is added, samples are mixed vigorously to transfer analytes from sample to organic phase, and centrifuged to separate the layers. After centrifugation, the organic layer is transferred and evaporated to dryness. A reconstitution solution is added to the evaporated extract, the container is sealed and mixed to solubilize the analytes and provide an injection matrix compatible with the LC method.

Advantages of LLE include low cost of materials and high selectivity, the potential to concentrate analytes while performing extensive cleanup of matrix, including very effective removal of phospholipids [1,23]. Disadvantages are the complexity of the process, slow throughput with manual LLE, high labor costs, a need for skilled labor, longer and more complex method development, and difficulty in automating with 96-well format [1]. Additionally, as polar compounds are poorly extracted, metabolites may have low recovery or require glucuronide/sulfate hydrolysis prior to LLE [1]. Incomplete separation of the layers—formation of an emulsion—occurs occasionally and breaking the emulsion by freezing or adding salt can be problematic. A clean, reproducible, transfer of the organic layer to another container for evaporation is labor intensive and technique dependent, therefore a major source of variability. At a minimum, a chemical hood and an evaporation manifold are needed. For higher throughput, a mechanical mixer or shaker is useful.

3.4 COMMERCIAL MEDIA—PHOSPHOLIPID REMOVAL SAMPLE PREPARATION PROTOCOL (PLR)

Commercially available media that retains both precipitated proteins and phospholipids but not analytes is a relatively recent development (2008). PRL media are available in 96-well format and in cartridges from several vendors, including Hybrid-SPE (Sigma-Aldrich), Ostro (Waters Co.), Captiva ND (Agilent), Phree (Phenomenex), and Maestro (SPEware Co.). The goal of adding robustness to PPT through removal of phospholipids while at the same time retaining the advantages of speed, simplicity, and high throughput drove the commercialization of these products [24]. Ligands in the plate bed bind either the polar phosphate group or the nonpolar fatty acids to remove > 90% of serum phospholipids. For example, zirconia coated silica particles in the Hybrid-SPE plate bind to the phosphate moiety through a Lewis acid:base functionality [23,24]. Differences between these products in the ability to remove polar lysophospholipids as well as more nonpolar phospholipids, in losses of analyte to nonspecific binding, and in recovery of acidic, neutral, and basic analytes have been described [23].

As with PPT, IS and serum/plasma can be mixed in the plate, followed by addition of the precipitating reagent and further mixing. Vacuum or positive pressure is applied, with retention of precipitated proteins and phospholipids in the plate, and flow through of the filtrate containing analytes to a collection plate. Constraints on the nature of the precipitation reagent and on the ratio of serum to precipitant volumes are specific to each vendor [23,24].

The advantages and disadvantages of PLR are the same as those of PPT—but with a major reduction in matrix effect, improvement in robustness, and increased cost.

3.5 COMMERCIAL MEDIA—SUPPORTED LIQUID EXTRACTION SAMPLE PREPARATION PROTOCOL (SLE OR SALL)

SLE or support assisted liquid–liquid extraction (SALL) is a means of immobilizing and therefore greatly facilitating automation of LLE [1]. With SLE the aqueous sample is added to the cartridge or plate and spreads out as small droplets widely dispersed in the bed of finely milled diatomaceous earth particles. When immiscible organic solvent is applied, nonpolar analytes partition from the polar aqueous sample with high efficiency in to the solvent. The technique has applicability to a wide range of analytes and can be effective at removing phospholipids. SLE plates and cartridges are available from a number of vendors, including Agilent, Biotage, Merck-Millipore, Thermo-Fisher, and United Chemical Technologies.

Sample, IS and a buffer are mixed. The maximum volumes of sample and aqueous diluent that can be used are dictated by the bed mass of the diatomaceous earth. For example, the Biotage ISOLUTE SLE+ plate with a 400 mg bed allows a maximum load volume of 400 μL [25]. A 1:1 dilution of sample with buffer is recommended, limiting the theoretical maximum sample volume to 200 μL [25]. Higher ratios of sample to buffer may be feasible but recovery experiments are necessary for optimization. The sample:buffer mixture is transferred to the SLE plate or cartridge, low vacuum is applied, and unlike SPE, the entire sample is absorbed by the bed. There is a 5 min wait while the sample distributes throughout the SLE bed. Then a volume of water-immiscible organic solvent, determined by the bed mass, is used for elution. With the 400 mg bed mass example—two applications of 900 µL would be used—the eluent flows through by gravity feed and analytes partition into the eluent which is captured in tubes or plates [25]. As SLE elution solvents are immiscible with water, evaporation of the eluate and reconstitution in a compatible injection solvent is necessary for analysis with reverse phase LC.

The advantages of SLE/SALL are ease of automation, particularly when compared to automation of LLE, effective removal of matrix, applicability to a wide range of analyte polarities, and simplicity compared to SPE [1]. The disadvantages are, like other commercial extraction media, increased cost compared to DIL, PPT, and LLE, requirement for a vacuum manifold and evaporator and limited concentrating capability (e.g., maximum sample volume 200–300 µL, if reconstituted in 100 µL, at most a 2.5–3.0 fold concentration).

3.6 COMMERCIAL MEDIA—SOLID PHASE EXTRACTION SAMPLE PREPARATION PROTOCOL (SPE)

SPE, whether manual, automated on a liquid handler or online, is essentially a low resolution chromatographic process. Like LLE, SPE was in wide use for HPLC-UV and GC-MS methods prior to the advent of LC-MS/MS. There is abundant literature on SPE sample preparation for LC-MS/MS as well as application notes and extensive support from SPE media vendors [1,26–29]. SPE chemistries for use with aqueous matrices are categorized as ion-exchange, reverse-phase, HILIC, or mixed-mode. Reverse phase SPE is less selective than mixed-mode SPE or LLE and is primarily useful for removing salts and polar matrix components. Nonpolar wash solutions that would remove neutral interferences from reverse phase SPE will also wash analytes to waste, as the only retention mechanism is adsorption to the stationary reverse-phase.

In contrast, mixed-mode SPE becomes highly selective by including an anion or cation exchange moiety in the same bed with the reverse-phase component (nonpolar polymer or C18 bonded to silica). This dual functionality is a powerful tool for removing matrix, because charged analytes can be retained with the ion exchange moiety while matrix is removed from the reverse phase with nonpolar wash solutions.

A vacuum or positive pressure manifold for cartridges or plates is necessary to perform SPE. Positive pressure moves fluids through the SPE bed more reliably than does vacuum. SPE plates with a small bed mass and hold-up volume (e.g., Waters µElution plate) can be eluted with < 0.5 mL of methanol or acetonitrile such that evaporation is optional [30]. But most SPE protocols require evaporation of water immiscible organic elution solvents so the analytes can be reconstituted in a smaller volume of a solvent:water mixture that is compatible with reverse-phase LC. The first steps for SPE are to mix IS with sample and an application buffer at a pH that maximizes retention of analytes on the stationary phase. Classically, the SPE bed requires conditioning with methanol or acetonitrile to activate the stationary phase. Then the bed is equilibrated with aqueous application buffer.

Most vendors now offer SPE media with a polymer acting as both the structural support and as the nonpolar functionality of the stationary phase, an alternative to C18 bonded to silica. Polymer SPE does not necessarily require conditioning. This can save both solvent and time. Comparing reproducibility and recovery with and without preconditioning is a good precaution. Another advantage of polymer based SPE is that drying of the bed does not adversely affect analyte retention, unlike silica-based SPE.

The diluted sample is loaded onto the cartridge/plate, with attention to the flow rates recommended by the SPE vendor. Analytes adsorb to the stationary phase and the liquid sample flows to waste. In general, slower flow is better with SPE to allow sufficient time for equilibration throughout the bed and interaction with all retention mechanisms. One or more wash solutions to remove matrix and exogenous interferences are applied and eluted through the SPE bed. After washing, the SPE bed is dried with air or nitrogen to remove residual water and solvent. The waste container is replaced with a collection container and elution solvent is applied to flush analytes from the stationary phase into the collection vessel. The eluate is evaporated and a reconstitution solution is added, the containers are sealed, mixed, and introduced to the LC-MS/MS.

The chief advantages of SPE are the capability to concentrate analytes and remove matrix, although SPE that is not specifically designed to remove phospholipids from serum can be less selective than LLE or SLE in this regard. SPE is relatively easy to automate. Disadvantages include cost, complexity of method development and production process, and the longer time required compared to DIL, PPT, PLR, or SLE. SPE has some degree of parallel processing and is often less technically demanding than LLE, so handling large numbers of samples may be easier with SPE than LLE.

3.7 ONLINE SPE

Online extraction traps analytes as they flow through an extraction column inline with the LC system while flushing matrix to waste. Then the mobile phase, and typically flow direction, are switched to elute analytes to either an analytical column for separation or directly to the MS/MS [31]. Samples are applied to the extraction column through standard LC flow or using turbulent flow LC (TFLC). With TFLC, the flow characteristics and retention mechanisms are fundamentally different than with standard LC [32]. With large (50–100 mm) particles, high flow rates (5–8 mL/min) and short, narrow columns (1 × 50 mm), the laminar flow of standard LC becomes instead a highly chaotic, turbulent, plug flow with faster mass transfer and therefore higher efficiency. Of significance for small molecule analysis, slower diffusing macromolecules (proteins) are excluded from, while small molecules can still interact with, the stationary phase, allowing for direct injection of high protein matrices, such as serum [32]. Therefore, only minimal sample preparation—dilution with IS or PPT for release from binding proteins—is necessary with TFLC before introducing samples to the LC-MS/MS. TFLC is commercially available from Thermofisher Scientific with Turboflow columns (patented in 1997).

The Gerstel MPS automates SPE by integrating a sample preparation robot into the autosampler, rather than by using LC flows to perform SPE. One sample is extracted with SPE while the previous sample is running on the LC-MS/MS. Throughput is therefore constrained by the minimum time in which an SPE can be performed or the length of the LC run, whichever is longer.

Additional LC pumps, switching valves, and software beyond a basic LC system are necessary to control flows and timing for all online extraction [31,32]. Simpler in-house developed systems are lower cost while more complex, commercial systems are significantly more expensive and, with LC multiplexing, have much higher throughput [3,4].

The advantages of online SPE are the minimal hands-on time required per sample, the potential for direct sample injection with matrix removal online, capability for high-throughput, and compatibility with thermally labile analytes and small sample volumes (no evaporation is required and the entire sample can be injected and concentrated online). The disadvantages are the greater complexity of LC plumbing and software and therefore additional expertise required for development and maintenance, higher cost of equipment and per sample, fewer options for improving selectivity than with offline SPE or LLE, and increased risk for carryover.

4 EVALUATION OF SAMPLE PREPARATION PROTOCOLS

4.1 EVALUATING CHROMATOGRAPHY

The first test of any sample preparation protocol is comparing the chromatography of an extracted biological matrix sample to that of neat standards in injection solvent. It is fairly common to discover that an extracted sample has symptoms of insufficient matrix removal and/or interferences, such as short term and/or long term noise/elevated baseline, peak asymmetry, decreased peak efficiency (wider peaks), or interfering peaks.

Problems with peak shape and Rt for early eluting peaks may be related to composition of the injection solvent matrix (see Section 6). Problems with baselines and interfering matrix peaks indicate more matrix removal and/or LC optimization is needed. Isobaric interfering peaks from structurally similar compounds, such as testosterone and dehydroepiandrosterone (DHEA) will usually require separation by LC rather than by low resolution sample preparation techniques.

4.2 EVALUATING SIGNAL TO NOISE (S/N) AT THE LLOQ

If the chromatography is acceptable, the next question to address is S/N at the LLOQ. The Clinical Laboratory Standards Institute document, Liquid Chromatography-Mass Spectrometry Methods; Approved Guideline, C62-A (CLSI C62-A) recommends a desirable S/N at the LLOQ of ≥ 20 and an absolute minimum S/N of 10 [33].

Although manipulation of LC and MS/MS parameters can significantly improve S/N—a primary goal of an optimized sample preparation protocol is to deliver the desired S/N at the LLOQ. If several fold improvements in S/N are necessary then a lower dilution or sample preparation protocols that can concentrate analytes—LLE and SPE, to a lesser extent SLE, should be considered. If the desired S/N at the LLOQ appears to be within reach, performing the quantitative and qualitative matrix effect experiments described in the next section can help to guide optimization efforts.

4.3 EVALUATING MATRIX EFFECT (ME)

A limitation of the ionization methods used in LC-MS/MS is their susceptibility to ME [34,35]. Although MEs on ionization are both negative (ionization suppression) and positive (ionization enhancement)—most of the discussion about deleterious effects on quantification references ion suppression [13–22]. There is a rich literature on ion suppression but the simplest explanation—competition between analytes and matrix for a limited amount of charge—is sufficient for the purposes of this discussion [34,35]. ESI is more prone to MEs than is APCI, but it is safest to assume that any clinical LC-MS/MS sample preparation protocol and ionization type will have some degree of ion suppression [34,35]. As mentioned previously, phospholipids as a class are an important contributor to ion suppression in serum or plasma samples, as are salts and strongly polar compounds, phthalates and other environmental contaminants, and high drug concentrations, in any matrix [13–16].

When serum samples are processed with a PPT protocol and analyzed with gradient reverse-phase LC, there is a predictable pattern of ion suppression [17–22]. There is strong ion suppression in the void volume from salts and polar compounds that are not retained on the LC column, followed by ion suppression in the first few minutes of the run from more polar lysophospholipids and later in the run from more nonpolar phospholipids [17–22]. The LC stationary phase, mobile phases, gradient, and the precipitation reagents selected modify ME, but unless phospholipids are depleted, some version of this ion suppression profile will occur.

Furthermore, after multiple injections of PPT samples, phospholipids are strongly retained and accumulate on C18 stationary phases [17–22]. Multiple investigators have described difficulty removing phospholipids from C18 with an acetonitrile wash at the end of each injection or at the end of the batch [17–22]. Washing with isopropanol or a solvent mixture appears to remove phospholipids more effectively than washing with 100% acetonitrile [17–22]. A frequently observed phenomenon in production is a constant bleed of phospholipid signal and ion suppression from inadequately washed C18 columns throughout the course of each LC injection cycle [17–22].

Ignoring ME during development of a sample preparation protocol is an easy mistake because it is a negative interference, essentially invisible. There are no interfering peaks—there is simply less signal because fewer analyte molecules become ionized in the source. A stable label IS that coelutes with the analyte can usually compensate for ME, but not always [36–41]. The extraction recovery and ME experiments described here are an essential tool for any clinical LC-MS/MS laboratory.

Ion suppression is an MRM specific phenomenon. Thus all analytes and all transitions in an SRM method should be tested for ME. Significant ion suppression could be missed if only one or two compounds are tested as surrogates for the other analytes in the method [16,17].

Four experiments to assess ME are presented in this chapter, a quantitative spiking experiment, a qualitative postcolumn infusion experiment, direct detection of phospholipids, and a matrix mixing experiment. Our laboratory uses the quantitative and qualitative experiments for evaluation and validation of all methods, direct detection when optimizing to reduce phospholipids, and matrix mixing to validate dilution matrices, rather than to evaluate ion suppression.

4.3.1 Quantitative Matrix Effect Experiment

What is characterized here as the quantitative ME experiment was described by Matuszewski in 2003 [42]. The design of this experiment differentiates between loss of analytes during extraction from biological sample matrix versus the effects of residual biological sample matrix on ionization. The same mass of analyte is spiked into biological samples before and after extraction and into injection solvent [42]. Extracts and spiked injection solvents are analyzed on the LC-MS/MS, and peak areas are compared. In summary, the set up and calculations for tubes designated A, B, and C is:

> A tube: Spike neat standard and IS in to the injection solvent. This tube is defined as 100% recovery and 0% ME.
>
> B tube: Spike neat standard and IS in to blank biological sample matrix (spike *before* extraction), then extract.
>
> C tube: Extract blank biological sample matrix, then spike with neat standard and IS (spike *after* extraction).
>
> Mean peak areas for replicates are calculated (means are a, b, and c) and their ratios are defined as:
> ME = c/a

Extraction recovery = b/c
Overall process efficiency = b/a

CLSI C62 recommends testing five native matrix samples with this protocol [33]. Our laboratory typically selects 10 patient samples with abnormal appearance—samples that are icteric, hemolyzed, lipemic, cloudy, unusual colors, or collected from patients with renal or hepatic failure. Triplicates are spiked at low (lower third of the AMR) and high (upper third of the AMR) concentrations. One precaution is to make sure neither the composition of the biological sample matrix nor the injection solvent is significantly altered by the solvent of the spiking solution.

C62 recommendations for acceptance criteria are comparison of calculated matrix bias (100% ME) to total allowable error, ME for an analyte to ME of the IS, and calculating the %CV of peak areas between samples to determine ME contribution to imprecision [33]. If between sample and/or IS peak area CV is >15%, further optimization of sample preparation is advised.

A quick graphic impression of between sample variability caused by ME can be seen in the plot of IS peak areas versus injection number. Reviewing this plot with each batch is a powerful tool to confirm that the variability of ME between samples remains acceptable, once the assay is validated and has been moved to production.

C62 does not list a minimum or a desirable range for percentage extraction recovery, but the spike before and after extraction experiment is useful to interrogate insufficient recovery from transfer, loading, mixing, wash, and elution steps of complex protocols, such as LLE, SPE, SLE, and PRL or losses from any protocol caused by nonspecific adsorption and transfers between containers. Acceptable S/N at the LLOQ is the true measure of acceptable recovery. Grant and Rappold in their short course at the Mass Spectrometry Applications in the Clinical Laboratory (MSACL) meeting recommend first optimizing for reproducibility between different patient samples, and with that achieved, then increasing recovery if necessary [43].

4.3.2 Qualitative Postcolumn Matrix Effect Experiment

The qualitative postcolumn infusion experiment visualizes ME over the time course of the chromatographic run [44]. A semi-quantitative assessment of ME can be made, but the best value is the chronologic characterization. It may be possible to modify Rt of the analyte relative to the pattern of ion suppression for a net increase in S/N and robustness [44].

The design of this experiment is to establish a constant MRM signal for both analytes and ISs by infusing neat standards in solvent postcolumn. Injecting a sample without matrix (e.g., water) will demonstrate a baseline profile for comparison with the deflections from ME seen when extracted biological matrix samples are injected (Fig. 3.2). As with the quantitative spiking ME experiment—the more abnormal the native matrix samples used, the better. The primary challenge with this experiment is adjusting the concentrations and infusion rate of the neat standards so the signals are in the midrange of counts per second (cps). Significant deflections can be missed when the infusion signal is too high and fluctuations from noise/baseline drift can make interpretation difficult when signal is too low. Although the goal is to see no loss of signal at analyte Rt, it is common to see some decrease but with closely paralleled deflections between stable labeled ISs and analytes.

4.3.3 Phospholipid Direct Detection Experiment

Building an MS/MS method that detects phospholipids directly is a more straightforward, but less comprehensive means than postcolumn infusion to detect the time course of ion suppression for serum

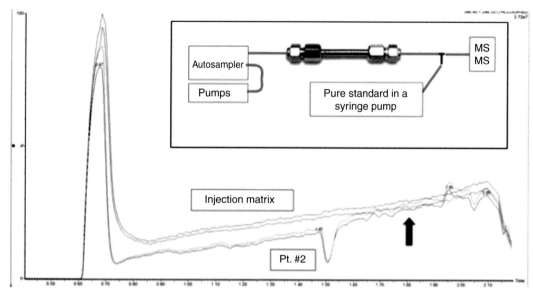

FIGURE 3.2 Post Column Infusion to Assess Matrix Effect

A schematic of the plumbing for post column infusion is inset. LC flow is from the pumps through the autosampler and column to the MSMS. Neat standard is introduced to the LC stream between the column and the MSMS through a syringe and T connection fitting.

Extracted ion chromatograms (XIC) for MRMs detecting 11-Nor-9-Carboxy-delta-9-tertrahydrocannabinol (THC-COOH) and deuterium labeled IS are shown overlaid for an injection of blank injection solvent (no matrix effect) and an injection of an extracted urine sample negative for THC-COOH. The *arrow* denotes the expected retention time (Rt) for THC-COOH. Ion suppression is seen prior to the THC-COOH Rt in the Pt. 2 XIC, but suppression is minimal and similar for analyte and IS at the THC-COOH Rt. Urine was extracted with mixed-mode strong anion exchange SPE.

samples. Our laboratory has found this to be a complementary technique to postcolumn infusion and easier to integrate into a development workflow. Several options have been described, one popular method uses collision induced dissociation (CID) in the source and MRMs 184/184 *m/z* and 104/104 *m/z* to detect fragments common to both late and early eluting phospholipids [18,19,21].

Glycerophospholipids, such as phosphatidyl choline, have a 3-carbon glycerol backbone that may be esterified with two fatty acids and a phosphate group. Lysophospholipids are a subgroup of the glycerophospholipid family with one of the hydroxyl groups on the three carbon glycerol backbone remaining unesterified, that is, containing only one fatty acid and a phosphate group. Predictably, lysophospholipids are relatively polar and an earlier eluting source of ion suppression compared to later eluting glycerophospholipids that have two fatty acid chains. Detecting both classes is informative as changes in LC methods and sample preparation may affect them differently [18,19,21].

Goals for direct phospholipid detection are an MS/MS acquisition that is quick to set up, minimizes duty cycle, and broadly detects both early and late eluting species. Xia and Jemal compared three MS/MS acquisition modes for detecting phospholipids and found that programing one MRM to represent each of the major phospholipid groups had greater selectivity, producing well-defined, quantifiable peaks, as compared to more comprehensive or generic acquisition strategies (e.g., precursor/neutral loss scans or in-source CID with m/z 184/184 and 104/104) [18]. Our experience evaluating PRL media supports this recommendation—we used in positive mode the MRMs m/z 524/184, 496/184 (lysophospholipids) and 704/184, 758/184, 786/184, 806/184 (other glycerophospholipids).

4.3.4 Matrix Mixing Experiment

C62 recommends matrix admixing experiments that are described in detail in the CLSI document EP-07 [33]. For example, matrix A is mixed with matrix B in the percentage ratios 100:0, 75:25, 50:50, 25:75, and 0:100. If ME is significantly different between the two matrices, observed concentrations will differ from expected. This protocol is appealing because it is simple, but it provides less information that the other three experiments.

4.4 EVALUATING METHOD PERFORMANCE

A lack of robustness in sample preparation can be a contributor to imprecision, inaccuracy, nonlinearity, and unacceptable limits of quantitation [1]. Except for S/N at the LLOQ, establishing the extent to which sample preparation causes unacceptable method performance, rather than or in addition to, contributions from suboptimal LC or MS/MS conditions, may not be obvious. Switching from an analog to a stable isotope labeled, coeluting IS usually improves precision. Changing to an IS with mass at least +3 m/z above the analyte m/z to correct nonlinearity from high analyte contributions to IS signal is also a simple fix. The quantitative ME and recovery spiking experiment is more time consuming but is valuable for interrogating sample preparation as a possible source of unacceptable method performance.

LLOQ can be improved by concentrating analytes to a greater degree, increasing extraction recoveries, and decreasing ion suppression by optimizing LLE, SLE, and SPE protocols. With a fully optimized method, the variance inherent to manual pipetting of sample and IS may ultimately be the limiting factor for precision.

4.5 EVALUATING PRACTICALITY

It is the norm to underestimate the amount of time consumed and potential for error in the repetitive sample sorting, labeling, sealing/unsealing, reracking of extraction/injection containers, and in the transfer of liquids between containers that is inherent to manual sample preparation for LC-MS/MS. Every sorting, racking, transfer, sealing, and labeling step that can be eliminated by creative attention to the process not only reduces labor costs, but also reduces the risk of sample misidentification, ergonomic injuries, lowers consumable costs, and may decrease variance from losses of analyte due to nonspecific adsorption to containers and transfer processes. A reduction in the number of transfers between containers, by pipetting or eluting directly into the injection vial or plate rather than in to an interim container can be very efficient but the compromise may be pipetting of much smaller volumes of sample, IS and elution solvent. Low volume pipetting can increase imprecision and evaporation of small volumes of organic solvents during processing is an additional risk. It is important to assess in a robust way the potential effects on precision and recovery from reductions in pipetting or elution volumes.

Tremendous increases in productivity are possible by adoption of 96-well plates for sample preparation instead of tubes/vials [3,4]. However, it is significantly easier to mislocate samples when manually pipetting small, colorless, liquid volumes into 96-well plates instead of tubes. Light boxes and other pipetting aids can facilitate manual addition of samples to plates, but using automated liquid handling, when feasible, is the best option. Using automated liquid handlers for LC-MS/MS sample preparation not only improves precision and reduces labor costs, but also saves time spent sorting samples and reduces sample misidentifications through positive identification with barcodes.

4.6 EVALUATING ROBUSTNESS

The definition of method robustness varies widely between laboratories. In some settings—a column lifetime of several hundred injections is a desirable trade-off for a faster, simpler sample preparation protocol. In contrast—another laboratory may routinely perform extensive, sometimes automated, sample clean up and define >10,000 injections as the minimum acceptable lifetime for LC columns.

In one setting, cleaning of the MS/MS interface (section of the MS/MS under vacuum between the source and the quadrupole rail) on a monthly or even weekly basis is accepted as the unavoidable outcome of high volume testing with minimal sample cleanup. To another laboratory—this additional labor, loss of instrument time, and operational unpredictability would be too problematic. Significant cost and labor would be invested to develop and use in production a sample preparation protocol that removes more matrix. Sufficient sample cleanup that introduces less matrix to the LC-MS/MS can yield an MS/MS interface that requires only scheduled 6 month or annual maintenance and a quadrupole rail that never needs cleaning.

Fit for purpose, sample preparation is defined here as a protocol that delivers acceptable method performance and removes sufficient matrix to confer the desired, predictable interval of good LC column and MS/MS performance.

Few laboratories, if any, can devote the resources to test two different sample preparation protocols for the same analyte on two different instruments over weeks and months to compare method performance, column lifetime, and MS/MS response as measures of robustness. Therefore, the "measure" of method robustness is usually an impression or a series of anecdotes, rather than quantitative data.

Process metrics that may be helpful to quantify robustness include:

- Tracking peak areas for a representative IS or LLOQ calibrator. Record peak area(s) by date and evaluate for trends, shifts, time course, and against action limits as an indicator of LC-MS/MS sensitivity.
- Number of sample repeats required because of ion suppression (low IS peak area relative to calibrators) or MRM ratio failures or interfering peaks.
- Number of run interruptions from over-pressure or leaks.
- Number of batch failures from unacceptable QC.
- Number of batch failures from unacceptable calibration.
- Average number and range of injections/column and guard column.
- Length of time after cleaning of the MS/MS interface to the next time that cleaning is needed to restore sensitivity.
- Number of tubing, fitting, inline frit, rotor seal, column changes because of leaks or over-pressure from clogging.
- Number and duration of delays in turn-around time because of LC-MS/MS instrument down-time.
- Negative variance of cost/reportable test, caused by frequent repeats, batch failures, delays, instrument downtime.

The quality of a sample preparation protocol can influence chromatographic robustness and the amount of time needed for data review, although good LC maintenance and optimization of LC, MS/MS, and peak integration algorithm parameters play a significant role as well. Metrics to quantify chromatographic robustness include:

- Number of samples/batch requiring manual peak integration.
- Number of batches with baseline signal above threshold (e.g., > 500 cps).
- Number of batches with S/N at the LLOQ ≤ 20.
- Number of batches with unacceptable resolution for critical peak pairs (Rs ≤ 2.0).
- Number of batches with average peak width and/or peak asymmetry above threshold (e.g., >10 s peak width).
- Number of adjustments required in MS/MS acquisition method and data analysis method windows because of Rt shifts caused by column degradation.

5 COMPARISON OF SAMPLE PREPARATION PROTOCOLS

Commonly used sample preparation protocols are characterized in Table 3.1 by relative cost and complexity, capability for concentration of analytes, and matrix removal. A major advantage of LC-MS/MS as a measurement technique are the many options for adjusting LC conditions, MS/MS parameters or the sample preparation protocol to compensate for limitations in one of the other phases of the analysis. Thus the choice of sample preparation protocol or complexity of optimization required should always take into account LC and MS/MS functionality.

As limitations in personnel, LC-MS/MS expertise, and operating budget are common to most clinical laboratories implementing LC-MS/MS, the first decision about sample preparation is often a consequence of practicality rather than chemistry—whether or not the simple sample preparation protocols—DIL and PPT—will suffice.

Table 3.1 Simplified Comparison of LC-MS/MS Sample Preparation Types

Sample Preparation Protocol	Analyte Dilution (D) or Concentration (C) Possible	Relative Cost	Relative Complexity	Relative Matrix Removal
Dilution (DIL)	D	Low	Simple	Less
Protein precipitation (PPT)	D	Low	Simple	Least
Liquid–liquid extraction (LLE)	D or C	Low	Complex	More
Phospholipid removal (LPR)	D	High	Moderately complex	More, selective[a]
Supported liquid extraction (SLE)	D or C (moderate)	High	Moderately complex	More
Solid phase extraction (SPE)	D or C	High	Complex	More
Online SPE/Turboflow	D or C	High	Complex	More

[a]Only phospholipids are removed, other matrix components are not depleted.

The sensitivity of the MS/MS is a first consideration. Purchasing the most sensitive MS/MS the laboratory can afford has many advantages. Sample preparation protocols, such as DIL, PPT, and PRL that do not concentrate analytes are more likely to achieve the desired LLOQ with a higher end MS/MS. The performance of a more sensitive MS/MS may take longer to degrade because less matrix is necessarily introduced with each injection.

Assessing the feasibility of simple sample preparation must take into account the nature of the analyte(s), sample type, throughput requirements, and required quantitation limits. With exceptions, sample preparation for endogenous analytes, such as serum steroid hormones with picomolar quantitation limits requires complex protocols—LLE, SPE, or SLE—that can concentrate analytes and deplete matrix. In contrast, drugs and metabolites at higher concentrations are often amenable to DIL, PPT, or PRL, even with a less sensitive MS/MS. Therefore, an initial method development task is to define on-column detection limits and the LLOQ needed.

If the desired LLOQ can be met with simple techniques, sample type is the next parameter to consider. DIL is appropriate for low protein matrices, such as urine whereas PPT or PRL are appropriate for high protein matrices, such as serum/plasma/whole blood. PPT with manual pipetting may be a reasonable choice with small sample numbers (<100/day). With higher test volumes (hundreds to thousands of samples per day), the demand for greater robustness and reliable, high throughput may justify the increased cost of PRL and automated liquid handling. PPT is the sample preparation protocol with the most risk for matrix contamination of the LC-MS/MS, but the choice of PPT protocol can make a difference in robustness. It may be possible to ameliorate the greater ion suppression seen with DIL and PPT by increasing the length of the LC run, but this will reduce throughput.

If a technique with greater selectivity, concentrating capability, and matrix removal is needed, the polarity (log P, log D), acidity/basicity (pKa), and thermostability of the analyte(s) are factors to consider. Neutral, nonpolar compounds, such as steroid hormones, have good analyte recovery and matrix removal with LLE and SLE [43,45,46]. Although reverse-phase SPE offers good retention for steroids, removal of serum phospholipids is challenging because the strong organic solvent washes that remove lipophilic matrix will also wash neutral steroids from the reverse phase, with subsequent low analyte recovery in the elution solvent. Neutral steroids are not retained by ion exchange mechanisms but steroid mixed mode SPE with a focus on ion exchange interaction by the matrix, rather than the analyte, has been described. Instead of adjusting pH to retain and then elute analytes, pH manipulation is used to minimize matrix retention by ion exchange mechanisms during loading and washing steps and maximize matrix retention on the ion exchanger when eluting neutrals [47–48].

For weakly acidic/basic molecules, mixed-mode ion exchange SPE can be highly selective because pH can be adjusted to retain analytes in the charged form with the ion exchange moiety while aggressively removing lipophilic matrix from the reverse phase with nonpolar washes. Then a pH shift converts analytes to their uncharged form for elution in organic solvent.

Thermally labile analytes may not tolerate heating during the evaporation step needed with LLE, SLE, and traditional, offline SPE, in which case online extraction or µElution SPE are options to consider.

When the decision is between LLE and other concentrating techniques (SLE or SPE)—lower cost materials favor LLE, but the labor required may be unfavorable. LLE also does not have good recovery for highly polar or permanently charged analytes.

If all of the complex techniques appear to be an option, there is no better way for new users to compare the hands-on practicality of LLE, SLE, and SPE then with a short experiment in their own laboratory to estimate precision, S/N at the LLOQ, and selectivity with patient samples. Vendors of commercial extraction media will provide onsite application support and loans of extraction and evaporation manifolds.

Clinical laboratories implementing LC-MS/MS often launch with a single method, low test volumes, manual pipetting, and DIL or PPT sample preparation. With time, as test volumes and test menu expand, there is a need for automation and more sophisticated sample preparation. Introducing more complex sample preparation protocols can be done incrementally and without a large capitol investment. A relatively low cost investment in an evaporator and a vacuum or positive pressure manifold can enable a laboratory to perform high quality manual LLE, SPE, or SLE.

Scaling up for sample preparation of larger numbers involves more capital investment with key decisions being the switch from tubes/vials to 96-well format, purchasing automated liquid handlers, and/or implementing online extraction.

Except for LLE, all of the sample preparation techniques described here can be done easily in 96-well format. The difficulty of thoroughly mixing immiscible liquids in 96-well plates and the risk of cross-contamination between wells means that automating LLE can be challenging and requires extensive validation.

Throughput, precision, error rates, and labor productivity can be improved with automated liquid handling. Smaller, sophisticated liquid handling automation for LC-MS/MS sample preparation has become available in the last few years, but the cost of robots that can move sample from tube to plate with excellent pipetting precision (<3%CV), using barcode readers, liquid level sensing, clot detection, and multiple, independently spanning pipetting channels remains high (easily >$100,000 for 8 pipet channels).

Online extraction options range from simpler, in-house designs that need one additional LC pump and six-port switching valve, to commercial systems with sophisticated automation that are capable of high throughput LC multiplexing, such as Symbiosis from Spark-Holland and TurboFlow from Thermo-Scientific [3,49]. The commercial systems require significant capitol investment and personnel with high throughput LC-MS/MS expertise are a distinct advantage for assay development and support during production. Commercial online extraction systems and LC multiplexing are more common in commercial reference laboratories where thousands or tens of thousands of samples are tested per day with LC-MS/MS, rather than in hospital laboratories testing hundreds of samples per day [49].

The decision to set up offline SPE versus an in-house developed online SPE is usually influenced by the extent of LC expertise present in the laboratory. Offline SPE does not require LC sophistication or additional LC components and a broader range of wash solutions can be used to improve selectivity and remove matrix. Online SPE, once set up and validated, requires significantly less hands-on time per sample than does offline SPE, but without LC multiplexing, may have slower throughput than an LC system that does not include online extraction.

Cost/sample is low and relatively easy to calculate for DIL and PPT, although cost/reportable test may be higher than anticipated when there is significant instrument malfunction from matrix contamination. Complex techniques that are more expensive on a cost/sample basis but can reduce matrix effects—online SPE and TurboFlow extraction or automation with liquid handling of PRL, SPE, LLE, or SLE—may prove in the long run to have a favorable cost/reportable test compared to DIL or PPT because of robust method performance, and reliable instrument operations [49].

6 OPTIMIZATION OF SAMPLE PREPARATION PROTOCOLS

Two topics that are applicable generically to sample preparation protocols are optimization of sample preparation in 96-well format and optimization of the LC injection matrix.

6.1 GENERIC—OPTIMIZING FOR 96-WELL FORMAT

Switching from manual sample preparation with tubes to 96-well plates with automated liquid handling can deliver exponential improvements in sample preparation productivity. For implementation of automated liquid handling however, a significant investment of time is usually necessary, typically months rather than weeks, to learn software and optimize robotic liquid handling and automated extraction protocols.

Precautions to keep in mind when implementing 96-well format include:

- The risk of sample misidentification with manual pipetting in to 96-well format.
- Awareness of the high potential for cross-contamination between wells during mixing, elution, and evaporation steps [43,50]. Cross-contamination experiments with a checkerboard pattern of high and low/blank wells should be routine after any protocol modification.
- The shape of wells can effect cross-contamination, mixing efficacy, minimum pipettable volumes, and pipetting accuracy. Variants are square and round wells, conical bottom, flat bottom, round bottom wells and with glass inserts. Square wells may have a larger volume for the same height, but round wells have lower potential for cross-contamination [43,51]. Plate heights range from standard microtiter to deep well-plates that can hold 0.5, 1, and 2 mL.
- The material of the plate, that is, polypropylene, polystyrene, polyethylene, or glass inserts, can affect recovery, nonspecific adsorption, baseline noise, interfering peaks, and pipetting accuracy [52]. All polypropylenes are not the same, a change of plate vendors can affect recovery [52].
- Adhesive plate seals are quick and easy to use but the adhesive can build up on, clog and/ or contaminate autosampler needles or act as a vehicle for carryover from the needle [53]. Alternatives are seals with adhesive surrounding but not over the wells, aluminum foil heat sealed to the plate, and plastic cap mats that indent into and seal each well.

6.2 GENERIC—OPTIMIZING LC INJECTION SOLVENT COMPOSITION

Considerations for the LC injection matrix are analyte solubility and effects on LC peak shape and Rt.

6.2.1 Analyte Solubility

The right proportions of organic solvent and water in the LC injection matrix (e.g., 50:50 mobile phase A:mobile phase B) are needed to solubilize analytes and prevent nonspecific adsorption to the container [54]. Solubility and nonspecific adsorption can be analyte dependent [log P/log D], container material dependent, and are affected not only by the percentage but also by the type of organic solvent present, for example, methanol versus acetonitrile [55,56]. Studies of highly nonpolar analytes, such as 11-Nor-9-Carboxy-delta-9-tertrahydrocannabinol (THC-COOH) provide some of the best examples of how surface chemistry plays a role in extraction recovery and precision [57–59].

6.2.2 LC Peak Shape and Rt

A sufficiently low organic solvent content in the injection matrix is necessary to achieve focusing of analytes in a narrow band at the head of a reverse phase LC column with no distortion of peak shape [54]. For an isocratic reverse phase LC method—the percentage of organic solvent in the injection matrix should be ≤ the %B in the mobile phase, for example, 90% of the mobile phase %B [54]. If the organic content in the injection matrix is too high—early eluting peaks can be too wide, asymmetrical, and with shifts in Rt because the analytes are more strongly partitioned into the mobile phase and insufficiently retained on the stationary phase [54].

For gradient analysis—a higher organic content of the injection matrix can be tolerated [54]. The injection solvent organic content, the type of organic solvent, injection volume, polarity of the analyte(s), polarity of the column stationary phase, and the %B at the start of the gradient can all affect column focusing in gradient analysis [54].

The best organic content for analyte solubility may be too high for optimum peak shape, therefore the solvent mix that delivers maximum extraction recovery while maintaining acceptable peak shape may need to be determined experimentally.

6.3 OPTIMIZING SPE

6.3.1 Stationary Phase

Selecting a stationary phase, bed mass and vendor are the first steps. Select the smallest possible bed mass that has enough capacity to achieve the LLOQ. Smaller bed mass equates to lower solvent volumes for elution, therefore lower cost, faster evaporation, and the potential to elute directly into an injection vial instead of a tube (avoiding a transfer step). There are many SPE vendors to choose from and the nature of the application support offered is a factor to consider in selecting a vendor.

The nature of the analyte will guide the choice of stationary phase, for example, mixed-mode cation exchange for basic compounds and mixed-mode anion exchange for acidic compounds. Using an SPE stationary phase with different selectivity than the LC column stationary phase is good practice for improving specificity [54]. Strong (permanently charged) ion exchange is used for weak acids or bases and weak ion exchange for permanently charged compounds.

6.3.2 Loading

For avidly bound analytes, protein precipitation before SPE can improve recovery, but the pH and polarity of the supernatant must be compatible with, or be adjusted to promote, analyte retention on the SPE stationary phase. With mixed mode SPE for acidic analytes, the application buffer pH should be at least 2 units > pKa, and for basic compounds, at least 2 units < pKa so the majority of the analyte is in the charged form for retention by the ion exchanger [1,63].

6.3.3 Washing

Wash solutions for mixed-mode SPE need to remove polar and intermediate polarity matrix and maintain the correct pH for analyte retention to the ion exchanger (e.g., with relatively polar acidic or basic water, methanol, acetonitrile mixtures) and then effectively remove lipids and other neutral matrix (with methanol, acetonitrile, ethyl acetate, hexane, dichloromethane, MTBE, etc. and their mixtures). For example, a strong anion exchange mixed mode SPE protocol for marijuana metabolite in urine used a polar, basic wash (1:22:77 vol:vol of NH_4OH:acetonitrile:H_2O) followed by an intermediate polarity organic solvent—ethyl acetate wash [60]. An alternative is to retain interfering substances on the stationary phase while analytes are eluted [1].

Good selectivity is more challenging with serum than with urine because depleting phospholipids while retaining analyte is complicated by phospholipids having both charged and nonpolar moieties. Several investigators have examined phospholipid removal with SPE systematically [61,62]. Chambers and coworkers took advantage of the greater solubility of phospholipids in methanol versus acetonitrile to optimize extraction of charged analytes with mixed mode ion exchange SPE [61]. Basified methanol as a wash solution and basified acetonitrile as the elution solvent reduced the phospholipid content in the final eluate [61].

6.3.4 Elution

When eluents contain highly volatile agents, such as organic solvents, glacial acetic acid, or ammonium hydroxide—advance preparation and storage of eluent should be validated against daily preparation. Changes in the eluent composition from evaporation can occur during storage, potentially decreasing recovery and selectivity. With mixed mode SPE, elution occurs by shifting the pH such that analytes are no longer charged and use of a nonpolar organic solvent to release analytes from both the ion exchange and reverse phase retention mechanisms. In the THC-COOH strong anion exchange mixed mode SPE example, the eluent was 1:30:69 (vol:vol) of glacial acetic acid:ethyl acetate:hexane [60]. Adjusting the polarity index of the elution solvent by changing the proportions of relatively polar (e.g., ethyl acetate) to relatively nonpolar (e.g., hexane) solvent may be useful to improve recovery or reduce interference [43]. Applying elution solvent in two steps, rather than the same total volume in one step, may increase recovery [63]. If recovery is lower than expected it may be feasible to quantify analytes in the eluates from loading, washing, and multiple elution steps (if eluate matrices are compatible with LC injection) to determine whether analytes were not retained during loading, were lost in wash steps or were never eluted [43].

6.3.5 Evaporation and Reconstitution

Thermally unstable or volatile analytes may be lost during the evaporation step. The time and temperature for evaporation should be validated and defined in the procedure. For example, it is common to add acidified methanol to the evaporation container so that amphetamines will become positively charged and less volatile during evaporation [64]. Precautions on correct adjustment of the height of the evaporator and flow rate of nitrogen, as well as testing for cross-contamination during evaporation, are important aspects to validate and include in the standard operating procedure [65].

6.4 OPTIMIZING PPT

Optimization parameters for PPT include the precipitating reagent or reagents, ratio of sample to reagent, mixing time, and the time for postmixing hold. Several authors have investigated the efficacy of different PPT reagents for protein removal and matrix effect [5–7]. Similar results were found for protein removal, but evaluations of ion suppression were more variable between investigators, both in methods (analyte, mobile, and stationary phases, experimental design) and in conclusions [5–7]. This observation supports the guidance that ME should be assessed by each laboratory for all clinical LC-MS/MS methods (CLSI C62-A and CAP Chemistry and Toxicology checklist).

Acetonitrile as a precipitating agent was found to remove more serum protein than did methanol [5–7]. Polson and coworkers found that both methanol and acetonitrile PPT had highly variable ion suppression depending on mobile phase composition but acetonitrile had somewhat less ion suppression than methanol with neutral mobile phases (68–83% vs. 86–92% respectively) [6].

However, they found $ZnSO_4$ and trichloroacetic acid (TCA) to be as effective at removing protein as the organic solvents and TCA had significantly less ion suppression (<10%) than did PPT with organic solvents [6]. Use of an organic solvent precipitating agent, in contrast to acid or $ZnSO_4$, needs to be evaluated for the effect on chromatography because of the high organic content (60–75%) of the injection matrix. The Rt and peak shape of early eluting analytes may be negatively affected. A divert valve is always recommended, but absolutely necessary when using $ZnSO_4$ to direct flow of the void volume to waste and prevent clogging and contamination of the MS/MS interface [6]. The stability of the LC column should be evaluated if using TCA as exposure to low pH in multiple injections may degrade chromatography [6].

Combining PPT reagents of different functionalities may be a mechanism for improving robustness. PPT with methanol/$ZnSO_4$ is a commonly used clinical sample preparation protocol for measuring immunosuppressants in whole blood [9,10], an application for which turn-around time and therefore instrument robustness is critical. Our laboratory found for serum 25-hydroxy vitamin D that PPT using both acetonitrile and methanol/$ZnSO_4$ precipitants had more reproducible recovery across the reportable range and yielded longer maintenance-free intervals for the MS/MS than did PPT with acetonitrile alone.

The most common recommendation for ratio of precipitation reagent to serum is 3:1 [5–7]. Uniform mixing for PPT samples in racks/plates can be problematic because of the difficulty in suspending a heterogenous solid phase (precipitate) in liquid constrained within an array of narrow diameter columns (autosamper vials in racks or 96-well plates on a multimixer). Optimizing the mixing time and protocol, the maximum number of samples mixed at one time, the postmixing hold before centrifugation, and defining the maximum batch size is advisable to insure reproducible recovery, avoid evaporation of organic PPT reagents or particulate from delayed precipitation [43]. The supernatant from solvent precipitation can be transferred to a clean container and evaporated in order to resuspend analytes in an injection matrix with lower organic content, but lengthy evaporation times and higher temperatures are necessary because of the high aqueous content [50].

6.5 OPTIMIZING DIL

Optimizing a DIL protocol consists of finding the correct dilution factor to achieve the desired LLOQ and verifying analyte solubility in the diluent while maintaining compatibility of the diluent with the LC method. As with PPT, assessment of ME early in the development process for DIL is good practice to determine whether adjustments in the LC or MS/MS method or selection of a more selective sample preparation protocol will be necessary to reduce ion suppression [66].

When analyzing a multicomponent panel with variable ionization characteristics and polarities, such as opiates or benzodiazepines, it is common to find 10 fold or greater differences in ionization efficiency. The right dilution for the least sensitive analyte may cause over concentration, detector saturation, and nonlinearity for analyte(s) with better sensitivity. To overcome this dilemma, the MS/MS collision energy for more sensitive compounds can be adjusted to decrease response (detuning) such that all analytes can be measured across the desired reportable range using one dilution.

6.6 **OPTIMIZING LLE**

The first goal is promoting partition of analytes from the sample to the immiscible organic extraction solvent by adjusting the pH and ionic strength of the extraction buffer. For acidic analytes, the buffer pH should be at least 2 units less than the pKa, and for basic compounds, the buffer pH should be at least 2 units higher than the pKa so the majority of the analyte is in the uncharged form and hence, more soluble in the organic phase [50]. High salt concentrations, or high ionic strength buffer, can be used to decrease the solubility of nonpolar analytes in the sample and improve partitioning into the organic phase [1].

The proportion of sample to extraction solvent effects extraction efficiency. Grant and Rappold in their MSACL short course recommend a ratio of 10:1 for solvent:sample [43].

The polarity index of the solvents used, their proportions, and the pH of the aqueous phase can all affect partitioning of analytes and phospholipids between the phases [23]. The more nonpolar the organic solvent, the less phospholipids are extracted, but adding a more polar solvent may be necessary for analyte recovery [23,43,50]. As the proportion of more polar solvent, such as ethyl acetate is increased in the organic phase, the concentration of extracted phospholipids will also increase [23,43,50]. Rappold et al., tested solvents and solvent mixtures with polarity indices ranging from 0 (hexane), 2.5 (MTBE), 4.4 (ethyl acetate) to 5.8 (acetonitrile), and recommended a calculated polarity index <3 to minimize extraction of phospholipids [23].

Optimization of mixing time with assessment of both analyte and phospholipid recovery can be useful as extraction of the two species may occur in different time frames [1]. After centrifugation, freezing can make clean separation of the layers more reproducible [3]. As mentioned earlier, it is important to check for sufficient mixing across the entire plate and for cross-contamination between wells when using 96-well plates for LLE [43,50]. The issues discussed for evaporation of SPE eluates are the same as for evaporation of the LLE organic layer.

6.7 **OPTIMIZING PLR**

The optimal precipitation reagent, maximum sample volume, and serum:reagent ratios to use with PRL media varies with the format and chemistry that different vendors have devised to remove phospholipids. For example, Sigma-Aldrich recommends use of Hybrid SPE with a 3:1 ratio of precipitant:serum with either 1% formic acid in acetonitrile for neutral or acidic compounds or 1% ammonium formate in methanol for basic compounds, unstable metabolites, or analytes with poor solubility in acetonitrile [67]. The inclusion of 1% formic acid in the acetonitrile is important for good recovery of acidic compounds with Hybrid SPE [67]. Further reagent optimization for the analyte(s) and LC method of interest may be productive, but comparing the PRL from different vendors has been more worthwhile in our experience because PRL typically has fewer parameters to modify than SPE, SLE, or LLE.

6.8 **OPTIMIZING SLE**

Modification of the loading buffer is the key for making a wide range of analytes amenable to extraction with SLE. For neutral, nonpolar analytes the sample can be diluted with water [25]. Weak acids or bases can be extracted by shifting the loading buffer pH to neutralize analyte charge [25]. Like LLE, acidic compounds are extracted at acid pH at least 2 pH units below the pKa and basic compounds at basic pH at least 2 pH units above the pKa [1,25,50] to promote the uncharged form. Highly polar and permanently charged compounds can be extracted by adding a volatile ion-pairing agent to the loading buffer [68]. Recovery of strongly protein bound analytes can be improved by adding an agent to the loading buffer that disrupts binding, such as acid, base, water miscible organic solvent (e.g., isopropanol) or $ZnSO_4$ [25].

The concentrating capability of SLE is nominally limited by the maximum loading volume for a given SLE bed mass and the recommended 1:1 proportion of sample to loading buffer. However, good recovery may be possible by adjusting the ratio and applying more sample with either less loading buffer or more concentrated loading buffer, particularly with neutral, nonpolar analytes [25].

As with LLE, the polarity index and pH of the elution solvent affects the recovery of both analytes and phospholipids. The less polar the elution solvent, the less phospholipids will be extracted, for example, a hexane or MTBE eluate will have lower concentrations of phospholipids than an ethyl acetate eluate [23,69]. However, the efficacy of analyte elution depends on the solubility of the analyte in the elution solvent. More polar analytes need a more polar eluent to increase recovery, for example, 10:90 ethyl acetate:hexane [23]. The trade-off is an increased extraction of phospholipids [23,69].

6.9 OPTIMIZING ONLINE EXTRACTION

There is an extensive literature on the optimization of online extraction using different stationary phases, single versus dual columns, use of denaturing SPE, and strategies to reduce carryover. Details are beyond the scope of this chapter, readers are referred to several comprehensive reviews [31,32].

7 SUMMARY

The fit for purpose approach requires an appreciation for the practical demands, as well as the chemistry and materials science, of sample preparation for LC-MS/MS in the clinical laboratory. Taking robustness into account during method development and validation is more time-consuming, but in the long run saves time and money once a procedure is in production. The amount of effort needed to evaluate fitness for purpose can be balanced against the risk of compromised quality and production instability that are all too common with insufficiently optimized sample preparation protocols. The rewards for the additional work should be the most robust method at the lowest cost/reportable test.

REFERENCES

[1] Bylda C, Thiele R, Kobold U, et al. Recent advances in sample preparation techniques to overcome difficulties encountered during quantitative analysis of small molecules from biofluids using LC-MS/MS. Analyst 2014;139:2265–76.

[2] Baecher S, Geyer R, Lehmann C. Absorptive chemistry based extraction for LC-MS/MS analysis of small molecule analytes from biological fluids—an application for 25-hydroxyvitamin D. Clin Chem Lab Med 2014;52:363–71.

[3] Grant RP. High throughput automated LC-MS/MS analysis of endogenous small molecule biomarkers. Clin Lab Med 2011;31:429–41.

[4] Grebe SKG, Singh RJ. LC-MS/MS in the clinical laboratory—where to from here? Clin Biochem Rev 2011;32:5–31.

[5] Blanchard J. Evaluation of the relative efficacy of various techniques for deproteinizing plasma samples prior to high-performance liquid chromatographic analysis. J Chromatogr 1981;226:455–60.

[6] Polson C, Sarkar P, Incledon B, et al. Optimization of protein precipitation based upon effectiveness of protein removal and ionization effect in liquid chromatography–tandem mass spectrometry. J Chromatogr B 2003;785:263–75.

[7] Souverain S, Rudaz S, Veuthey JL. Protein precipitation for the analysis of a drug cocktail in plasma by LC-ECI-MS. J Pharm and Biomed Anal 2004;35:913–20.

[8] Thibeault D, Caron N, Djiana R, et al. Development and optimization of simplified LC–MS/MS quantification of 25-hydroxyvitamin D using protein precipitation combined with on-line solid phase extraction (SPE). J Chromatogr B 2012;883:120–7.

[9] Buchwald A, Winkler K, Epting T. Validation of an LC-MS/MS method to determine five immunosuppressants with deuterated internal standards including MPA. Clin Pharmacol 2012;12:2–11.

[10] Annesley TM, Clayton L. Simple extraction protocol for analysis of immunosuppressant drugs in whole blood. Clin Chem 2004;50:1845–8.

[11] Breaud AR, Harlan R, Kozak M. A rapid and reliable method for the quantitation of tricyclic antidepressants in serum using HPLC-MS/MS. Clin Biochem 2009;42:1300–7.

[12] Kuhn J, Knabbe C. Fully validated method for rapid and simultaneous measurement of six antiepileptic drugs in serum and plasma using ultra-performance liquid chromatography–electrospray ionization tandem mass spectrometry. Talanta 2013;110:71–80.

[13] Hall TG, Smukste I, Bresciano KR et al. Identifying and overcoming matrix effects in drug discovery and development, tandem mass spectrometry—applications and principles. In: Prasain J, editor. InTech; 2012. Available from: http://www.intechopen.com/books/tandem-mass-spectrometry-applications-and-principles/identifying-and-overcoming-matrix-effects-in-drug-discovery-and-development

[14] Trufelli H, Pierangela P, Faiglini G, et al. An overview of matrix effects in liquid chromatography-mass spectrometry. Mass Spectrom Rev 2011;30:491–509.

[15] Eeckhaut AV, Lanckmans K, Sarre S, et al. Validation of bioanalytical LC–MS/MS assays: evaluation of matrix effects. J Chromatogr B 2009;877:2198–207.

[16] Furey A, Moriarty M, Bane V, et al. Ion suppression: a critical review on causes, evaluation, prevention and applications. Talanta 2013;115:104–22.

[17] Côté C, Bergeron A, Mess J-N, et al. Matrix effect elimination during LC–MS/MS bioanalytical method development. Bioanalysis 2014;1:1243–57.

[18] Xia Y-Q, Jemal M. Phospholipids in liquid chromatography/mass spectrometry bioanalysis: comparison of three tandem mass spectrometric techniques for monitoring plasma phospholipids, the effect of mobile phase composition on phospholipids elution and the association of phospholipids with matrix effects. Rapid Commun Mass Spectrom 2009;23:2125–38.

[19] Ismaiel OA, Halquist MS, Elmamly MY, et al. Monitoring phospholipids for assessment of ion enhancement and ion suppression in ESI and APCI LC/MS/MS for chlorpheniramine in human plasma and the importance of multiple source matrix effect evaluations. J Chromatogr B 2008;875:333–43.

[20] Wu ST, Schoener D, Jemal M. Plasma phospholipids implicated in the matrix effect observed in liquid chromatography/tandem mass spectrometry bioanalysis: evaluation of the use of colloidal silica in combination with divalent or trivalent cations for the selective removal of phospholipids from plasma. Rapid Commun Mass Spectrom 2008;22:2873–81.

[21] Ismaiel OA, Zhang T, Jenkins RG. Investigation of endogenous blood plasma phospholipids, cholesterol and glycerides that contribute to matrix effects in bioanalysis by liquid chromatography/mass spectrometry. J Chromatogr B 2010;878:3303–16.

[22] Xinghua Guo, Ernst, Lankmayr. Phospholipid-based matrix effects in LC–MS bioanalysis. Bioanalysis 2011;3(4):349–52.

[23] Rappold B, Holland P, Grant R. The phospholipid fix: quantitative measurement and analytical solutions for phospholipid depletion. Conference abstracts and proceedings of the fifty-eighth American Society for mass spectrometry. Salt Lake City: UT; 2010. Poster 411/2018.

[24] Neveille D, Houghton R, Garrett S. Efficacy of plasma phospholipid removal during sample preparation and subsequent retention under typical UHPLC conditions. Bioanalysis 2012;4:795–807.

[25] ISOLUTE ® SLE+ User Guide, Biotage. Available from: http://www.biotage.com/literature/download/sle_user_guide_web.pdf?ref=http%3A%2F%2Fwww.biotage.com%2Fsearch%3Fq%3DUser%2520guide%2520SLE; 2014.

[26] Feng J, Lanqing W, Dai I, et al. Simultaneous determination of multiple drugs of abuse and relevant metabolites in urine by LC-MS-MS. J Anal Toxicol 2007;31:359–68.

[27] Burrai L, Nieddu M, Trignano C. LC-MS/MS analysis of acetaminophen and caffeine in amniotic fluid. Anal Methods 2015;7:405–10.

[28] Patela DS, Sharmab N, Patela MC. LC–MS/MS assay for olanzapine in human plasma and its application to a bioequivalence study. Acta Pharmaceutica Sinica B 2012;2:481–94.

[29] Marin SJ, McMillin GA. LC-MS/MS analysis of 13 benzodiazepines and metabolites in urine, serum, plasma, and meconium. Methods Mol Biol 2010;603:89–105.

[30] Wang R, Wang X, Liang C, et al. Direct determination of diazepam and its glucuronide metabolites in human whole blood by μElution solid phase extraction and liquid chromatography-tandem mass spectrometry. Forensic Sci Int 2013;233:304–11.

[31] Kuklenyik Z, Calafat AM, Barr JR, et al. Design of online solid phase extraction-liquid chromatography-tandem mass spectrometry (SPE-LC-MS/MS) hyphenated systems for quantitative analysis of small organic compounds in biological matrices. J Sep Sci 2011;34:3606–18.

[32] Couchman L. Turbulent flow chromatography in bioanalysis, a review. Biomed Chromatogr 2012;26:892–905.

[33] CLSI liquid chromatography-mass spectrometry methods; approved guideline. CLSI document C62-A. Wayne, PA: Clinical and Laboratory Standards Institute; 2014.

[34] Annesley TM. Ion suppression in mass spectrometry. Clin Chem 2003;49:1041–4.

[35] Vogeser M, Seger C. Pitfalls associated with the use of liquid chromatography-tandem mass spectrometry in the clinical laboratory. Clin Chem 2010;56:1234–44.

[36] Wieling J. LC-MS-MS experiences with internal standards. Chromatographia 2002;55:S107–13.

[37] Liang HR, Foltz RL, Meng M, et al. Ionization enhancement in atmospheric pressure chemical ionization and suppression in electrospray ionization between target drugs and stable-isotope-labeled internal standards in quantitative liquid chromatography/tandem mass spectrometry. Rapid Commun Mass Spectrom 2003;17:2815–21.

[38] Lindegardh N, Annerberg A, White NJ, et al. Development and validation of a liquid chromatographic-tandem mass spectrometric method for determination of piperaquine in plasma Stable isotope labeled internal standard does not always compensate for matrix effects. J Chromatogr B 2008;862:227–36.

[39] Wang S, Cyronak M, Yang E. Does a stable isotopically labeled internal standard always correct analyte response? A matrix effect study on a LC/MS/MS method for the determination of carvedilol enantiomers in human plasma. J Pharm Biomed Anal 2007;43:701–7.

[40] Tana A, Hussaina S, Musukub A, et al. Internal standard response variations during incurred sample analysis by LC–MS/MS: case by case trouble-shooting. J Chromatogr B 2009;877:3201–9.

[41] Jemal M, Schuster A, Whigan DB. Liquid chromatography/tandem mass spectrometry methods for quantitation of mevalonic acid in human plasma and urine: method validation, demonstration of using a surrogate analyte, and demonstration of unacceptable matrix effect in spite of use of a stable isotope analog internal standard. Rapid Commun Mass Spectrom 2003;17:1723–34.

[42] Matuszewski BK, Constanzer ML, Chavez-Eng CM. Strategies for the assessment of matrix effect in quantitative bioanalytical methods based on HPLC-MS/MS. Anal Chem 2003;5:3019–30.

[43] Grant R, Rappold B. Short course handbook, Development and validation of quantitative LC-MS/MS assays for use in clinical diagnostics. San Diego, CA: Mass spectrometry applications in the clinical laboratory (MSACL); 2013.

[44] Bonfiglio R, King RC, Olah TV, et al. The effects of sample preparation methods on the variability of the electrospray ionization response for model drug compounds. Rapid Commun Mass Spectrom 1999;13:1175–85.

[45] Kushnir MM, Rockwood AL, Bergquist J. Liquid chromatography-tandem mass spectrometry applications in endocrinology. Mass Spectrom Rev 2010;29:480–502.

[46] Kushnir MM, Rockwood AL, Roberts WL, et al. Liquid chromatography tandem mass spectrometry for analysis of steroids in clinical laboratories. Clin Biochem 2011;44:77–88.

[47] Foley D, Keevil B, Calton L. Analysis of serum testosterone and androstenedione for clinical research using either manual or automated extraction. American Association for Clinical Chemistry Annual Meeting, Abstract A-415; 2014.

[48] Keevil BG, Hawley J, Foley D, et al. A highly sensitive method for aldosterone analysis using LC-MS/MS. Endocrine Society Annual Meeting. Poster FRI-367; 2015.

[49] Vogeser M, Kirchhoff F. Progress in automation of LC_MS in laboratory medicine. Clin Biochem 2011;44:4–13.

[50] Li G, Aubry A-F. Best practices in biological sample preparation for LC-MS bioanalysis. In: Li W, Zhang J, Tse FL, editors. Handbook of LC-MS bioanalysis, best practices, experimental protocols, and regulations. Hoboken, NJ: John Wiley & Sons, Inc; 2013.

[51] Hill HM, Smith GT. Evaluation and elimination of carryover and/or contamination in LC-MS bioanalysis. In: Li W, Zhang J, Tse FL, editors. Handbook of LC-MS bioanalysis, best practices, experimental protocols, and regulations. Hoboken, NJ: John Wiley & Sons, Inc; 2013.

[52] Pawula M, Hawthorne G, Smith GT, et al. Best practice in biological sample collection, processing, and storage for LC-MS in bioanalysis of drugs. In: Li W, Zhang J, Tse FL, editors. Handbook of LC-MS bioanalysis, best practices, experimental protocols, and regulations. Hoboken, NJ: John Wiley & Sons, Inc; 2013.

[53] Razavi A, Bayer D. US patents 20,040,096,622, US 20,030,077,440 A1. Multi well plate with self sealing advantages; 2003; 2004.

[54] Unger S, Weng N. Best practice in liquid chromatography for LC-MS bioanalysis. In: Li W, Zhang J, Tse FL, editors. Handbook of LC-MS bioanalysis, best practices, experimental protocols, and regulations. Hoboken, NJ: John Wiley & Sons, Inc; 2013.

[55] Silvester S, Zang F. Overcoming non-specific adsorption issues for AZD9164 in human urine samples: consideration of bioanalytical and metabolite identification procedures. J Chromatogr B 2012;893–894:134–43.

[56] Rodila R, Kim GE, Fan L, et al. HPLC-MS/MS determination of a hardly soluble drug in human urine through drug-albumin binding assisted dissolution. J Chromatogr B 2008;872:128–32.

[57] Stout PR, Horn CK, Lesser DR. Loss of THCCOOH from urine specimens stored in polypropylene and polyethylene containers at different temperatures. J Anal Toxicol 2000;24:567–71.

[58] Roth KDW, Siegel NA, Johnson RW, et al. Investigation of the effects of solution composition and container material type on the loss of 11-nor-delta9-THC-9-carboxylic acid. J Anal Toxicol 1996;20:291–300.

[59] Jamerson MH, McCue JJ, Klette KL. Urine pH, container composition, exposure time influence adsorptive loss of 11-nor-delta9-tetrahydrocannabinol-9-carboxylic acid. J Anal Toxicol 2005;29(7):627–31.

[60] Hochrein H, Akin J, Stone J, et al. Optimization and validation of cannabinoid metabolite confirmation in urine using LC-MS/MS and biotage EVOLUTE EXPRESS AX SPE cartridges. Abstracts and proceedings of the seventh mass spectrometry applications in the clinical laboratory. San Diego, 2015. Poster #15; 2015.

[61] Chambers E, Wagrowski-Diehl DM, Ailing L. Systematic and comprehensive strategy for reducing matrix effects in LC/MS/MS analyses. J Chromatogr B 2007;852:22–34.

[62] Lahaie M, Mess JN, Futado M, et al. Elimination of LC-MS/MS matrix effect due to phospholipids using specific solid phase extraction elution conditions. Bioanalysis 2010;2:1011–21.

[63] McDonald PD, Bouvier ESP, editors. Solid phase extraction applications guide and bibliography: a resource for sample preparation methods development. 6th ed. Milford, MA: Waters Corporation; 2001.

[64] Lee M-R, Yu S-C, Lin C-L, et al. Solid phase extraction in amphetamine and methamphetamine analysis of urine. J Anal Toxicol 1997;21:278–82.

[65] Clark ZD. Three strategies to stop a pervasive high-throughput sample preparation quality problem you likely didn't know you had. Abstracts and proceedings of the sixth mass spectrometry applications in the clinical laboratory. San Diego. Poster # 2.a; 2014.

[66] Clark ZD. Diluting & shooting yourself in the foot: complications with sample-to-sample variations in signal suppression. Abstracts and proceedings of the fifth mass spectrometry applications in the clinical laboratory. San Diego. Podium presentation; 2013.

[67] Instructions & Troubleshooting for Hybrid SPE®—Phosphlipid Ultra cartridge. Bellefonte, PA Sigma-Aldrich Co. Available from: https://www.sigmaaldrich.com/content/dam/sigma-aldrich/docs/Supelco/Data-sheet/1/T710124.pdf; 2010.

[68] Scheidweiler KB, Desrosiers NA, Huestis MA. Simultaneous quantification of free and glucuronidated cannabinoids in human urine by liquid chromatography tandem mass spectrometry. Clin Chim Acta 2012;413:1839–47.

[69] Cao H, Jiang H, Fast D. Impact of extraction conditions on matrix effect and recovery efficiency by supported liquid extraction in quantitative LC-MS/MS. Conference abstracts and proceedings of the fifty-ninth American Society for mass spectrometry. Denver, CO 2011. Poster 152/485; 2011.

VALIDATION, QUALITY CONTROL, AND COMPLIANCE PRACTICE FOR MASS SPECTROMETRY ASSAYS IN THE CLINICAL LABORATORY

D.F. Stickle*, U. Garg**

*Department of Pathology, Jefferson University Hospital, Philadelphia, PA, United States;
**Department of Pathology and Laboratory Medicine, Children's Mercy Hospital, Kansas City, MO, United States

1 INTRODUCTION

Mass spectrometry was once considered to be too specialized and expensive technology for routine use in clinical laboratories. Whereas gas chromatography mass spectrometry (GC-MS) has been in use for several decades in specialized clinical laboratory services, liquid chromatography tandem-mass spectrometry (LC-MS/MS) has been gaining in prevalence for a wide and growing variety of uses in general clinical laboratories [1–6]. While GC-MS is used mostly in toxicology laboratories, LC-MS/MS is being used in a wider array of fields, such as toxicology, therapeutic drug monitoring, endocrinology, biochemical genetics, and newborn screening [7–13]. For common analytes (e.g., 25-OH vitamin D, immunosuppressants), a primary driving force for adoption of mass spectrometry has been substantial cost savings over commercial immunoassays; for many other analytes, the driving force has been the ability to produce analytically superior results, or results that cannot efficiently be produced by any other current means (e.g., steroid analysis, toxicology, and metabolic screening). In all cases, these are laboratory-developed tests (LDTs). And, in all cases, development of an assay with suitable performance characteristics is a major investment in time and cost. Such investment is substantial even if only to reproduce and validate an assay for which a method exists in the literature. There is overlap, certainly, between development and validation, but a validation exercise to fully document the performance characteristics of the method is an essential step for bringing an assay online for production of patient results [14–16]. Validation includes numerous elements according to regulatory and professional standards. In this chapter, we will describe the essential elements of GC-MS and LC-MS/MS assay validation, with additional discussion of elements of quality control (QC) and regulatory compliance that apply to mass spectrometry LDTs.

2 ELEMENTS OF ASSAY VALIDATION ACCORDING TO CLIA, CAP, AND FDA GUIDELINES

There are hundreds of variables and parameters that affect the performance of a mass spectrometry assay. Method development seeks to optimize conditions so as to lead to an assay that is successful for its intended use. Although there is almost always the possibility that some refinement in operating parameters could lead to performance improvement in some way, there is some point in development at which it is judged that an assay is suitable for its intended purpose. Once an assay is developed to this stage, it must be fully characterized in such a way as to document that its performance can meet requirements of intended use, and, additionally, that its performance has been documented with respect to requirements of professional practice and/or accreditation standards. This exercise is referred to as validation, which takes place once all aspects of the method have been stipulated and prescribed.

There are multiple sources of guidelines that delineate elements and/or requirements of validation. Our objective is to discuss major resources and to highlight certain elements of practice. In the United States, essential sources of guidance for clinical laboratories include at a minimum regulations associated with the College of American Pathologists (CAP) accreditation standards for chemistry and toxicology [17] and the Clinical Laboratory Improvements Amendments (CLIA) [18,19] within the US Code of Federal Regulations. Professional practice and diagnostic industry guidelines, such as Clinical and Laboratory Standards Institute (CLSI) documents on mass spectrometry [20,21], and the US Food and Drug Administration (FDA) Guidance for Industry on bioanalytical method validation [22] are additional resources. Generally, CAP and CLIA guidelines specify elements of performance characteristics which must be documented in validation, but they do not specify how such demonstration should be made, or by what criteria they should be deemed as acceptable. The FDA guidelines, however, are a succinct source of information regarding standards of procedures and evaluation whereby validation of an assay can be made, and these standards are widely accepted as minimum standards for validation. In this light we will first summarize elements of validation according to CLIA regulations, and expand on methods and criteria for acceptable results according to the FDA document. For simplicity, we assume that all methods under consideration utilize the ratio of analyte signal to internal standard signal as the primary measure, and that all analyses are processed in batch forms.

CLIA regulations (Subpart K—Quality System for Nonwaived Testing) [19] specify requirements for establishment and verification of performance specifications for methods developed in-house. These include characterization of the elements shown in Table 4.1

Each of these elements is reasonably straightforward to interpret with respect to accepted definitions. Later, each of these elements is discussed in terms of design of validation and acceptance criteria according to guidelines.

Table 4.1 CLIA Verification of Performance Elements
1. Accuracy
2. Precision
3. Analytical sensitivity
4. Analytical specificity to include interfering substances
5. Reportable range of test results for the test system
6. Reference intervals (normal values)
7. Any other performance characteristic required for test performance

Accuracy refers to the ability of mean test results to reproduce the concentrations of known standards. According to the FDA guidelines, accuracy should be determined using at least five determinations per concentration, with a minimum of three concentrations utilized covering the expected range of patient samples. Acceptance criteria are that the mean value should be within 15% of the nominal value except at the lower limit of quantitation (LLOQ), where deviation ≤ 20% is allowable.

Accuracy of an assay can be determined by various means. One approach to evaluate accuracy is to compare the results of the method in development with the results from a well-established reference method. Another method of assessing accuracy is to compare the measured concentrations with the known concentrations of a certified material. When these approaches are not available or possible, a means of determination of recovery to evaluate accuracy is used. This is performed by spiking the analyte of interest in the analyte-free blank matrix, and comparing the measured values with the target values. Ideally, commercially available matrix-matched certified standards/materials should be used when available.

Precision refers to statistics of reproducibility, namely, the closeness of values surrounding the mean. Precision studies are generally conducted using materials intended to be used in routine QC, provided that they are of the same biological matrix as the intended samples. This is typically characterized as the coefficient of variation (CV) (%CV, standard deviation/mean ×100%), assuming that precision data are normally distributed. According to the FDA guidelines, precision should be assessed using at least five determinations per concentration, with a minimum of three concentrations utilized within the range of expected patient concentrations. Acceptance criteria are that CVs should not exceed 15% of the CV except for the LLOQ, where CV < +20% is allowable. Precision assessment should be made for within-run/intrabatch studies, as well as for between-run (interbatch) studies over time. Interbatch studies are typically conducted across users and across days. CLSI guidelines are available for evaluating and verifying precision [23,24].

The FDA guidelines also recommend that both accuracy and precision analyses be performed on diluted specimens having initial concentrations that are above the linear range of the assay. Dilutions of such specimens should ideally occur in a matrix-matched diluent, however, it may be possible to show consistent acceptability of results in simpler matrices.

Note that it is often the case in the literature that results of precision studies are reported as mean ± CV, with only the implied premise that the input data for such calculations are normally distributed, that is, the calculated CVs are meaningful with respect to probabilities of results according to the characteristics of a normal distribution. While the premise is almost certainly correct for most precision studies data, it is advisable nonetheless, in validation studies to verify/document that this is the case by analysis of data distributions, for example, by assessment using a normality probability plot [25].

Analytical sensitivity refers to limits of detection and/or quantitation. For most clinical assays, the lower limit of quantitation (LLOQ) and the upper limit of quantitation (ULOQ) will be most important, as they will define the linear range of the assay. LLOQ can be determined by analyzing samples with decreasing concentrations of the analyte. LLOQ is the lowest concentration at which acceptable precision and accuracy can be obtained. According to FDA guidelines, the lowest calibration standard should be accepted as the LLOQ if the following criteria are met: the response at that concentration should be at least 5 times of the zero blank response; calculated concentration should not deviate more than 20% from the nominal assigned value of the standard; precision (CV) should not exceed 20% at LLOQ. For ULOQ, FDA recommends that this be the concentration of the highest calibration standard, assuming accuracy of mean deviation of < 15% of the nominal assigned value, and CV of less than

15%. It is recommended that at least five replicate measurements be used to establish these properties for LLOQ and ULOQ.

Analytical specificity or selectivity refers to the characterization of cross-reactivities of the assay with unintended targets. Ideally, the effects of the presence of nontarget compounds on quantitation of results should be zero. However, this is often not possible to demonstrate definitively, and therefore a wide range of potential interferences should be evaluated. FDA guidelines recommend testing for interference using blank samples obtained from at least six sources near the LLOQ. Additionally, potential interference should be investigated for expected medications within the likely subject population. This investigation should include samples from patients receiving these medications, as metabolites may be interferents even when parent drugs are not. Thus, the range of the experimentation needed for these studies may be quite large. As compared to other techniques, such as immunoassays and spectrophotometry, mass spectrometry assays generally are relatively more specific and less prone to interferences.

Reportable range, also called analytical measurement range (AMR) or linearity, refers to a quantitative range of concentrations of analyte over which a valid quantitative result may be obtained without dilution. As noted above, generally this will be within the limits of the lowest and highest standards used to produce a calibration curve. Reportable range can be calculated by analyzing 4–6 concentrations covering the whole range. Measured values are compared with target values within allowable error. CLSI guidelines are available for evaluating linearity [26].

Reference intervals refers to the range of values found in a normal or reference population of subjects. For endogenous compounds, by convention, a reference range is usually set to encompass the central 95% of results for a normal or reference population [27]. For new analytes for which there are no predicate assays available, establishment of a reference range may be an extensive undertaking. How subjects are selected as normal is a necessary first step. The number of subjects needed may depend on preliminary results, for example, whether the normal range can be established as parametric (i.e., data are normally distributed, for which fewer data points are required) or not (for which data must be sufficient in number to be able to resolve the 2.5th and 97.5th percentiles). For analytes for which predicate assays are available, *method comparison* studies can assist in determination and/or transference of reference intervals. If a reference interval exists from another method, then correlation studies indicating high correlation may be able to provide sufficient data either to corroborate or to modify the reference range associated with the predicate method. CLSI guidelines are available for establishing and verifying reference ranges [28].

Verification of other performance characteristics referred to in the CLIA regulations represents potentially an open-ended list of elements to consider. It is recommended that these elements should include those of the FDA guidelines not specifically mentioned earlier. These elements are as follows:

Recovery should be assessed and documented by comparison of results of signal for analyte and internal standard from spiked matrix samples versus spiked unextracted solvent solutions. As stated in FDA guidelines, recovery of the analyte need not be 100%; the assessment should be able to demonstrate that the recovery is consistent and document the extent of variation recovery among different spike matrix samples. It is recommended that recovery experiments should be conducted using three concentrations across the expected range of patient samples by comparison of results to those from unextracted standards that represent 100% recovery. For assays with low recovery, it is important to demonstrate that accuracy and precision are not affected, particularly near the LLOQ.

Calibration studies are more part of method development than of validation per se, as all results for accuracy and precision will depend critically on calibration. It should be documented how exactly calibration is achieved; specifically, the dependent variable (typically, the ratio of integrals of analyte to internal standard), the number and composition of calibration samples, the frequency of calibration, the nature of data reduction for calibration (e.g., whether weighting of different concentrations is applied; whether a fixed zero intercept is applied), and criteria for acceptance of the calibration curve. It is common to run a calibration curve with every analytical run, however, it has been demonstrated for some assays this may not be necessary for good performance [29].

FDA guidelines recommend that calibrators should be made in the same biological matrix as the samples and that they correspond to the expected range of sample analyte concentrations. It is recommended that the curve should consist of a blank sample (matrix sample without analyte or internal standard), a zero sample (matrix sample with internal standard only), and at least six nonzero samples (matrix samples having both analyte and internal standard) including one at LLOQ. For method validation, it is recommended that experiments should involve a minimum of six runs conducted over separate days, with at least four concentrations (including LLOQ, low, medium, and high) analyzed in duplicate in each run. Obviously, these experiments correspond to those used to establish accuracy with respect to criteria for acceptability of results.

Stability of specimens should be documented in validation studies. These should include evaluation of stability of specimens at different stages in the process of analysis (preanalysis, during an analysis cycle, and postanalysis), anticipating the variations that may occur in routine operations of collection, shipping, storage, processing, and reanalysis. The FDA guidelines recommend evaluation in a number of categories mentioned later, using at least three replicates at low and high analyte concentrations, with comparison made to calibrator or QC samples; acceptable results should be at least within 15% of nominal concentrations. *Freeze and thaw stability* should be evaluated for a minimum of three cycles. *Bench top stability* should be evaluated at room temperature for a period of time to exceed the characteristic processing time. *Long term stability* should be evaluated under storage conditions for a period to exceed that characteristic of normal sample handling. *Stock solution stability* should also be evaluated so as to establish appropriate storage intervals. *Processed sample stability,* for example, the stability of extracted samples, should also be evaluated.

Incurred sample reanalysis (ISR) is recommended by FDA guidelines as a component of validation studies. ISR involves repeat analysis of patient samples across runs on different days. Such analyses are viewed as providing "critical support" for the accuracy and precision studies previously established using calibrator and QC samples. Essentially, incurred sample reanalysis should reproduce each sample's prior data according to preestablished acceptable degrees of accuracy and precision.

Although the general intent of the CLIA regulations is easily understood, it is important to note that there are few exact definitions of elements, and, more importantly, there are few specifications with respect to how these elements should be evaluated. Nor, correspondingly, are there always associated criteria within the CLIA regulations whereby results of characterization of elements are deemed to be acceptable. Meeting the true intent of these regulations, then, should include not just characterization of results but also specification by the director in documentation of the exact definition of the element (e.g., how specifically analytical sensitivity was defined, how a normal population was defined, etc.), a delineation of the design of the evaluation, and a delineation of the criteria used to determine that the performance characteristics for each element were deemed to be acceptable with respect to intended use. Per above, meeting FDA guideline recommendations as described earlier should be considered a minimum standard for validation.

2.1 CAP INSPECTION CHECKLIST ELEMENTS RELATED TO MASS SPECTROMETRY ASSAY VALIDATION

CAP inspection checklists are directly related to documentation of method validation. Many aspects of validation are in general checklists (e.g., "Are procedures adequate?"). There are currently a small number of entries specifically targeted to operation of mass spectrometry assays in the chemistry and toxicology checklist [17]. These are as follows:

1. There are documented procedures for operation and calibration of the mass spectrometer.
   This question relates specifically to tuning of the mass spectrometer itself. The specification is that the mass spectrometers are tuned each day of patient testing, or that tuning occurs according to manufacturer's recommendations. In GC-MS, mass spectrometer is typically tuned everyday using autotune. In LC-MS/MS, tuning generally occurs on the schedule of preventive maintenance as recommended and performed by the manufacturer of the instrument. Records of tuning should be maintained, as well as documentation of criteria whereby tuning procedures are deemed to be acceptable.

2. There are identification criteria for single stage mass spectrometry (i.e., GC-MS, LC-MS) that are in compliance with recommendations.
   This question applies to single stage MS, such as GC-MS or LC-MS, and indicates that there should be specific criteria for results analysis with respect to compound identification by ion ratios or ion ratios plus retention time. For quantitative analyses, specifications state that the assay should have full documentation of validation. Typically three ion ratios are used for GC-MS, and two ion ratios for LC-MS assays. However, for some analytes there may be only few characteristic ions, and two to three ratios may not be possible. In this situation, one ion ratio of two characteristic ions may be acceptable if another identifying characteristic, for example, retention time, is used in analyte identification.

3. The identification criterion for tandem mass spectrometry (MS/MS) are validated and documented.
   This question specifies simply that for tandem mass spectrometry using multiple reaction monitoring, there is at least one transition monitored for the internal standard and another for the analyte. Retention time is also generally used in identification criterion.

4. There is documentation of assessment of matrix effects in LC-MS test development.
   Matrix effects refers to the influence of coeluting compounds to alter ionization of the target analyte [30–32]. This question specifies that evaluation of matrix effects on ionization must be part of the documentation of assay validation. Moreover, the question specifies that matrix effects analysis must be repeated during periodic revalidations of the assay. As noted in the CAP inspection checklist, two methods are common in making this assessment: postcolumn infusion and spiked matrix analyses [30]. In postcolumn infusion experiments, blank matrix injected as sample is mixed postcolumn with a constant rate of infusion of analyte in mobile phase solvent. Assessment is to determine whether the ionization of analyte is affected, particularly at the known retention time for the analyte in patient samples. A second, simpler method to assess matrix effects is to evaluate the difference in ion currents/counts between analyte spiked into blank matrix versus mobile phase solution. The second method is most useful and reliable when analyte recovery is high.
   CAP commentary for this checklist item is unusual in that it makes specific recommendations regarding how investigation should be conducted. The checklist item states that at least 10

different sources of matrix should be tested for matrix effects. Average ion suppression/enhancement should not exceed 25%, and %CV of the ion suppression/enhancement should not exceed 15%; otherwise, data to demonstrate that assay accuracy are not adversely affected must be part of the assay validation documentation, with the recommendation that validation of accuracy under this circumstance should be by assessment of variation of the calibration curve slope according to whether calibrators are made up within mobile phase solution versus within a series of blank sample matrices.

5. The laboratory LC-MS assay procedure includes an evaluation for possible ion suppression or enhancement in patient samples during routine testing.

This checklist item specifies that there should be routine monitoring of the signal intensity of internal standards. Aberrant signal intensity relative to standards, QC or other samples can indicate either an error in sample processing or signal interference in a single patient sample that may be due to unusual components of the matrix. Monitoring of signal intensity of internal standard across samples is often referred to as a "metric plot." Determination of aberrance of signal intensity for a given sample is generally by a fixed criterion of being outside of the bounds of a certain percentage of the average intersample signal intensity (e.g., ±25%) or on signal/noise ratio. Samples not meeting acceptance criteria should be repeated, or investigated further by means, specified in procedures.

2.2 CONSIDERATIONS REGARDING USE OF DRIED BLOODSPOTS (DBS) AS A SAMPLE TYPE

Blood and urine are the most commonly used sample types for mass spectrometry assays. However, there has been recently a great expansion in literature regarding use of dried bloodspots (DBS) as a sample type. For this reason, specific commentary on method validation for DBS is warranted. Interest in utilization of DBS has been driven largely by the interest in pharmaceutical development to utilize DBS in clinical trials, in part as a matter of great potential for operational cost savings [33,34]. Development of LC-MS/MS technology and assays has made this possible [35]. Longstanding use of DBS in newborn screening has demonstrated outstanding success of use of DBS in semi-quantitative analyses [36–38]. For quantitation, intersample variation in hematocrit as a preanalytical variable is cited as a common complication for some quantitative assays [39,40]. Recently, use of potassium measurement has been demonstrated as a means to estimate and correct for variation in hematocrit [41–43].

In the context of increased interest and use of DBS, current draft versions of the FDA guidelines for method validation include discussion of DBS, noting that a comprehensive validation of use of DBS is essential. The guidelines specify that validation should address the following elements: effects of temperature in storage and handling temperature, homogeneity of sample spots, effects of varying hematocrit, stability, carryover, and reproducibility including incurred sample reanalysis. Studies showing correlation with traditional samples should also be conducted.

A number of nuances concerning DBS should be noted. First, there is a significant accretion of red cells at the perimeter of filter paper DBS [44]. In validation studies, this property can affect apparent net recovery of red cell-associated analytes as calculated from the punch area relative to total DBS area [44]. In general, it is therefore advisable that DBS assays should use DBS standards [45]. Second, whereas hematocrit can affect volume per DBS area, it is not the sole interpatient variable affecting this parameter [46]. Generally, DBS volume per area is increased with hematocrit for filter paper cards; in contrast, noncellulosic collection cards show the reverse relationship, with DBS volume per area decreasing with hematocrit [46]. Ultimately, whole DBS excision after volumetric application of blood is likely to be the solution to the counter the influences of interpatient variation in bloodspot spreading properties on quantitative analyses. To this end, a novel combined card plus capillary device for controlled volume application of blood to filter paper has recently been described. Last, it should be noted that there are numerous options that have been investigated with respect to how internal standard is introduced into the analysis, such as pretreatment of cards with internal standard [47]. These various options are likely to be explored in DBS method development and validation. There are numerous resources that specifically address validation of DBS assays [48–52].

3 ELEMENTS OF QUALITY CONTROL

It is commonplace for routine QC to follow some subset of rules established by Westgard [53] to determine acceptability of a run, based on statistical parameters of previous runs of quality controlled materials. Routine QC may be less than that conducted during validation. For quantitative assays, CAP regulatory requirements are for at least two levels of control per day. FDA recommends three controls: low, mid, and high. The low control should be within 3 times of the LLOQ, mid control in the midrange, and high control approaching the high end of the AMR. Depending on availability of assayed control materials, laboratories may choose to run either assayed or unassayed control materials.

Procedures must be in place for establishment of statistical parameters for use of control materials. Once the statistical parameters are established (means and standard deviations (SD) of replicates), the acceptability of controls is evaluated using some combination thereof (e.g., mean $\pm$ 2 or 3 SD, or mean $\pm$ 20%) within and across runs using various statistical rules (e.g., Westgard rules). Theoretically, the application of such rules reduces the probability of release of inaccurate results on patient samples.

There are no CLIA rules on acceptability of controls. According to FDA guidelines, three controls should be run in duplicate and at least 67% (four out of six) of the QC samples should be within 15% of their respective nominal (theoretical) values; 33% of the QC samples (not all replicates at the same concentration) can be outside the $\pm$15% of the nominal value.

In practice, there is a widened scope of what might be considered QC for mass spectrometry assays that go beyond the measurement of designated QC samples. For example, metric plots as discussed earlier are used as a form of QC applied to individual samples. There may be additional single-sample control elements, such as specifications for ratios of qualifier ions. General QC may include specifications of operating conditions on start-up (e.g., column pressures, baselines, minimum peak heights) compared to historical readings. Such items may be part of a daily "start-up" checklist for running of the assay.

4 REGULATORY COMPLIANCE

Most laboratories that introduce mass spectrometry will be doing so in the framework of existing accreditation of the laboratory at-large. The requirements for regulatory compliance will thus be in common with other laboratory sections and familiar. It is important that documentation for the new assay includes all of the elements needed for inspection of nonwaived, high-complexity testing in general.

In particular, CAP and CLIA regulations specify necessity of written procedure manuals for all aspects of activities needed to successfully perform the assay. These are documents that would already have been produced along with a method validation. These include specimen requirements (patient preparation; specimen collection, labeling, storage, preservation, transportation, processing; criteria for specimen acceptability, and rejection), test calculations and interpretation of results, calibration and calibration verification procedures, control procedures, and maintenance procedures.

4.1 PROFICIENCY TESTING

CLIA and CAP regulations require proficiency testing for accreditation of laboratories. Generally, proficiency programs are available for common analytes by subscription, with challenges having a fixed schedule and number of specimens so as to meet regulatory requirements. However, for many analytes that may be measured by mass spectrometry, there may be no formal proficiency testing programs available. In such cases, alternative means of proficiency testing must be established. These include measurement of assayed materials, comparison of split sample measurements with other laboratories, or alternative measurement method comparisons. In such cases, it will be the responsibility of the laboratory director to specify frequency of testing and acceptance criteria.

4.2 COMPETENCY ASSESSMENT

The Center for Medicare and Medicare Services (CMS) and CAP have relatively recently established a rather elaborate set of requirements to provide evidence of competency of personnel performing any nonwaived form of patient testing in the laboratory [54,55]. These requirements as follows:

1. Direct observations of routine patient test performance, including patient preparation, if applicable, specimen handling, processing, and testing.
2. Monitoring the recording and reporting of test results.
3. Review of intermediate test results or worksheets, QC records, proficiency testing results, and preventive maintenance records.
4. Direct observations of performance of instrument maintenance and function checks.
5. Assessment of test performance through testing previously analyzed specimens, internal blind testing samples, or external proficiency testing samples.
6. Assessment of problem solving skills.

Competency assessment, which includes the six procedures, must be performed for testing personnel for each test that the individual is approved by the laboratory director to perform. All tests performed simultaneously on the same testing platform may be combined as long as there are no unique aspects, problems or procedures associated with any test on the testing platform. However, any test with unique aspects, problems or procedures within the same testing platform should be assessed separately to ensure that staff maintain their competency to report test results promptly, accurately, and proficiently [54]. Laboratories will need to design means of completion and maintain records of completion for each of these requirements, and do so separately for each of the assays performed by mass spectrometry. Evaluation and documentation of competency of personnel responsible for testing is required at least semi-annually during the first year of testing, and at least annually thereafter.

4.3 PROPOSED NEW FEDERAL REGULATIONS FOR LABORATORY-DEVELOPED TESTS

Recently, the FDA issued a draft proposal regarding regulatory oversight of LDTs, effectively announcing their intent to require registration of all LDTs with the FDA [56,57]. Virtually all mass spectrometry assays in clinical use would fall in the category of LDTs.

The draft guidance document states that CLIA governs the accreditation, inspection, and certification process for laboratories, and CLIA requirements address the laboratory's testing process. However, according to the FDA, accreditors do not evaluate test validation under CLIA prior to marketing nor do they assess the clinical validity of a LDT (i.e., the accuracy with which the test identifies, measures, or predicts the presence or absence of a clinical condition, or predisposition in a patient). Under the FD&C Act [58], the FDA asserts that it is obliged to assure both the analytical validity (e.g., analytical specificity and sensitivity, accuracy, and precision) and clinical validity of diagnostic tests through its premarket clearance or approval process.

Meeting FDA registration requirements for LDTs would significantly increase the regulatory burden of laboratories providing mass spectrometry analyses. This action by the FDA has been opposed by professional societies, citing the quality oversight already provided by CLIA regulations and accreditation requirements [59–61]. This issue is currently unresolved, but it is important for laboratories to be aware of proposed changes in regulations, and to follow their development in order to be prepared to face an increased level of regulation for LDTs.

5 ADDITIONAL RESOURCES FROM THE LITERATURE

There are numerous additional resources in primary literature and elsewhere that provide guidance on method validation. The European Medicines Agency's "guideline on bioanalytical method validation" [62] parallels the US FDA guidance to industry [57]. The CLSI documents on mass spectrometry [20,21], although focused primarily on method development, contain excellent and carefully considered discussion of the myriad facets of mass spectrometry that pertain also to method validation. Numerous review articles provide guidance on method validation [9,63–67]. Instrument manufacturers also provide guidance; for instance, a document from Agilent Technologies [68] is an excellent example of such resources. Professional societies, such as the Royal Society of Chemistry (RSC) have produced valuable documents concerning method development and validation [69]. Moreover, instructions to authors from various journals (e.g., Clinical Chemistry [70]) may provide specific instructions and guidance regarding elements of validation needed for publication of new methods; specific examples of papers presenting method validations correspondingly can also be useful (e.g., Annesley and Clayton [71]). In all cases of formal guidance documents, it is likely that each will evolve with periodic review and updates as advances in mass spectrometry instrumentation and techniques take hold [72,73].

6 SUMMARY

Accredited clinical laboratories must produce and maintain documents pertaining to validation of in-house derived mass spectrometry methods. At a minimum, validation elements should include those included in CLIA and CAP regulations. There is judgment to be applied as to how such studies are conducted and what criteria are applied for acceptance. FDA guidance concerning method development is widely used as a standard of reference for means and criteria of acceptance of validation studies. Numerous publications provide additional resources for reference.

It is important to note that, even with availability of numerous sources of guidance, many elements of practice require discretion, judgment, and specification by the laboratory director regarding means of evaluation and criteria for acceptance of performance characteristics, for example, whether imprecision is within limits compatible with intended use; whether the extent of investigation of possible interferents is deemed adequate; and etc. There is great overlap of elements of validation with elements in method development, and it is useful therefore to have method validation in mind during all phases of method development. Postvalidation, there are additional elements of documentation that are needed to meet regulatory and accreditation requirements (e.g., maintenance and personnel records). Last, laboratory directors should be aware of the possibility that mass spectrometry assays in the United States may soon be subject to regulatory oversight by the FDA, which circumstance will likely greatly increase the amount of documentation required to maintain or bring these assays online to produce patient results.

REFERENCES

[1] Wu AH, French D. Implementation of liquid chromatography/mass spectrometry into the clinical laboratory. Clin Chim Acta 2013;420:4–10.
[2] Grebe SK, Singh RJ. LC-MS/MS in the clinical laboratory—where to from here? Clin Biochem Rev 2011;32(1):5–31.
[3] Kushnir MM, Rockwood AL, Bergquist J. LC-MS/MS in clinical laboratories. Bioanalysis 2013;5(1):5–6.
[4] Leung KS, Fong BM. LC-MS/MS in the routine clinical laboratory: has its time come? Anal Bioanal Chem 2014;406(9–10):2289–301.
[5] van den Ouweland JM, Kema IP. The role of liquid chromatography-tandem mass spectrometry in the clinical laboratory. J Chromatogr B Analyt Technol Biomed Life Sci 2012;883–884:18–32.
[6] Strathmann FG, Hoofnagle AN. Current and future applications of mass spectrometry to the clinical laboratory. Am J Clin Pathol 2011;136(4):609–16.
[7] Viette V, Fathi M, Rudaz S, Hochstrasser D, Veuthey JL. Current role of liquid chromatography coupled to mass spectrometry in clinical toxicology screening methods. Clin Chem Lab Med 2011;49(7):1091–103.
[8] Peters FT. Recent advances of liquid chromatography-(tandem) mass spectrometry in clinical and forensic toxicology. Clin Biochem 2011;44(1):54–65.
[9] Bozovic A, Kulasingam V. Quantitative mass spectrometry-based assay development and validation: from small molecules to proteins. Clin Biochem 2013;46(6):444–55.
[10] Carvalho VM. The coming of age of liquid chromatography coupled to tandem mass spectrometry in the endocrinology laboratory. J Chromatogr B Analyt Technol Biomed Life Sci 2012;883–884:50–8.
[11] Kushnir MM, Rockwood AL, Bergquist J. Liquid chromatography-tandem mass spectrometry applications in endocrinology. Mass Spectrom Rev 2010;29(3):480–502.
[12] Kushnir MM, Rockwood AL, Roberts WL, Yue B, Bergquist J, Meikle AW. Liquid chromatography tandem mass spectrometry for analysis of steroids in clinical laboratories. Clin Biochem 2011;44(1):77–88.
[13] Lehotay DC, Hall P, Lepage J, Eichhorst JC, Etter ML, Greenberg CR. LC-MS/MS progress in newborn screening. Clin Biochem 2011;44(1):21–31.
[14] Ackermans MT, Endert E. LC-MS/MS in endocrinology: what is the profit of the last 5 years? Bioanalysis 2014;6(1):43–57.
[15] Vogeser M, Seger C. Pitfalls associated with the use of liquid chromatography-tandem mass spectrometry in the clinical laboratory. Clin Chem 2010;56(8):1234–44.

[16] Seger C. Usage and limitations of liquid chromatography-tandem mass spectrometry (LC-MS/MS) in clinical routine laboratories. Wien Med Wochenschr 2012;162(21–22):499–504.

[17] College of American Pathologists (CAP). CAP accreditation program: chemistry and toxicology checklist. Northfield, IL: College of American Pathologists; 2015.

[18] Centers for Medicare and Medicaid Services (CMS). Clinical Laboratory Improvement Amendments (CLIA). Available from: https://www.cms.gov/Regulations-and-Guidance/Legislation/CLIA/index.html?redirect=/clia/

[19] Centers for Medicare and Medicaid Services (CMS). CLIA Regulations, Title 42, Chapter IV, Subchapter G, Part 493. Subpart K—Quality System for Nonwaived Testing; 2015. Available from: www.ecfr.gov

[20] Clinical and Laboratory Standards Institute (CLSI). Mass spectrometry in the clinical laboratory: general principles and guidance; Approved guideline (C50-A). Wayne, PA: Clinical and Laboratory Standards Institute; 2007.

[21] Clinical and Laboratory Standards Institute (CLSI). Liquid chromatography-mass spectrometry methods; approved guideline (C62-A). Wayne, PA: Clinical and Laboratory Standards Institute (CLSI); 2014.

[22] U.S. Department of Health and Human Services, Food and Drug Administration, Center for Drug Evaluation and Research (CDER), Center for Veterinary Medicine (CVM). Guidance for industry: bioanalytical method validation. Draft guidance. 2013; Available from: http://www.fda.gov/downloads/drugs/guidancecompliance-regulatoryinformation/guidances/ucm368107.pdf

[23] Clinical and Laboratory Standards Institute (CLSI). Evaluation of precision of quantitative measurement procedures; Approved guideline—3rd ed. (EP05-A3). Wayne, PA: Clinical and Laboratory Standards Institute (CLSI); 2014.

[24] Clinical and Laboratory Standards Institute (CLSI). User verification of precision and estimation of bias; Approved guideline—3rd ed. (EP15-A3). Wayne, PA: Clinical and Laboratory Standards Institute (CLSI); 2014.

[25] Weiss NA. Introductory statistics. 9th ed. New York: Addison-Wesley; 2012.

[26] Clinical and Laboratory Standards Institute (CLSI). Evaluation of the linearity of quantitative measurement procedures: a statistical approach; Approved guideline (EP06-A). Wayne, PA: Clinical and Laboratory Standards Institute (CLSI); 2003.

[27] Solberg HE. Establishment and use of reference values. In: Burtis CA, Ashwood ER, Bruns DE, editors. Tietz textbook of clinical chemistry and molecular diagnostics. 5th ed. St. Louis, MO: Elsevier Health Sciences; 2011. p. 425–48.

[28] Clinical and Laboratory Standards Institute (CLSI). Defining, establishing, and verifying reference intervals in the clinical laboratory; Approved guideline—3rd ed. (EP28-A3C). Wayne, PA: Clinical and Laboratory Standards Institute (CLSI); 2010.

[29] Rule GS, Rockwood AL. Alternative for reducing calibration standard use in mass spectrometry. Clin Chem 2015;61(2):431–3.

[30] Hall TG, Smukste I, Bresciano KR, Wang Y, McKearn D, Savage RF. Identifying and overcoming matrix effects in drug discovery and development. In: Prasain, JK, editor. Tandem mass spectrometry—applications and principles. InTech; 2012. p. 389–421. Available from: http://cdn.intechopen.com/pdfs-wm/29012.pdf

[31] Annesley TM. Ion suppression in mass spectrometry. Clin Chem 2003;49(7):1041–4.

[32] Van Eeckhaut A, Lanckmans K, Sarre S, Smolders I, Michotte Y. Validation of bioanalytical LC-MS/MS assays: evaluation of matrix effects. J Chromatogr B Analyt Technol Biomed Life Sci 2009;877(23):2198–207.

[33] Spooner N. Dried blood spot sampling for quantitative bioanalysis: time for a revolution? Bioanalysis 2010;2(11):1781.

[34] Timmerman P, White S, Globig S, Ludtke S, Brunet L, Smeraglia J. EBF recommendation on the validation of bioanalytical methods for dried blood spots. Bioanalysis 2011;3(14):1567–75.

[35] De Kesel PM, Sadones N, Capiau S, Lambert WE, Stove CP. Hemato-critical issues in quantitative analysis of dried blood spots: challenges and solutions. Bioanalysis 2013;5(16):2023–41.

[36] Chace DH, De Jesus VR, Spitzer AR. Clinical chemistry and dried blood spots: increasing laboratory utilization by improved understanding of quantitative challenges. Bioanalysis 2014;6(21):2791–4.

[37] Hannon WH, Therrell BL. Overview of the history and applications of dried blood samples. In: Li W, Lee MS, editors. Dried blood spots: applications and techniques. USA: John Wiley & Sons, Inc; 2013. p. 3–15.

[38] Clinical and Laboratory Standards Institute (CLSI). Newborn screening by tandem mass spectrometry; Approved Guideline (NBS04-A). Wayne, PA: Clinical and Laboratory Standards Institute (CLSI); 2010.

[39] Wilhelm AJ, den Burger JC, Swart EL. Therapeutic drug monitoring by dried blood spot: progress to date and future directions. Clin Pharmacokinet 2014;53(11):961–73.

[40] Timmerman P, White S, Cobb Z, de Vries R, Thomas E, van Baar B. European Bioanalysis Forum. Update of the EBF recommendation for the use of DBS in regulated bioanalysis integrating the conclusions from the EBF DBS-microsampling consortium. Bioanalysis 2013;5(17):2129–36.

[41] Capiau S, Stove VV, Lambert WE, Stove CP. Prediction of the hematocrit of dried blood spots via potassium measurement on a routine clinical chemistry analyzer. Anal Chem 2013;85(1):404–10.

[42] De Kesel PM, Capiau S, Stove VV, Lambert WE, Stove CP. Potassium-based algorithm allows correction for the hematocrit bias in quantitative analysis of caffeine and its major metabolite in dried blood spots. Anal Bioanal Chem 2014;406(26):6749–55.

[43] den Burger JC, Wilhelm AJ, Chahbouni AC, Vos RM, Sinjewel A, Swart EL. Haematocrit corrected analysis of creatinine in dried blood spots through potassium measurement. Anal Bioanal Chem 2015;407(2):621–7.

[44] El-Hajjar DF, Swanson KH, Landmark JD, Stickle DF. Validation of use of annular once-punched filter paper bloodspot samples for repeat lead testing. Clin Chim Acta 2007;377(1–2):179–84.

[45] Clinical and Laboratory Standards Institute (CLSI). Measurement procedures for the determination of lead concentrations in blood and urine—Approved guideline, 2nd ed. (C40-A2). Wayne, PA: Clinical and Laboratory Standards Institute (CLSI); 2013.

[46] McCloskey LJ, Yoo JH, Stickle DF. Interpatient distributions of bloodspot area per fixed volume of application: comparison between filter paper and non-cellulose dried matrix spotting cards. Clin Chim Acta 2014;437:187–90.

[47] van Baar BL, Verhaeghe T, Heudi O, et al. Is addition in bioanalysis of DBS: results from the EBF DBS-microsampling consortium. Bioanalysis 2013;5(17):2137–45.

[48] Timmerman P, White S, Globig S, Ludtke S, Brunet L, Smeraglia J. EBF recommendation on the validation of bioanalytical methods for dried blood spots. Bioanalysis 2011;3(14):1567–75.

[49] Li W, Tse FL. Dried blood spot sampling in combination with LC-MS/MS for quantitative analysis of small molecules. Biomed Chromatogr 2010;24(1):49–65.

[50] Koster RA, Alffenaar JW, Botma R, Greijdanus B, Touw DJ, Uges DR, Kosterink JG. What is the right blood hematocrit preparation procedure for standards and quality control samples for dried blood spot analysis? Bioanalysis 2015;7(3):345–51.

[51] Jager NG, Rosing H, Schellens JH, Beijnen JH. Procedures and practices for the validation of bioanalytical methods using dried blood spots: a review. Bioanalysis 2014;6(18):2481–514.

[52] McDade TW. Development and validation of assay protocols for use with dried blood spot samples. Am J Hum Biol 2014;26(1):1–9.

[53] Klee GG, Westgard JO. Quality management. In: Burtis CA, Ashwood ER, Bruns DE, editors. Tietz textbook of clinical chemistry and molecular diagnostics. 5th ed. St. Louis, MO: Elsevier Health Sciences; 2011. p. 163–203.

[54] Centers for Medicare and Medicaid Services (CMS). What do i need to do to assess personnel competency? Available from: http://www.cms.gov/Regulations-and-Guidance/Legislation/CLIA/Downloads/CLIA_CompBrochure_508.pdf; 2012.

[55] College of American Pathologists (CAP). CLIA, CAP accreditation, and CAP competency assessment program comparison. Available from: http://www.cap.org/apps/docs/education/competency_assessment/competency_requirements_GEN55500.pdf; 2013.

[56] U.S. Food and Drug Administration (FDA). Laboratory developed tests. Available from: http://www.fda.gov/MedicalDevices/ProductsandMedicalProcedures/InVitroDiagnostics/ucm407296.htm; 2014.

[57] U.S. Food and Drug Administration (FDA). Draft guidance for industry, food and drug administration staff, and clinical laboratories—framework for regulatory oversight of laboratory developed tests (LDTs). Available from: http://www.fda.gov/downloads/MedicalDevices/DeviceRegulationandGuidance/GuidanceDocuments/UCM416685.pdf; 2014.

[58] U.S. Food and Drug Administration (FDA). Federal food, drug, and cosmetic act (FD&C Act). Available from: http://www.fda.gov/regulatoryinformation/legislation/federalfooddrugandcosmeticactfdcact/; 2015.

[59] American Association for Clinical Chemistry (AACC). Oversight of laboratory developed tests. Available from: https://www.aacc.org/~/media/files/position-statements/laboratorydevelopedtestspositionstatement2014.pdf?la=en; 2014.

[60] American Medical Association (AMA). Statement on FDA proposal regarding diagnostic testing. Available from: http://www.ama-assn.org/ama/pub/news/news/2014/2014-08-01-fda-proposal-diagnostic-testing.page; 2014.

[61] Clement PD, Tribe LH. Laboratory services, as the practice of medicine, cannot be regulated as medical devices. Available from: http://www.acla.com/wp-content/uploads/2015/01/Tribe-Clement-White-Paper-1-6-15.pdf; 2015.

[62] European Medicines Agency, Committee for Medicinal Products for Human Use (CHMP). Guideline on bioanalytical method validation. Available from: http://www.ema.europa.eu/docs/en_GB/document_library/Scientific_guideline/2011/08/WC500109686.pdf; 2011.

[63] Honour JW. Development and validation of a quantitative assay based on tandem mass spectrometry. Ann Clin Biochem 2011;48(Pt 2):97–111.

[64] Kruve A, Rebane R, Kipper K, et al. Tutorial review on validation of liquid chromatography-mass spectrometry methods: part I. Anal Chim Acta 2015;870:29–44.

[65] Kruve A, Rebane R, Kipper K, et al. Tutorial review on validation of liquid chromatography-mass spectrometry methods: part II. Anal Chim Acta 2015;870:8–28.

[66] Nowatzke W, Woolf E. Best practices during bioanalytical method validation for the characterization of assay reagents and the evaluation of analyte stability in assay standards, quality controls, and study samples. AAPS J 2007;9(2):E117–22.

[67] Scientific Working Group for Forensic Toxicology. Scientific Working Group for Forensic Toxicology (SWGTOX) standard practices for method validation in forensic toxicology. J Anal Toxicol 2013;37(7):452–74.

[68] Huber L. Validation of analytical methods. Available from: http://www.chem.agilent.com/Library/primers/Public/5990-5140EN.pdf; 2010.

[69] Sargent M, editor. Guide to achieving reliable quantitative LC-MS measurements: RSC Analytical Methods Committee. Available from: www.rsc.org/images/AMC%20LCMS%20Guide_tcm18-240030.pdf; 2013.

[70] Clinchem.org. Clinical chemistry: instructions for authors. Available from: http://www.clinchem.org/site/info_ar/info_authors.xhtml; 2015.

[71] Annesley TM, Clayton LT. Quantification of mycophenolic acid and glucuronide metabolite in human serum by HPLC-tandem mass spectrometry. Clin Chem 2005;51(5):872–7.

[72] Garofolo F, Michon J, Leclaire V, et al. US FDA/EMA harmonization of their bioanalytical guidance/guideline and activities of the Global Bioanalytical Consortium. Bioanalysis 2012;4(3):231–6.

[73] DeSilva B, Garofolo F, Rocci M, et al. 2012 white paper on recent issues in bioanalysis and alignment of multiple guidelines. Bioanalysis 2012;4(18):2213–26.

BEST PRACTICES FOR ROUTINE OPERATION OF CLINICAL MASS SPECTROMETRY ASSAYS

B. Rappold

Essential Testing, LLC, Collinsville, IL, United States

1 NEW ASSAY IMPLEMENTATION

Successful routine analysis of diagnostic assays is quite different from the undertakings of development and validation of those assays. While development and validation may require a degree of ingenuity and experimental design, routine mass spectrometric operations demand a high level of continuous effort. Rugged assays are as much designed as they are performed. The initiation of this performance occurs prior to the physical launch of the assay. Following validation, a number of key components should be addressed to ensure a smooth transition from the development/validation phase to the operational/utility phase, including the following components: staff training, material verification, redundancy testing, development of a standard operating procedure (SOP), and prelaunch walkthrough to verify transference from development to operations.

A documented procedure should be made during the validation of the assay. It is recommended that this SOP contain the following details, adopted from the Clinical Laboratory Improvement Amendment [1], as a minimum:

Specimen collection, processing, and rejection criteria requirements.
Limitations in methodologies, including interfering substances.
Step-by-step performance of the procedure.
Preparation of solutions, calibrators, controls, reagents, and other materials.
Calibration and calibration verification procedures.
Control procedures.
The reportable range of patient test results as established or verified through method performance specifications; reference or normal ranges, including relevant calculations/correction factors.
Result values requiring immediate physician/clinician notification.
Remedial action to be taken when calibration or control results fail to meet the laboratory's criteria for acceptability and description of the steps to be taken in the event that a test system becomes inoperable.
Specimen storage and preservation.
Criteria for the referral of specimens including procedures for specimen submission and handling.
Pertinent literature references.

Example system suitability test solution chromatograms.
A clear description of the methodology used, including LC gradient program, ions or precursor-product ion pairs and MS parameters.

The SOP should include sufficient detail to use both in testing and as a reference for operational personal. Expansion of certain items from the list mentioned earlier is included for further detail.

1.1 SPECIMEN COLLECTION, PROCESSING, AND REJECTION CRITERIA REQUIREMENTS

This element should refer to appropriate sample types and method of collection. If testing on blood-based matrices, the variety of anticoagulants or tube-type should be assessed in validation for acceptance. For example, interference in testosterone measurement was observed from separator gels used in blood collection tubes [2]. Additionally, clot activation factors used in sample draws were identified as a cause of falsely elevated testosterone measurement for a single analyte transition, while fluoride tubes yielded a lower recovery in testosterone [3]. In the analysis of methylmalonic acid, it was observed that citrate serum collection tubes resulted in an interference with a qualitative ion [4]. The results of validation experiments to define acceptable means of sample draws should be used to define appropriate tube types. Certain laboratories may not receive the source tube to enable visual identification of the sample type, but only receive a subaliquot of the sample. For assays wherein a certain sample type creates excessively erroneous results, reflex testing to establish the matrix type should be included in the SOP. Clarity in the manner in which the blood draw occurs (supine/upright) or the time or patient condition of sampling (first morning urine void, fasting, 24 h collections, etc.) should be explicit in the SOP. Analytically important notes should be attached to the specimen. As an example, urine metanephrines is typically performed on a 24 h urine collection with 30 mLs of 6 N hydrochloric acid added to stabilize the samples. One can easily observe that the functional pH of a 200 mL 24 h collection is quite different than that of a 2000 mL 24 h collection. The decreased pH in the first sample may exhibit a change in the recovery of urinary metanephrines, dependent on the extraction motif utilized. Restrictions on the quality of the sample (i.e., hemolysis, lipemia) should be indicated if they manifest an error.

A minimum sample volume should be designated for each procedure. Commonly, adequate sample to perform two analyses at a minimum, with additional excess volume required in pipetting, is acceptable. As laboratory test SOP's may be utilized to define procedures for patient sampling and sample triaging, consideration for the patient and realistic requirement for sampling should be included in this section. Requiring excessive sample testing volume may provoke the exclusion of a sample due to insufficient volume or may request that a phlebotomist to overly draw samples from neonatal/pediatric patients. In these cases, addressing insufficient sample volume may utilize a reflex for automatic dilution. These considerations should be applied on an assay by assay basis with thought toward the expected patient population and relevant medical decision points (i.e., data from assays in which elevations in relation to disease may not suffer from twofold dilution in sample analysis).

1.2 LIMITATIONS IN METHODOLOGIES, INCLUDING INTERFERING SUBSTANCES

A statement describing sample-based causes of error as determined in validation should be elucidated. Interfering substances can include both compounds which interfere with analysis as well as compounds which can alter the clinical interpretation. Take salivary cortisol measurements as an example. This test is normally performed immediately following overnight sleep by the patient; lack of adherence to that protocol will render results meaningless.

1.3 STEP-BY-STEP PERFORMANCE OF THE PROCEDURE

Each step of the analytical procedure should be comprehensive and clear. Important details concerning the handling of samples should be described in the beginning of the procedure. For example, vitamin K is photosensitive; adding this detail at the end of an SOP will not serve a technician following the procedure stepwise [5]. In most cases, the first step in a procedure would be to allow the calibrators, quality controls (QCs), and patient samples come to room temperature, unless specific temperature considerations are necessary, and mixed to homogenize the materials. The second step is typically pipetting of those samples, followed by the addition of the internal standard (third step). All volumes, reagents, hardware, and other materials should be defined as necessary and referenced to their origin.

1.4 PREPARATION OF SOLUTIONS, CALIBRATORS, CONTROLS, REAGENTS, AND OTHER MATERIALS

Details regarding the in-laboratory manufacturing of assay supplies is fundamental in ensuring assay success. Explicit instructions on solution preparation is highly recommended; newly hired staff may not be aware of what scientists consider "primary knowledge." All experienced laboratory staff have witnessed or heard telling of errors, such as adding organic to buffered solutions, precipitating the salts to the bottom of the bottle. Or possibly the bench technician who added concentrated nitric acid to a basic solution in pH adjustment. Consider the preparation of a 90:10 water:methanol mobile phase. A proper habit is to individually measure the water and the methanol in separate graduated cylinders and combining in a discrete vessel. However, the practice of adding 900 mLs to a 1 L graduated cylinder and bringing the volume up to 1 L with methanol is not uncommon in laboratories.

Solutions and materials for the test should be described with manufacturer and, if available, an acceptable (Chapter 4) alternative. Anecdotally, the years 2008 and 2009 saw a dramatic decrease in the availability of acetonitrile, causing some laboratories to revalidate their methods with alternative mobile phases (such as methanol) [6]. As the validation procedure for diagnostic assays should take considerable time, being prepared for such events can lead to better turn-around-time for patient analysis. It is recommended that, at a minimum, alternative products for common materials, such as mobile phases, SPE materials or dimensions, glass tubes, and even LC columns be addressed in validation and included in the SOP.

1.5 SPECIMEN STORAGE AND PRESERVATION

The short term housing of patient specimens should be designated in regards to temperature conditions and length of time. Causes of reanalysis of patient samples, such as required dilution, random errors related to individual sample preparation or batch failure, are not uncommon. Validation studies to establish limits of analyte stability should include evaluation of multiday storage at various conditions including room temperature, refrigerated (2–8°C), standard freezer (0 to −20°C), and ultra-freezer (−60 to −80°C). If possible, storage up to 4 weeks following initial analysis is recommended as a minimum. In general out-patient testing, clinicians and physicians may not review reports immediately upon receipt from the laboratory. Preanalytical error, such as missed ordering or misinterpretation of the order, can be followed up and corrected if the sample is stored rather than discarded following testing [7].

Additional considerations for items related to calibration of the assay and QC procedures shall be further addressed later in this chapter.

Access to an SOP initiates the personnel training phase of assay implementation. Individuals who are designated to perform the test should be able to locate, prepare, and manage all materials associated with testing. A thorough review and documentation of review of the SOP is recommended. Following SOP review, the operational staff should witness the validating technicians/scientist(s) perform an example batch preparation. A draft of the SOP should be compared while in-process, allowing for corrections and clarifications to be made permanent in an SOP revision. Operational staff should then perform the assay procedure under the supervision of a validating scientist to ensure transference of the necessary skill set, particularly in those steps which require specific timing or special care.

Verification of the assay's performance in the operational setting should include appropriate content to mirror expected patient sample analysis. This evaluation should emulate the reality of the assay expectations, which is to perform accurately and precisely over numerous changes in columns, calibrator lots, reagent preparations, multiple liquid chromatography-mass spectrometry (LC-MS/MS) systems, and most importantly, time. As a consequence, all new materials (including calibrators) should be prepared by the operational staff for comparison to those materials prepared in validation, using real patient samples and appropriate statistical comparison to gauge success.

This verification batch is performed simultaneously (or as nearly as possible) by both the validation group and the operations group using materials prepared by each respective group. A batch of at least 40 patient samples spanning the analytical range, possibly those specimens used in reference interval generation/reference interval verification and 10 replicates of each QC material is utilized. A batch prepared in this manner allows for the appraisal of the following factors: reagent and material selectivity, calibration curve linearity, intraassay imprecision, and interassay comparison. The results of the batches should be that all materials are free of interfering or contaminating substances originating from reagent preparation (as measured by the response of the analyte in extracted blank matrix), the calibration curves exhibit r-values of > 0.995 (indicating good linearity), the precision of the 10 replicates is approximate to that as determined during validation of intraassay imprecision. Comparison of patient samples yields a Deming regression slope of 0.9–1.1 and an R-value greater than 0.9. Deming regression, as opposed to linear least squares regression, is recommended here as batches prepared by both the validation and the operational staff should exhibit imprecision (representing error in both the x- and y-axis of the comparison). Linear least squares would be an acceptable method if the validation assay has been determined to be free of error, as would be the case in reference method procedures developed in-house.

This evaluation can elucidate common forms of assay translation errors, such as incorrect salt correction factors for lyophilized materials during calibrator preparation, deviations in sample extraction, or imprecision in pipetting. A number of these errors shall be addressed further later in the chapter.

2 SYSTEM SUITABILITY STANDARDS

System suitability test (SST) solutions are perhaps the most useful tool MS users have in routine operations. These solutions provide daily and longitudinal checks for the performance of the analytical system. As users of LC-MS platforms are certainly aware, these instruments can be politely described as fickle. SST's allow for the early detection of possible errors prior to the start of batches with samples, preventing the loss of time, effort, and materials in a wasted analysis. Instituting and monitoring SST performance over time also allows for the enhancement of preventative maintenance scheduling and the continuation of operational analyses.

A basic definition for SST solutions has been described in the *Bioanalytical Method Validation Guidance for Good Laboratory Practice* by the FDA [8] as a reference standard solution used to assess system performance. For clarification, this solution shall be further described with choice of solvent and concentration. It should be noted that this solution should include both the analyte(s) and the internal standard(s) for the assay. Certain recommendations related to the analysis of pharmaceutical products endorses the use of a generic SST solution which does not have the analytes of interest included, but instead has compounds which are both readily available and stable [9]. In environments where assays have short lifespans relative to diagnostic tests, such as drug metabolism/pharmacokinetic testing, this approach affords expediency. In diagnostic testing, however, it is appropriate to gauge the capability of a platform to measure the analyte of interest with the actual compound of measure.

The choice of solvent should be made under conditional restrictions of stability and injection volume. Long-term stability of the analyte in the SST solution is preferred to enable long-term evaluation of the LC-MS/MS system characteristics. The SST solvent of choice need not be biological matrix; indeed, neat highly pure solvent may prevent adsorptive loss and degradation of the analyte(s) and internal standard(s) as well as obviating variable matrix effects and should thus be chosen over extracted or neat biological material. The choice of solvent must also be selected relative to injection volume, in that the injected solvent does not induce breakthrough of the analytes of interest on the head of the column. For example, an SST solution of pure methanol may provide acceptable long-term stability for the hydrophilic analytes analyzed via reverse-phase LC. Yet on injection, the analytes are detected at the solvent front as they are not properly retained on the column's stationary phase material. Thus, a balance must be found between stability and injection volume.

It is recommended that evaluation of the long-term stability of SST solvents be undertaken at the early stages of method development. This experiment can be easily performed by spiking the same concentration into various solvents of choice and storing them at both long-term storage conditions (i.e., ultra-freezer or similar) as well as the temperature of the autosampler expected to be utilized in operations. The actual concentration should be determined by an approximation of response in the mass spectrometer based on provisional signal-to-noise calculations. The concentration of a final SST solution may not match this initial concentration; this determination supports stability only. Following weeks or months (dependent on the difficulty and duration of method development), preparations of fresh solutions at the provisional concentration and comparison of the response of the stored versus the fresh allows for an approximation of stability. This experiment can be included as a separate validation exercise during or prior to final validation of the assay.

The concentration of an operational SST solution should be that of the *extracted* lower limit of quantitation (i.e., the lowest calibrator or lowest concentration reported by the assay). The emphasis on the extracted lower limit of quantitation should take into account not only the expected dilution or concentration factor of the sample preparation process, but also the mean recovery as established in validation of the assay (Chapter 4). The reasoning is such that a review of the response of a high level sample may not adequately demonstrate changes in the background noise of an assay. Often times, peak review is performed with the *y*-axis related to the highest peak in the chromatogram. Cursory review of a highly abundant peak may mask interfering peaks or noise changes at the lower levels of measured response. Additionally, daily expectations for analytical performance should include the lower limit of quantification; if a neat, un-extracted solution (representing the appropriate preparative recovery) does not yield an acceptable response, extracted calibrators, patient samples, and QCs certainly will not. Thus, evaluation of the SST at this level would indicate the need to perform maintenance, troubleshoot sensitivity, or to check operating conditions of the platform.

An example of an SST solution preparation scheme for the analysis of plasma metanephrines by solid phase extraction (SPE) with an LLOQ of 25 pg/mL is as follows. It was determined during validation that the SPE process results in an average recovery of approximately 75%. Furthermore, this SPE procedure concentrates the samples twofold from the original sample volume. Therefore, the SST solution shall be prepared at an approximate concentration of 37.5 pg/mL (25 pg/mL × twofold concentration × 75% recovery).

Inclusion of the internal standard at a concentration representing a moderate response is recommended to confirm that the peak of interest in the analyte transition(s) is indeed that of the analyte. Drifts in mass accuracy or resolution or contamination during mobile phase preparation may contribute an isobaric interference which closely resembles the analyte of interest, providing a false positive signal. Alignment of the retention times of the analyte and internal standard can alleviate such concerns as well as providing an additional check for changes in sensitivity and ensuring that the lack of response in the analyte transition is indeed related to instrument function and not trivial events, such as injecting air from an empty SST vial! The response of the internal standard should be in the middle of the dynamic range of the mass spectrometric detection to ensure that imprecision in response over time is not affected by detector blinding or source saturation.

There are additional pragmatic details which should be considered in the preparation of SST solutions. A solution which exhibits exceptional stability for the analyte of interest may be deleterious to the internal standard. This often occurs with deuterium labeling and highly acidic solvents. For example, catecholamines (epinephrine, norepinephrine, and dopamine) are stable in 1–6 N HCl and with the addition of sodium metabisulfite [10]. Commercially available deuterated internal standards, however, exhibit in-solution hydrogen-deuterium exchange at very low pH. A balance between stability of the internal standard and analyte should be achieved in a working SST solution. For assays in which there are known isobaric interferences, particularly in endogenous analyte assays, adding those interferences ensures appropriate chromatographic resolution prior to sample analysis. When isomers demonstrate similar precursor and product ions, degraded chromatographic performance can adversely impact analysis by coeluting peaks, changing the noise levels for the analyte of interest, and even misidentifying peaks on data review. Examples of isobaric pairs frequently observed in clinical LC-MS/MS are shown in Table 5.1. An example of this is shown in the analysis of reverse triiodothryonine (rT3) by LC-MS/MS. The SST for this assay includes the circulating isomer, tri-iodithryonine (T3). Fig. 5.1A shows an acceptable SST which indicates a system capable of delivering quality data and an unacceptable SST (Fig. 5.1B), indicative of column performance degradation where in the T3 begins to tail into the rT3 peak. This example is also useful in pointing out the clinical nature of building the SST solutions. In the normal population, T3 circulates at a far higher concentration that rT3 [11]. Sufficient resolution should be shown between these two similar molecules to ensure that tailing T3 peaks at a higher abundance do not interfere with the quantitation of rT3.

Frequency of SST analysis should be determined by the frequency of actual batches for that assay. At a minimum, SST injections should be made the same day in which samples are analyzed. Additionally, review of SST injections should be made sufficiently before the arrival of prepared samples to

Table 5.1 Examples of Isobaric Pairs Warranting Inclusion in System Suitability Test Solutions

Analyte	Isomer/Isobar	MW
Methylmalonic acid	Succinic acid	118.1
Triiodothyronine	Reverse-triiodothyronine	650.3
Testosterone	Dehydroepiandrosterone	288.2
Cortisone	Prednisolone	360.2
Morphine	Oxymorphone	285.1
Codeine	Hydrocodone	299.1
Alloisoleucine	Isoleucine/Leucine	132.1

MW, Molecular weight.

allow the most lengthy and common maintenance procedure, which is venting of a mass spectrometer and cleaning of the interface and front end optics. This exercise can take anywhere from 1 to 4 h, dependent on the platform and thoroughness of cleaning. Assays which are performed on a daily basis may benefit from scheduling of SST samples to be analyzed immediately following the completion of a batch of patient samples, allowing review of the SST data without waiting for the injection cycle to complete.

The number of SST injections to be made is preferably which are injected following an equilibration blank. As a general rule of thumb for LC-MS/MS assays, the first injection is discarded as the LC system is not in dynamic equilibrium. This is important for batches as well; all samples should be treated identically. The first sample is the only analysis which does not have a sample prior (hence the name, "first"). Using a blank or equilibration sample is recommended for consistent data across a batch. Three replicates demonstrate reliability in the data collected without expending additional analytical time for little value. Factors for review of the SST include retention time (relative to previous SST solutions), peak height (or signal to noise), peak width, peak asymmetry, and if applicable, relative resolution between isobaric interferences.

3 VERIFICATION OF NEW MATERIALS

Calibration materials and QCs must be periodically replaced due to either stability or to exhaustion of initial supplies. Calibration standards underpin the entirety of assay performance; errors in preparation of these materials results in erroneous data generation in the analysis of patient samples. QC's are equally important as they are directly representative of the longitudinal performance of the assay, linking the original validation studies, including reference interval generation and comparisons to certified standard reference material or a reference method procedure, to current patient results. Special care and attention should be paid to the production and qualification of these important assay components.

As a general rule, preparing calibrators and QCs at isolated times using independent solutions is preferred. Noncoincidental expiration of batches of QC's and calibrators is a mechanism by which the materials can be independently justified for clinical use. The timing of new material preparation should

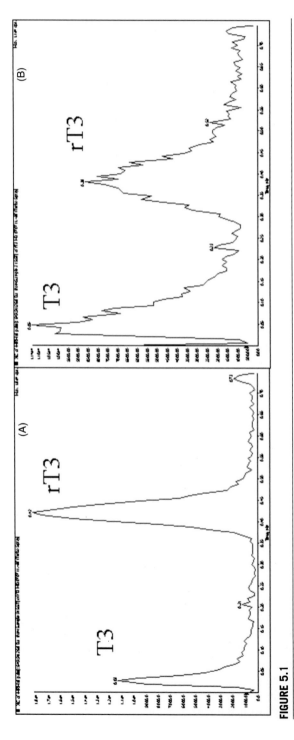

FIGURE 5.1

Example SST chromatograms demonstrating acceptable column performance as indicated by resolution between reverse triiodothryonine and an isomer, triiodothyronine in the (A) and unacceptable column performance in (B)

be such that neither QC's nor calibrators expire or are consumed at similar times, creating uninterrupted traceability. If both lots of standards or QCs should run out or unexpectedly reach expiration, a laboratory then has minimal internal resources in which to verify a new series of materials. When prepared at distinct times, an active lot of QC materials can be used to ensure a new lot of calibrators are provisionally acceptable; or a working lot of standard material can be used to generate acceptance ranges for new QC's.

3.1 CALIBRATORS

In preparation of new materials, traceability is the fundamental objective [12]. Each component used should be linked to the laboratory's validation studies as well as external internationally recognized standards. Before studies to address the precision or accuracy of new calibration standards or QC's are considered, the laboratory should ensure that fundamental aspects of metrology are in place. Primary stock solutions, from which further dilutions into calibrators shall be performed, should have a certificate of analysis appropriate for the measurand. Anecdotally, simple qualifying techniques, such as Karl-Fischer analysis, are insufficient to assign true purity of the analyte. Sufficient evidence of acceptable purity checks should be performed to provide an unequivocally accurate starting material. If lyophilized powder is to be used for initial starting material, the laboratory should have certified balances of sufficient sensitivity to accurately weigh the masses required. The laboratory should not accept a vendor's claim of "X mg per vial" as true unless that vial is accompanied by a certificate of analysis which details the qualification of the balance used in the weighing at the vendor's laboratory. Additionally, using a balance capable of weighing 1–300 g is unsuitable for the weighing of 0.5 g. Basic principles of measurement error, significant digits and metrology, addressed in thorough detail elsewhere, should be followed [13].

Accurate transfer of solutions should always be performed by Class A volumetric glassware. If a dilution scheme requires the use of volumes not typically found in standard Class A volumes (the author has not yet seen, e.g., a 829 µL Class A volumetric pipette), adjustment of the dilution scheme should be modified to encompass volumes necessary to achieve a traceable measure. In certain instances where Class A volumetric glassware is not an option, a well-calibrated manual pipette used in reverse pipetting mode can be used. It is recommended that this pipette be checked for accurate and precise delivery (preferably checked by the person preparing the solutions) on a certified balance to ensure its acceptable operation prior to use in making calibration materials.

Once laboratory materials are ready, the matrix used for calibrator preparation should be checked for possible interferences or contamination. If, for instance, charcoal stripped serum or plasma is used to generate calibrators for endogenous substances, the charcoal stripping process may be ineffective in completely removing the analytes of interest. One can imagine a case wherein one lot of charcoal stripped serum donated by mostly females is absent of testosterone yet another lot of serum donated by 20–28 year old males still exhibits testosterone even after three rounds of charcoal stripping. Certain exogenous compounds suffer from this difficulty, as thorough screening of donor material and/or self-reporting of drug consumption (legal or otherwise) is not error free [14]. Preparing of calibrators only to discover a positive bias in the concentrations due to extant content, and the lengthy investigations of cause, can be easily avoided by prescreening matrix prior to calibration preparation.

A thorough reading of the certificate of analysis provided with the material can also prevent improper calibration preparation. Lyophilized materials can often be sold as a salt, commonly HCl or Na.

A correction for the salt weight in gravimetric preparations should be instituted and documented in the process. Depending on the molecular weight of the compound and the molecular weight and number of salts, a proportional bias that is within acceptance criteria can be perpetuated, creating a systematic error in the assay.

Prior to preparing materials, general good laboratory practices should be followed, including allowing all solvents and materials to come to room temperature and weighings performed on a vibration resistant table. Fortification of material into matrices which exhibit protein binding (i.e., blood-based) should be allowed to reach equilibrium before further pipetting. Additionally, the volume of spiking organic content into blood-based matrices should be such that precipitation of the matrix, and a subsequent volume change due to a change in partial density, does not occur. Generally, less than 5% total volume of organic solvent will not induce protein precipitation. Adding a relatively large volume of a methanol solution to a volume of plasma can quickly precipitate the proteins of the matrix. These precipitated proteins may also hinder further aliquoting as they can obstruct the tip of pipettes.

Confirmation of a new lot of calibration materials should be performed by comparison to the previous lot of calibration materials as well as against the results of patient samples. With appropriate planning, this can be performed in the context of normal operations by reserving positions in a batch for the new calibrators. It is recommended that the new calibrators be analyzed in triplicate, at a minimum, for this experiment. This allows for a small precision measurement which can be used to exclude outliers. It is also recommended that the previous lot of calibrator materials be analyzed in duplicate in the same batch, ensuring that the extremities of the calibration curves, or those values most likely to exhibit random variation in a linear least squares regression, be deemed acceptable prior to assigning acceptance for new calibration standards (see Section 5 for calibrator acceptance).

To perform the comparison of calibrators, a procedure similar to that described in method comparison and bias guidelines from the Clinical and Laboratory Standards Institute is performed [15]. A batch is prepared with the following materials: duplicate original calibration curves, current lot of quality controls, triplicates (at minimum) of the new calibration curve, and preferably 40 patient samples. For a 96-well plate and calibration schemes with eight standard solutions, this is approximately 96 samples. The patient samples should be chosen to represent values throughout the analytically measurable range of the assay. Samples are analyzed and data are reviewed and processed under two conditions. First, the original calibration materials are set as standards and used to generate a calibration curve, which should exhibit an r-value greater than 0.99 in linear least squares regression. QCs are then assessed to ensure appropriate analytical performance; if unacceptable, investigation into the cause of failure is initiated and the batch repeated following corrective action. Then the mean concentrations of the new calibrators (measured as unknowns) are assessed and compared to the expected. Values which are within 15% (or the chosen criteria to meet analytical performance expectations, whichever is less) are considered provisionally acceptable. The results for the patient samples are then computed and retained. The original calibration curve is then removed from calculations, using the triplicates of the new calibration lot, all points included, to generate a calibration curve. Again QC's should be checked for acceptance, wherein failure renders the concentration of the new calibration standards disputable. As a reminder, these QC's are from an existing lot with values ascribed by the previous lot of calibration standards. If QC's fall within accepted ranges, the 40 patient samples are then recalculated against the new calibration curve. These results are then compared to the patient concentrations generated from the original calibration curve as a Bland-Altman plot. Acceptance criteria for this patient comparison should be undertaken judiciously, using knowledge of the medical decision points and acceptable error in analysis. In general, the Bland-Altman plot should indicate no systemic bias; all points should be

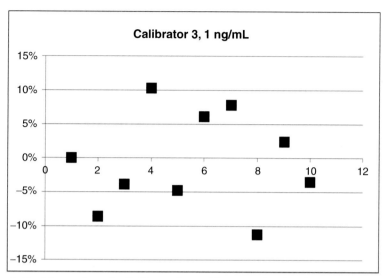

FIGURE 5.2 Longitudinal Comparison of a Calibrator Across Nine Lots of Calibration Preparation Verification Indicating Random Scatter of Bias From Lot-to-Lot Comparisons

scattered at random around the 0% line, with a mean inaccuracy less than the validated mean interassay inaccuracy. It should be noted that the expected failure should only result in a fixed bias, assuming that all previous experimental criteria here are followed before patient comparison.

It is also recommended that historical calibration curve comparisons be retained and evaluated for longitudinal drift. Over the course of multiple calibration curve preparations, each new lot may compare favorably to the previous lot but may drift unacceptably from the original calibration preparation (those calibrators utilized in the original generation of reference intervals and medical decision points). Reviewing historical calibration verification bias over the course of many lots of materials may reveal a change over time which would not be immediately discernible in the standard course of calibration comparison. This data may be expressed as a scatter plot with the bias from expected on the *y*-axis (as determined via the comparison described previously) with each replicate as a data point. Limits of acceptability should be defined as precision on the *y*-axis (equal to the interassay imprecision as determined in validation) as well as a visual observation of scatter around the 0% bias line. An example of this is shown in Fig. 5.2. Plots which exhibit consistent biases (either positive or negative) indicate longitudinal drift, which may be due to degradation or loss by adsorption of top stock materials used in preparation.

3.2 QUALITY CONTROLS

QCs are often more easily prepared and qualified than calibrator materials. For the vast majority of LC-MS/MS assays, QC's are utilized to determine the precision of the method, not the intrinsic accuracy of the test. Hence, the goal of QC's is simply *approximate* concentrations with *longitudinal* precision measurements. The longitudinal aspect of this requires the most planning—it is recommended that at least 20 distinct days be used to assess the variation of measurement (precision) across all experimental conditions. Indeed, the greater the truly random variation which occurs in extraction, analysis, and comparison to distinct calibration curves, the more confident one can be in the range of precision generated in QC schemes. However, purposefully inducing variation may not assist in this evaluation. QC's should be treated as patient samples in every aspect. "Special" batches prepared just for the generation of QC precision is quite different from that of typical patient samples and is thus discouraged.

The appropriate number of discrete concentrations of QC's is dependent on the use of the assay. In a test where in clinically normal versus abnormal is clearly differentiated by a single medical decision point, having greater than two QCs does not necessarily increase the quality of the data. For example, an assay is utilized to measure the presence of argininosuccinic acid in the assessment of inborn errors of metabolism. This compound is generally considered to be in measurable quantities only when a patient presents with argininosuccinate lyase deficiency [16]. Thus the QC scheme can appropriately consist of a weakly positive sample (low QC) and a strongly positive sample (high QC), while the negative/blank/carryover sample serves to illustrate false positive signals. Note that contamination in all samples may not necessarily report patient samples as false positives as the calibration curves should correct for such signal as an increased intercept but such signals may cause confusion on data review. As mass spectrometric assays frequently utilize isotopically labeled internal standards for analysis, recovery of the patient samples through the assay is indicated by IS recovery and the QC's demonstrate precise measurement. Distributing more than two QC's throughout the analytically measurable range serves only to lengthen turn-around time of the assay with additional sampling, documentation, and data review.

However, analytes which are compared to numerous reference intervals or in a plethora of disparate diagnostic workups, additional QC samples may be appropriate. Numerous biomarkers have reference intervals specific for gender, age, and ethnicity; determining the QC policy for these assays should link the medically relevant decision points to the number and levels of QC's. Consider the case of dehydroepiandrosterone measurement which has both age specific reference intervals and Tanner stage specific reference intervals, each of which are further stratified by sex. Additionally, there are reference intervals for menopausal status and pre- and postmenarche [17]. Imagine a QC procedure in which there is a normal and an abnormal QC for each of these ranges. Given 5 tanner stages and 11 age groups across 2 genders and 4 menstrual statuses (across only females, logically), 36 reference intervals would yield 72 quality controls. A laboratory undertaking such an arduous QC plan would spend most of their time assaying controls and rarely test a specimen. In these cases, QC's should be placed in regions of the calibration curve distinct from expected calibration standard values or at medical decision points considered most important. As the potential variety of QC schemes in these instances can be nearly limitless, direct recommendations shall not be made.

The matrix of choice for QCs is a subject of interesting discussion. It is often recommended that the QC matrix match that of the test matrix, commonly meaning if the assay is to be performed in serum, the QCs should be prepared in serum as well. It should be considered that "matrix matching" can also be more than just the type of the matrix to be assayed, but relevant to the clinical population to be sampled. Consider that the population a vendor would utilize for whole blood donations from that of organ transplant patients, particularly in protein binding effects or in that distribution of many immunosuppressants is primarily intracellular [18,19]. Thus, for the analysis of immunosuppressant compounds, the most closely "matrix matched" QCs would not be random blood donors. As a separate example, normal sex hormone binding globulin (SHBG) ranges differ dramatically between males (10–57 nmol/L) and females (18–144 nmol/L, nonpregnant) [20]. Assessment of serum estradiol measurement is most frequently utilized in evaluations of hypogonadism, oligo-amenorrhea, monitoring of hormone replacement therapy, inference of follicle development for invitro fertilization, and studies related to bone fracture risk in postmenopausal states for females (granted use in males for evaluating feminization and in diagnosing estrogen–producing neoplasms is also performed, though to a lesser degree). As analyte binding and subsequent internal standard equilibration play a critical role in efficient and precise extractions, utilizing female-donor matrix for QCs may be a better representation of the assay's performance rather than male donors. This concept is similar to the most desirable specimen type as recommended in the Clinical and Laboratory Standards Institute (CLSI) Linearity of Quantification procedure (EP6-A) [21].

QC preparation can be performed in a number of different ways. First, as described earlier, patient pools can be prepared from in-operation assays. These pools can be generated by annotating samples within the reference range (normal QC) and those samples outside the reference interval (abnormal QC) with closely related concentrations. If more than two QC's are required to suitably address the clinical population the assay serves, additional concentration-annotated pools may be sequestered from the testing population. If specimen materials in the abnormal range are unavailable, the "normal" pool may be fortified with the analyte to levels considered "abnormal" for the assay. A second means of QC generation is executed by purchase of blank matrix and subsequent fortification of the analytes. This can be useful for assays in which multiple analytes are being measured in single run and the incidence of elevations of all measurands in a single sample, much less multiple samples in which to generate the pool, is unreasonable. This would be appropriate "positive" QCs preparation for assays, such as amino acid analysis or a drug of abuse panel. Similar to preparation of calibrators, the spiking scheme should be developed in a manner that reduces the overall spiking solution volume relative to the sample pool so as not to substantially modify the makeup of the matrix. Excessive dilution of human matrix with neat solution is an important feature to avoid, as the QC is then representative not of the patient population, but of diluted patient samples. A third option is the purchase of commercial control materials. This approach has certain advantages, such as predefined approximate values and the possibility of interlaboratory comparison of concentrations without directly referring specimen. The makeup of these external materials should be extensively vetted by the laboratory to ensure that the matrix supplied (often lyophilized) is equivalent to real human samples. This may be performed during validation of the assay by the 5-level admixing scheme described in the validation Chapter 4 and in CLSI document EP7-A2, Interference Testing [22]. In this case, all the QC materials should be utilized for the mixing (i.e., the high QC with a low patient sample or pool; low QC with a high patient sample or pool).

Once QCs are prepared, samples should be subaliquoted to vials (ensuring the vessel material does not induce absorptive loss, as determined in validation) and stored at the most stable conditions available. The volume of the subaliquots can depend on the nature of the analyte. Compounds which exhibit degradation following a freeze-thaw cycle should be aliquoted for single use, while more stable compounds can be subaliquoted to volumes sufficient for multiple analyses. If the analyte(s) are amenable to the latter approach, it is recommended that the QC vials in use be stored in the same manner as patient samples.

4 SAMPLE ANALYSIS

Considerations for the actual analysis of samples should be addressed in this transference phase, as the occurrence of the testing, the style of calibration, and the necessary resources required for operational analysis must be defined prior to the testing of patient samples. Information critical to these determinations include the expected testing volume (per patient), the requirement for expedited testing (i.e., STAT testing), and the hands-on sample preparation time (to requisition appropriate human resources for testing). For low testing volume, it may be inefficient and cost-prohibitive to analyze a small number of samples in a single batch, as the costs of calibrators, controls, and reagents are fixed for each batch. Absent the need for accelerated results (as in STAT testing), samples may be retained until sufficient volume is reached to balance the costs of static resources with the dynamic test volume. In these cases, stability limits of patient samples determined in validation should be enforced, with margins of error accounting for reanalysis in the event of failed batches or repeats for interferences or dilution. If the laboratory receives samples from external locations, the expected transit time should be included in the determination of run frequency relative to sample stability.

The construction of batches for sample analysis is determined from sample volume/testing frequency as well as operational limits defined in validation. Some methods are sufficiently robust that calibration curves included in every batch of samples is not necessary. In these cases, calibration curves may be analyzed ad hoc, while QCs are measured with samples to determine acceptable performance. In this mode, calibration curves and a set of QC's would be analyzed at the beginning of a work-shift and verified for performance, with samples and additional QC's processed as needed. Assays capable of this calibration style are not typically associated with complex sample extraction motifs wherein internal standard recovery variance is expected. Crucial to this style of operations is reproducible internal standard pipetting and mass spectrometer response, such that the normalizing effect of the internal standard is appropriately utilized. In cases where batching samples with calibration and control materials are appropriate, the location and frequency of calibration standards and QCs should be determined a priori, either in a randomized fashion or as a set of procedures.

Randomization of an entire batch is, from a purely scientific perspective, is the most appropriate way of ordering samples. Arbitrary placement of calibrators and controls can elucidate intrabatch variation not related to sample preparation, as well as provide confidence that all extraction steps are performed precisely on all samples, regardless of origin. This, however, has practical implications which may make randomization of sample order undesirable. Prompt data reduction of calibration curves and QCs, analyzed prior to patient samples in a batch, can indicate insults to the analytical system. Carryover from patient samples with extremely elevated analyte levels could be ran prior to a calibration standard or QC, preventing suitable interrogation of the data.

Batch-based analysis of samples wherein the calibration materials and QC's are in dedicated locations can provide certain advantages. As stated previously, quantifying an initial calibration curve and QCs can lead to rapid appraisal of assay acceptance. In cases of initial failure, root-cause analysis can initiate without loss of instrument time and reagents as patient samples are unnecessarily acquired. Thus, it is recommended that the beginning of each batch consist of calibration standards, a blank with internal standard, a blank without internal standard, and a series of QCs. Provisional review of the data is identical to calibration curve generation and batch review as described in further detail in this chapter.

5 DATA REVIEW

Successful completion of data collection is inherently followed by data reduction. This may be among the most time consuming task of laboratory staff. Calibration curves must be generated and applied, quality assurance rules checked, QC samples verified, and individual samples inspected for appropriate data integration. Fortunately, computer software has streamlined many of these steps. Imagine performing integrations by hand on ink chart recorders for a 70-analyte acyl carnitine panel or linear least squares regression with pen, paper, and a slide rule. Despite the advantages computers and higher order processing have yielded, there remains flaws in the mechanisms by which peaks are chosen and integrated. In complex samples, isobaric peaks may be present, creating difficulties in automatic peak selection. Variations in baseline noise may result in peaks being integrated as though they are superimposed in the noise or above, but rarely consistently. As such, peaks should be manually reviewed by qualified scientific staff to ensure accuracy and precision rather than relying on inadequate algorithms to autoselect and integrate the correct peaks correctly. Improvements in peak integration algorithms combined with software capable of mathematical determinations of peak quality have been recently introduced [23,24]. These types of solutions can provide substantial improvements over the subjective nature of human data review, but are not yet widely implemented.

Data integration software packages, regardless of vendor, utilize algorithms to convert analog signal in the mass spectrometer (detected ions) to a digital signal observed on the computer screen. There are a number of options available to scientists in data review which may be arbitrarily used to the detriment of the quality of the data. The two most prevalent are the over-use of smoothing factors and the ability to manually integrate peaks of interest.

Mass spectrometry is a scanning instrument, capable of moving between mass ranges during data acquisition. Data points collected are averaged as counts-per-second, which can induce variations in the signal produced. The appearance of jagged peaks is not uncommon in MS data. Smoothing of such data can be performed to facilitate appropriate peak integration; however, over smoothing can result in the addition of a chromatographically resolved peak in the integrated peak area. In Fig. 5.3, an example of unsmoothed 17α-hydroxypregnenolone (raw data) is shown next to its smoothed counterpart. The obvious isobar which elutes after 17α-hydroxypregnenolone is included in the smoothed data, resulting in an overestimation of the peak area.

Manual integration of data involves a scientist selecting the beginning, end, and baseline level of a peak. This is a highly subjective process and should be avoided. Among five different scientists, one can expect five different manual integrations; additionally, if each scientist were asked to repeat the manual integration, they would most likely end up differing from the initial attempt. Software, while not being entirely accurate in peak selection, is capable of consistent integration. Manual manipulation of data can create errors as well as take up substantial time in data review.

The process of data review can be most easily arranged in two groups: overall batch review and individual sample review. The first primarily focuses on calibration curve acceptance, overall internal standard recovery and QC acceptance. Individual samples undergo inspection for appropriate integration of samples, checks against retention time alignment between the internal standard and the analyte, transition ratio analysis, and review of other features which may reveal the quality of the data.

It is recommended that data review initiate with a check of batch acceptance. Prior to review of actual samples, a check of the pressure traces of the batch (if available) can swiftly reveal unacceptable data analysis. Fluctuations in a pressure trace over the course of the batch may indicate a leak, lack of mobile phase, or a possible air bubble trapped in a check valve or pump head. Rather than expending time reviewing a batch in which data were acquired inappropriately, the scientist can begin a reinjection process. In the absence of historical pressure traces, briefly scanning the last sample of a batch can indicate an issue which presented earlier in the analysis, particularly since leaks, autosampler errors or loss of MS signal rarely self-corrects over the course of a batch and depleted mobile phase does

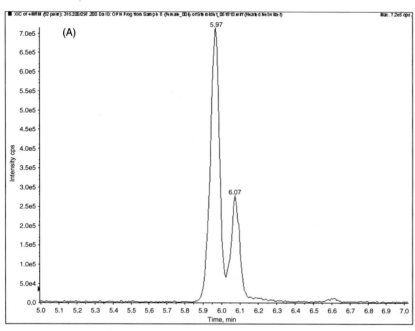

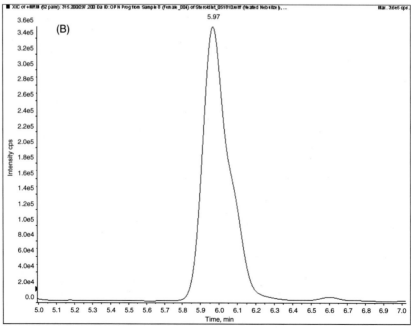

FIGURE 5.3

Chromatograms of unsmoothed 17α-hydroxypregnenolone (A) with an isobaric species eluting after the peak of interest. When excessively smoothed, the isobaric analyte response is included in the integrated peak area (B)

not prepare itself (note that small air bubbles trapped in LC check valves can be corrected over time). Alternatively, SST's ran at the end of a batch can establish appropriate instrument function with a longitudinally tracked solution. In cases where the end of batch SST is unacceptable, data review can be undertaken more judiciously and troubleshooting can be initiated.

Following confirmation that data were suitably acquired, a cursory review of internal standard recoveries is in order. Data reduction software provides access to internal standard recovery plots. Deviations from normal recovery may indicate a host of issues to trouble shoot, including instrument drift or charging, poor preparative recovery, failure to precisely aliquot internal standard in sample preparation, injection of incorrect samples from a discrete assay (if, for instance, the platform analyzes different assays or is multiplexed), unacceptable ionization suppression/enhancement from matrix effects, or source fouling.

Sample preparation-induced drift can be observed by a diverse range of origins related to the mode in which sample preparation occurs. It is advantageous to initially assess the retention time of the internal standards; the retention time and recovery of this peak should demonstrate consistency throughout the batch. Poor signal or deviations in the retention time of the IS indicates a possible issue with the sample, thus obviating further data review and initiating corrective action. Poor IS recovery can be a result of a number of errors in sample preparation or in analysis. Examples are listed in Table 5.2.

For example, the preparation of 96 samples for liquid–liquid extraction may have IS recovery aberrations as a function of interaction time between the sample and the extraction solvent. If individual tubes are used for each sample, differences in the residence time of the extraction solvent on the sample

Table 5.2 Common Sources of Reduced Internal Standard Recovery

Sample Preparation Technique	Possible Source of Error
Solid phase extraction	Lack of IS equilibration with sample
	Excessive positive pressure using vacuum manifold
	Improper wash solution or elution solution
	Inadequate IS volume pipetted
	Excessive temperature or gas flow on evaporation (if used)
	Incorrect reconstitution solution (if used)
Liquid–liquid extraction	Lack of IS equilibration with sample
	Wrong fraction captured
	Inadequate IS volume pipetted
	Excessive temperature or gas flow on evaporation (if used)
	Incorrect reconstitution solution (if used)
Protein precipitation	Inadequate IS volume pipetted
Analysis	Possible source of error
Liquid chromatography	Insufficient volume of injection
Mass spectrometry	Ionization suppression (including charge competition)

IS, Internal standards.

may occur between the 1st and the 96th sample. In solid phase extraction, the amount of time solvents or samples are allowed to pass through the bed may also result in IS recovery differences, particularly when vacuum or positive pressure is used to draw/push samples through the sorbent material. Validation experiments are commonly performed to address these issues; however, human interactions are not infallible and random errors do occur.

Charging may also induce IS response differences across a batch. Charging is, briefly, the buildup of materials on the ion optics which generates deviations in the electric fields, disturbing the trajectory of the ions of interest. This sort of drift can be commonly confirmed by performing a polarity switch (i.e., positive to negative and back to positive mode) across the mass spectrometer's optics while infusing the compounds of interest or a standard tuning solution (i.e., polypropylene glycol or polytyrosine). A substantial increase in sensitivity following a polarity switch is indicative of charging. Thorough cleaning of the corrupted ion path is in order once detected.

Ionization suppression, one form of matrix effects, can yield decreased IS responses. In rare cases, matrix effects can result in ionization enhancement, indicated as high recovery of the IS relative to calibrators, QC's, and other samples (in these instances, ensuring that the IS peak is pure and that an internal standard was not overly aliquoted is essential). Known matrix effect-causing molecules, such as phospholipids or dioctyl phthalates, should be accounted for in development [25]. Matrix effects of unknown origin may be related to an individual patient, as caused by a drug/ metabolite or associated with a disorder or dysfunction unaccounted for in interference studies, or an occurrence related only to that particular sample with no determinate cause of error.

Rarely, IS drift may be associated with deuterium labeled internal standards undergoing hydrogen-deuterium exchange in the source of the mass spectrometer or inconsistent deuterium scrambling in collisional dissociation, resulting in disparities in IS recovery over the course of an analytical run [26]. The constituents of the gas phase in the source will certainly change as more samples are introduced and as coeluting (but unobserved in the mass spectrometer) sample constituents are introduced to the ionization and interface regions. Drift of this type should be assessed during validation; remediation is selection of a more stable IS transition or a distinctly labeled IS all together. In conventional MS/MS applications, the neutral loss established for the analyte product ion of interest is typically the same neutral loss used to generate the product ion of the internal standard. Note that the actual mass lost may be different, dependent on the position of the labeling. It may be the case that an entirely different fragmentation pathway provides the most reliable normalization of analyte recovery in analysis [27].

Acceptable recovery of the internal standard should be addressed on an assay-by-assay basis, taking into consideration the clinical application of the assay, the calculated concentration of the sample, and guided by the medical decision points of the test.

After IS drift/deviations have been assessed, calibration curves should be generated. In many cases, MS assays include calibration materials with each analytical run. Thus, de novo curve generation is required for each batch. Data reduction software provided by mass spectrometer vendors can automatically perform linear least squares regression from expected calibrator values and the response ratios of the analyte and the internal standard. It is essential to ascribe the same fit (linear or quadratic) and the same weighting factors ($1/x$ or $1/x^2$, commonly) as was determined in validation. Fitting of calibration curves without criteria or validation is strongly discouraged; changes in the appropriateness of the fit should induce an investigation of cause rather than an attempt to mask a possible instrument or preparative error. As a reminder for all aspects of operations, validation studies objectively demonstrate the analytical claims; deviations from the protocols established in validation leads to unproven processes.

Standard practice for calibration curve fitting is to utilize all points which are within ± 15% of the expected value as back-calculated in the curve, with the exception of the lowest calibrator which is acceptable at ± 20% [8]. This difference in acceptance criteria for the distinct concentrations is due to the heteroscedastic nature of data most commonly generated by quantitative mass spectrometric assays. Briefly, the influence of noise and changes in baseline signal are greatest at the lowest peak response (i.e., the lowest calibrator). If, due to a substantial deviation in the accuracy of the lowest calibrator(s), those values are unacceptable, acceptance of the curve should be determined by both QC acceptance as well as impact to clinical decision making. For instance, imagine an assay with calibrators an LLOQ of 1 ng/mL and a second calibrator at 5 ng/mL, where for a single batch the 1 ng/mL calibrator reads at 0.5 ng/mL, or 50% accuracy in the calibration curve. In this same assay, values less than 3 ng/mL are considered clinically actionable or abnormal. The process and raw data should be reviewed to determine the source of the bias and the batch reassayed as the medically relevant decision point is obfuscated. Similarly, calibration standards which either are the concentrations of medical decision points or bracket medical decision points should not be discarded without a firm understanding of the reason for failure and impact on the patient results.

QC values should be assessed following completion of calibration curve generation. Minimum QC schemes utilize two samples, allowing for review of recovery and precision below and above a single medical decision points in the batch. These QC samples should also be used to monitor precision of analytical performance over time. QC schemes and their implementation have been discussed previously in Chapter 4. It is further recommended that a blank with internal standard (a true negative) made from calibration matrix be included in each batch as check for carryover and contamination resulting from internal standard solutions. Additionally, blank samples which contain no internal standard or analyte (a double blank) should be included to ensure isobaric contributions to the internal standard response have not occurred, or that contribution to the analyte transition observed in a blank-with-IS is due to the internal standard solution and not the analytical process. The placement of these samples is such that the true negative immediately follows the highest calibrator and the double blank succeeds the true negative. Review of contributions to the analyte trace in the true negative sample and the internal standard trace in the double blank sample should indicate no unexpected responses; failure of the criteria should prompt an investigation of contamination.

Following this general batch review, individual samples should be assessed. Each chromatogram for a patient must be reviewed to ensure appropriate integration of peaks as well as identification of possible errors in data acquisition. Characteristics for review include assessments of retention time reproducibility, retention time alignment with the internal standard, evidence of matrix effects, peak width, evaluation of interfering or isobaric peaks, and transition ratio analysis.

Retention time reproducibility and alignment of the retention times of analyte(s) and their respective internal standard(s) is an important feature for review. Data software packages for LC-MS/MS systems can calculate the retention times; this data can be quickly plotted to assess deviations from expected retention times generated from calibrators and QC's. Assessment of peak width at half height allows for normalization of chromatographic observations at concentrations within the dynamic range of the mass spectrometer. The analyte and the internal standard should, given adequate signal, exhibit a similar peak width at half height for peaks of high purity. Deviations in the peak width may indicate a coeluting isobaric interferent that does not appear as a distinct peak or inappropriate peak selection. Fig. 5.4 is a plot of the peak-width-at-half-height analyte/internal standard ratios for samples analyzed for cotinine. The wide distribution of ratios

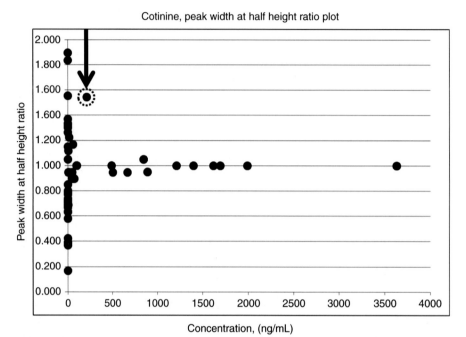

FIGURE 5.4 Example Peak Width at Half Height Plot for Cotinine, Including Standards, Quality Controls, and Human Samples

The *dashed circle* and *arrow* indicates a sample with an unexpected deviation from the predicted value of approximately 1.

at the low end of the concentration range is to be expected as the assay attempts to integrate noise as a peak, though all generated values are less than the LLOQ. However, one obvious nonzero sample is indicated by the dashed circle and arrow. The actual integrated chromatogram for that sample is shown in Fig. 5.5.

Quantitative assays typically utilize one precursor-product ion pair for quantitation (some assays will sum more than one transition together, while using less than one is yet unreported in the literature). However, many molecules analyzed via tandem mass spectrometry will in fact generate more than one fragment ion of interest. These other ions should not be discarded following their observation in developmental infusion experiments. They provide yet another avenue in which a mass spectrometric assay can produce selective analysis:transition ratios.

If the use of stable, isotopically labeled internal standards is considered mass spectrometry's greatest advantage by precisely normalizing the detection of the compounds of interest, transition ratios should be thought of as a not-too-distant second. Transition ratios are simply the comparison of the response of two fragment ions generated from the same precursor molecule. Collisional dissociation for the generation of product ions is inherently selective and reproducible. The comparison of two or more consistently created product ions increases that selectivity many fold over, allowing MS users additional confidence in the reporting of assay results.

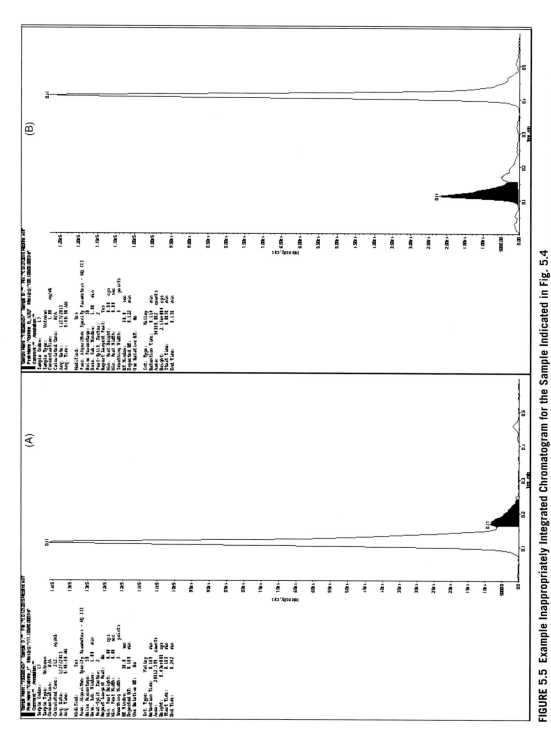

FIGURE 5.5 Example Inappropriately Integrated Chromatogram for the Sample Indicated in Fig. 5.4

The cotinine peak (A) was incorrectly chosen by the auto-integration algorithm as it does not align with the internal standard, cotinine-D4 (B).

Table 5.3 Observed Peak Areas for Quantifying and Qualifying Peak Area Ratios for Calibrators in Urinary Free Cortisol

	Cortisol, Urinary, Free			
Calibrator (ng/mL)	363.3 → 121.1 Peak Area	363.3 → 97.1 Peak Area	Transition Ratio	Number of Standard Deviation's From Mean
1	1476.6	464.4	0.315	2.23
2	2466.4	586.5	0.238	−0.31
5	6764.5	1657.6	0.245	−0.07
10	13904.8	3293.9	0.237	−0.34
25	33258.2	7893.6	0.237	−0.33
50	68573.2	15606.4	0.228	−0.65
100	129556.7	29952.1	0.231	−0.53

Calibrator 1 demonstrates an unacceptable ratio for use in transition ratio generation.

Transition ratios are performed by generating peak areas of a quantifying ion and a qualifying ion, wherein the quantifying ion is the precursor-product ion pair on which validation data were generated and patient results shall be released. These peak areas are divided according to the equation:

$$\text{Qualifying Ion Peak Area} \div \text{Quantifying Ion Peak Area}$$

The transition ratios generated from samples of known purity, such as the calibrators and QCs are used to generate the mean ratio for each batch, however, certain transition ratios should not be included. Quantifying ion peaks should have sufficient sensitivity to enable precise measurement at the lowest calibrator; this may not be true for the qualifying ion. Certain compounds may exhibit a single dominant product ion of high abundance while other ionized fragments are produced with low efficiency. In these cases, only standard and QC concentrations in which acceptable sensitivity is achieved for the qualifying product should be utilized. The criteria for acceptance of particular data points should be assessed during validation of the assay. For example, optimized product ions for cortisol in the analysis of urine free cortisol are shown in Table 5.3. The lowest calibrator (1 ng/mL) has sufficient sensitivity for accurate and precise quantification for the precursor-product ion pair of 363.3→121.1; the qualifying product ion of 97.1 does not have sufficient sensitivity to allow for a precise peak area to be calculated Fig. 5.6. In this case, the second calibrator of 2 ng/mL is utilized. This is further shown via the deviation of the transition ratio for that calibrator series in the batch observed in Table 5.3.

As relative sensitivity of a mass spectrometer may change over time, comparison of the transition ratio of the samples to that of known samples should be performed on an intrabatch basis. The criteria for acceptance should be sufficient to indicate deviations in the selectivity of the quantifying peak for the assay.

An example of the use of transition ratios is shown in Fig. 5.7 for urinary free cortisol. Note sample number 17, in which the deviation from the mean (generated from acceptable calibrators and QCs) is immediately apparent. Review of this sample indicated no integrated peak for the qualitative product ion and an approximate concentration of 1.3 ng/mL. This sample is considered acceptable as internal standard recovery is within normal range for the batch and the result is not considered to be clinically abnormal.

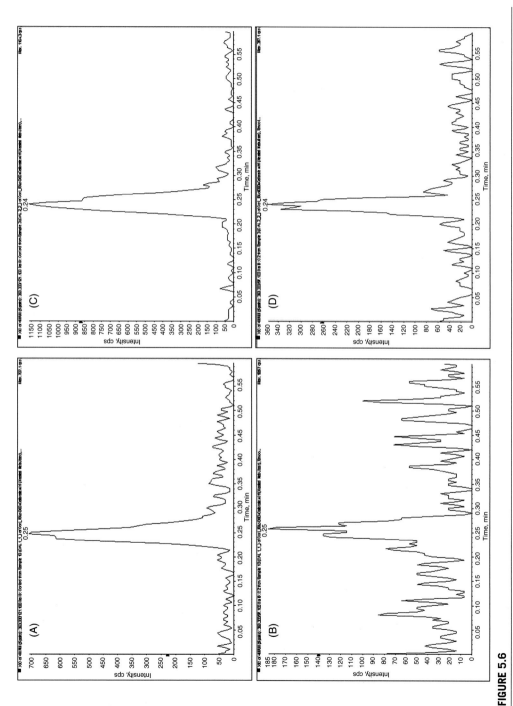

FIGURE 5.6

Example 1 ng/mL Calibrator Quantifying Ion (A) and Qualifying Ion (B) and 2 ng/mL Calibrator Quantifying Ion (C) and Qualifying Ion (D) for cortisol. The 1 ng/mL Qualifying Ion demonstrates insufficient sensitivity for precise evaluation of a transition ratio and is excluded in data generation.

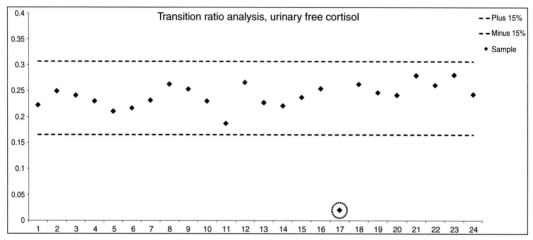

FIGURE 5.7 Example Transition Ratio Acceptance Plot Generated in Microsoft Excel Using Criteria of ± 15% of the Mean From Acceptable Calibrators for Urinary Free Cortisol

Refer to text for explanation of unacceptable transition ratio for sample 17.

In certain cases, a molecular ion does not offer up a second transition capable of sufficient sensitivity for practical use. This has been eloquently addressed by Kushnir, Rockwood, and colleagues, relying on the standard principles of product ion formation, specifically that product ions can be formed at different efficiencies for isobaric pairs, but those product ions are also formed at different optimal collisional dissociation energies [28]. In this case, the same fragment is monitored using distinct collision energies to generate a transition ratio. Though it is the same product ion, the relative effectiveness of the two different collisional dissociation rates yields fundamentally the same information as a transition ratio with two different neutral losses.

After data have been reviewed and determined to be successful, patient results can be readied for release. In the case of certain assays, results which require immediate medical care according to values identified in the SOP should be identified and expedited. Such examples of these tests can include immunosuppressant concentrations in the supra-therapeutic region or blood-based concentrations of drugs of abuse indicative of overdose. Cutoffs for these alert values should be documented in the laboratory's test SOP.

6 INSTRUMENT MAINTENANCE

Drafting and implementation of a preventative maintenance program for instrumentation is essential to robust operations of assays. The scheduling of such maintenance should focus on timely replacement or cleaning of components with a limited useful life. In some cases, such as valve or rotor seals or plungers on injection syringes, the useful life is dictated by number of times employed. Manufacturers typically provide information in instrument manuals regarding the lifespan and replacement procedure for these items. Other components are time-related, linked to environmental factors or are assay specific. For example, the mass accuracy of time of flight instruments may drift due to modifications in laboratory temperature or humidity, requiring frequent mass calibrations to ensure appropriate operation. Roughing pump oil is changed on a consistent schedule, such as 4, 6, or 12 months, depending on the brand of roughing pump used. Still other components have no set limit on either time or frequency of use, such as electrospray electrodes, in which degradation of the ability to generate charged solution can be related to a number of unconnected factors and should be replaced not as needed, but prior to expected failure.

6.1 LIQUID CHROMATOGRAPHY SYSTEM

Liquid chromatographs, compared to mass spectrometers, have far more user-accessible components. In a standard clinical laboratory, the LC system also requires the most care. General repair of LC components, exclusive of the column, relates to pump seals, rotor seals used in valves, tubing, and the autosampler. As each LC manufacturer has different specific components and recommended maintenance procedures, detailed steps for LC hardware maintenance will not be discussed here. However, following each maintenance procedure, planned or unplanned, SSTs should be performed to ensure the performance of the system prior to analysis of patient samples.

For maintenance of columns and guard columns, observations of pressure traces, SST's or samples within a batch can point toward deleterious column function. Normal operating conditions, pictorialized as a pressure trace, should be included in the SOP. This can be referenced against the pressure traces of the daily SST's or patient analysis. If the pressure trace is reading near the pressure limit of the LC system, an evaluation of the source of high pressure should be undertaken. In the case of high pressure being generated from the column or guard column (due to matter occluding the interstitial space of the packed beds or the frits at the head of the column), a replacement is necessary. Alternatively, washing the column with a bevy of mobile phases or reversing the column direction has been reported [29]. It should be noted that these processes are not always cost- or time-effective and may not resolve pressure issues; indeed, columns are consumables and should be treated as such [30].

When a chromatography column is replaced on the system, observations of increased retention times of the compound(s) due to increased activity are not uncommon. In these cases, passivation of the column material is required. Injection of repeated SST's, or even extracted patient samples is recommended until acceptable retention times are observed. This manner of passivation should be noted in ruggedness evaluations in validation of the assay, though it may appear across various lots of column material and thus not be observed in validation exercises. Prior to use in analytical runs, at a minimum, the SST chromatogram of the new column should be compared to that of the chromatogram described in the SOP. More stringent evaluations may include the reanalysis of calibrators, QC's, and patient samples for new column acceptance [31]. In this case, it is recommended that a previously analyzed batch be utilized without repreparation of the samples to exclude random error in sample preparation as a possible cause of new column failure. If analytes are determined to be unstable, and each batch contains appropriate calibration materials and QCs, SST's are sufficient evidence for new column acceptance.

Replacement of LC tubing, whether it is polyetheretherketone (PEEK) or stainless steel, is a periodic process. PEEK tubing will degrade over time, particularly if tetrahydrofuran is used as a mobile phase constituent, and is only useful up to a certain pressure (approximately 400 bar). Stainless steel tubing is typically more rugged, wherein replacement is required due to fittings and ferrules generating leaks. It should be noted that each column and valve manufacturer has preferred fittings, ferrules as well as depths of the column end-fitting. Recommendations for each LC component's preferred materials and fitting depth should be followed to prevent unnecessary dead volume.

Autosampler maintenance should be performed on a similar schedule to the LC systems. In addition to the replacement of consumable parts, such as injection syringes, valve or needle seals, syringe guides, etc., the autosampler should be calibrated to ensure proper needle position on autosampler vials or 96-well plates as described in the instrument's manual. Few sounds in the laboratory are more disturbing than the jolt that occurs when an autosampler attempts to punch through 4 cm of polypropylene as opposed to 2 mm of a silicone mat or 1 mm of sealing foil. Unfortunately, that sound is often followed by delays in analysis as well as expenses incurred from early replacement of previously sound autosampler components.

6.2 MASS SPECTROMETER MAINTENANCE

Planned or preventative maintenance for mass spectrometry equipment typically includes the following components: mass resolution and accuracy calibration, source cleaning, and ion optics. The frequency at which each of these needs to be performed is dependent on the instrumentation, the laboratory conditions, and the assays analyzed on a particular platform. Preservation of mass spectrometer performance is typically related to the front end of the mass spectrometer, a region which consists of the source and the interface. The frequency of maintenance on these components is broadly determined by the cleanliness of the LC stream reaching the mass spectrometer and the robustness of the source utilized. The cleanliness of the samples may have little to do in relation to the cleanliness of the source and interface if a bypass valve is used appropriately. Such a bypass valve directs waste materials away from the source, preventing fouling from materials in the samples or generated during sample preparation, as discussed in Chapter 3.

Cleaning of the source and interface region is MS vendor specific. For the specific steps of cleaning the MS, users should refer to the manufacturer or field engineers for instructions on cleaning as well as the details included in the laboratory's SOP on instrumentation. However, certain best practices should be followed. First, high purity, LC-MS grade solutions should be used for cleaning. Impure materials used for cleaning may leave residues on lenses, orifices, and other MS components which can create undesirable performance characteristics. Some commercial metal polishes are used to clean ion lenses or optics; these polishes may leave a hydrophobic residue (they are intended to create a luster after cleaning) which can thermally degrade, possibly interfering with gas phase ion generation, ion transmission, or stability. Rather than such polishes, very fine aluminum oxide powder, manufactured at a high purity, can be used to polish metal components of a mass spectrometer without leaving such a residue and is easily rinsed away. Detergents can be used with care, ensuring that the MS component is thoroughly rinsed to remove surfactants and salts. Lint free wipes should be considered for cleaning applications, as liberated fibers can obscure the orifice of mass spectrometers, resulting in minimal transmission of ions.

Mass spectrometer calibration is a considered distinct from assay calibration. In the MS sense, calibration is the accuracy of mass selection as well as the m/z width observed in the instrument, and may also be described as "tuning." The frequency of necessary tuning depends on a number of factors, including the vendor (and thus the quality of the quadrupoles), the type of mass spectrometer, the frequency of venting the instrument to clean, the ambient conditions of the laboratory, among other subtle mechanisms [32]. Some uses of mass spectrometry can quickly elucidate when tuning is required. For example, full scan instruments, such as time-of-flight or Fourier transform-based platforms capture all data within a required mass window. Dependent on the matrix, sample preparation scheme and the ionization mechanism, certain ions are perpetually present in analysis, allowing for speedy identification of drift in the m/z accuracy, similar to the use of separate solutions used in mass calibration of TOF instruments [33]. These ions can be of biological origin, such as phospholipids in blood-based matrices, or of chemical origin, such as polymers, phthalates/plasticizers or clusters of mobile phases (LC) or crystal matrix (MALDI). Such lock-mass ions can be interrogated for mass accuracy as well as for peak width (peak width is commonly used at half-height, though alternatives are also used).

Unlike full-scan instruments, quadrupoles or triple quadrupoles operating in single ion monitoring (SIM) or selected reaction monitoring (SRM) do not allow the observation of the m/z as a piece of useful data. These mass spectrometers lock the radio frequency/direct current values at a certain value preprogramed for a particular m/z. In effect, the mass spectrometer excludes ions which would be indicative of calibration drift. Thus, identification of changes in either mass accuracy or peak width must be established in a separate mode. Periodic checks of mass analyzing quadrupoles is encouraged for all MS instruments. This may be performed with the tuning solutions preferred by the mass spectrometer's manufacturer or can be performed in-house using the compounds to be analyzed.

In high resolution analysis, satisfactory mass accuracy is assay and platform independent. For example, certain TOF instruments are incapable of 50 ppm or better mass accuracy. Certain applications, such as microbial identification via MALDI-TOF may only require 100 ppm mass accuracy [34]. For the majority of MS/MS applications, acceptable mass inaccuracy is less than ± 0.1 m/z. Instrument resolution is also application and platform dependent. When coeluting isobars (but not isomers) are present in high resolution applications, sufficient differentiation should be obtained to correctly distinguish the compounds. In MS and MS/MS utility, unit resolution (defined here as peak-width at half mass) of 0.6–0.8 is acceptable. Some applications benefit from the use of lower mass resolution, such as immunosuppressant analysis. Observable gains in sensitivity are found in lower resolution as higher order, naturally occurring isotopes maintain a stable trajectory through the quadrupoles. In these instances, the resolution should be constrained to allow for adequate sensitivity of the molecules and no wider. Excessively poor mass resolution can result in falsely elevated signals due to compounds with similar m/z's being detected in the mass spectrometer.

7 QUALITY ASSURANCE FOR SPECIMEN ANALYSIS

Mass spectrometric platforms inherently involve a large degree of human interaction, from sample preparation to data review. For the actual analysis of samples, however, the level of automation is such that the systems are fundamentally self-sufficient. However, errors and shutdowns do occur, often times 5 min prior to leave for a long weekend or vacation! To prevent these events, quality assurance protocols may be introduced to the laboratory's workflow to ensure consistent instrument performance, even in the absence of human minders of the system.

These protocols are separated between the LC system and the mass spectrometer. For the chromatography platform, the following system components are recommended to be checked and documented with daily or with each analysis, as appropriate.

Leak checks and detection: Modern LC systems include leak detection sensors in the body of the LC pump. These are placed such that leaking purge valves, pump heads, check valves, and other elements of the pump are detected, as long as the leak is both persistent and does not evaporate prior to reaching the sensor. These sensors, however cannot, detect leaks at fitting junctures, columns, external switching valves, autosampler loops, or other elements external to the pump housing. Immediately following SST analysis (before possible solvent evaporation can occur), each connection should be inspected. Leaks, by virtue of their location or volume, can be difficult to detect. A folded lint free wipe can be touched to these difficult to reach or see locations and inspected for the presence of liquid. Successful acquisition of SST samples does not immediately indicate the absence of a leak. In this author's experience, leaks tend to increase over time as opposed to eventually self-resolving. Entrepreneurial scientists would do well to create a solvent which self-seals, similar to that in bike tires.

Pressure trace monitoring/ air bubbles: Current LC models are competent at degassing of mobile phases, particularly at lower flow rates. These systems, however, are incapable of removing large air bubbles from solvent lines. On occasion, air bubbles may get trapped at check valves or pumping components. Often times, these are discovered by observations of a jagged or saw-toothed LC pressure profile, as the trapped air bubble creates flow and thus pressure disturbances. Corrective action is often purging of the LC system, though occasionally cracking the fittings of the pump may also be in order while solvent is flowing. Checks for possible air bubbles should be performed by observation of SST pressure traces prior to analysis.

Loss of mobile phase: Calculation of the exact volume of mobile phase(s) necessary for an analytical run can be easily performed; some LC software will calculate this for the user. That volume, however, should not be considered the appropriate volume needed for analysis. Mobile phases should always be prepared in excess to prevent possible drying of the LC pumps. Documentation of acceptable levels of mobile phases should be included in a daily assay check.

Orientation of 96-well plates: Present day autosamplers are equipped to handle 96-well plates. Relative to vials, 96-well plates offer reduction in foot print with the added capability of higher order pipetting (manual 12-channel pipettes for aliquoting reagents in a single row are not uncommon) as well amenability to automation. A singular disadvantage of 96-well plates is the opportunity to turn one around. Vendors do supply well plates with corners removed on the "front" of the plate; however, this does not fully discourage laboratory staff from placing the front of the plate where the back would normally be. This event can be observed on data review (well position 1 is now well position 96; the calibrators will not be in order for linear least squares regression to be recognizable). After loading plates in to the autosampler, plates should be reinspected to ensure appropriate orientation and positioning.

A mass spectrometry system is a sensitive instrument. This sensitivity relates not just to the capability to measure at low levels, but also that a mass spectrometer is susceptible to environmental changes. Temperature and humidity can play a role in the effectiveness of MS analysis. In TOF instruments, the power supply of the instrument as well as the flight tube can be affected by fluctuations in the ambient conditions [31]. Environmental conditions should be controlled as much as possible to reduce the amount of tuning and calibration required.

Most MS platforms require pressurized gases for operation. Triple quadrupoles require inert gas, such as nitrogen or argon, for controlled collisional dissociation. LC-MS frequently utilizes nebulizing gas flow to encourage efficient electrospray formation as well as a drying gas to facilitate desolvation or complete evaporation in APCI. Ensuring adequate gas availability should be a routine check of the instrument on a daily basis. Most current mass spectrometer models have built in pressure regulators and monitors and will force a shutdown of the platform in the absence of sufficient gas availability.

The majority of large volume tanks are capable of refilling while still supplying gas to the instruments. Smaller gas cylinders, such as dewars, typically do not have this functionality and require a venting of the instrument prior to swapping of gas tanks. If a shut down and vent of the platform is necessary, SST samples should be used to confirm instrument performance after the new tank is in service.

Present day LC-MS/MS systems are currently available with the ability to provide staggered parallel multiplexed LC delivery into a single mass spectrometer, as previously discussed in Chapter 3. These platforms, while increasing throughput and efficiency, are also susceptible to cross-stream carryover. This occurs in two instances. The first is where the solvent from one stream resides in the tubing following a selector valve which rotates between the various LC systems. As a different LC stream is brought into line with the mass spectrometer, the remnant solvent (and any molecules contained therein) are introduced to the MS system at the beginning of the acquisition of the sample being introduced. For this reason, sufficient time to clear the volume of the tubing should be in place before the acquisition of the first peak to be detected. This allows for any bolus of material which may include an isobaric interfering species from the previous sample of another stream to contribute to the data being collected. Calculating the amount of time is performed by dividing the approximate volume of the tubing by the flow rate used. Appropriate excess time should be included to ensure that longitudinal diffusion of the material in the new sample stream does not increase background levels for the peak of interest.

The second source of cross-stream carryover is in the source of the mass spectrometer. All sources are equipped with exhaust to remove excess gas, both introduced via nebulizing gas or heating gas as well as produced from desolvation of liquid chromatography outputs. Some gas phase molecules can persist, however, increasing the background for subsequent samples. For example, methylmalonic acid (MMA) analysis serves two purposes: assessment of vitamin B12 deficiency or methylmalonic acidemia diagnosis/monitoring. Expected values for healthy, B12 sufficient individuals can be in the micromolar region (76–270 µM); individuals with methylmalonic acidemia can produce concentrations in the molar regime [35,36]. If a methylmalonic acidemia patient is analyzed on one stream, the subsequent sample of the next stream exhibits an increase in the background level of MMA, yielding a limited amount of quality data for that elevated-background sample.

Maintaining operating conditions in the laboratory is critical to proper mass spectrometer performance. This includes environmental considerations as well as consistent power supply. Laboratory temperature should be kept below 23°C to ensure that roughing pumps and turbo pumps do not overheat. Fluctuations in laboratory temperature should be avoided, particularly when using high resolution platforms. Humidity should be maintained at a low level, as condensation within cooled autosamplers can promote failures of the autosampler components through corrosion.

Power supply to MS equipment should be uninterrupted. At a minimum, the roughing or backing pumps for the MS unit should be placed on battery backup. Accelerated venting of the mass analysis region by a hard stop can result in seizure of the bearings in the analyzer pumping system. As a worst case, and particularly with older instruments, the vacuum within the mass spectrometer so far exceeds that of the backing pumps that the backing pump check valves will crack, releasing oil and oil vapor into the body of the mass spectrometer. Battery backups, further supported by a generator for long-term outages, allows laboratory staff to rest on evenings and weekends.

8 CONCLUSIONS

The growth of mass spectrometry in hospitals, clinics, and references laboratories must be mirrored by the growth in understanding of how the laboratory can facilitate rugged applications for clinical use. Success in the laboratory is not gauged by the volume of samples processed, but the quality in which results are produced, particularly long-term quality. Implementation of procedures to address calibration, QC, system checks, and data review and guided by fundamental laboratory principles as well as the mass spectrometry specific components outlined in this chapter, a successful clinical mass spectrometry operation is achievable.

REFERENCES

[1] Clinical Laboratory Improvement Amendments. 42.CFR.493.part 1251; 1988.
[2] Shi RZ, van Rossum HH, Bowen RA. Serum testosterone quantitation by liquid chromatography-tandem mass spectrometry: interference from blood collection tubes. Clin Biochem 2012;45:1706–9.
[3] Wang C, Shiraishi S, Leung A, Baravarian S, Hull L, Goh V, et al. Validation of a testosterone and dihydrotestosterone liquid chromatography tandem mass spectrometry assay: interference and comparison with established methods. Steroids 2008;73:1345–52.
[4] Kushnir MM, et al. Analysis of dicarboxylic acids by tandem mass spectrometry. High-throughput quantitative measurement of methylmalonic acid in serum, plasma, and urine. Clin Chem 2001;47(11):1993–2002.
[5] Jedlička A, Klimeš J. Determination of water-and fat-soluble vitamins in different matrices using high-performance liquid chromatography. Chem Papers 2005;59(3):202–22.
[6] Desai, Ankur M, et al. Acetonitrile shortage: use of isopropanol as an alternative elution system for ultra/high performance liquid chromatography. Anal Methods 2011;3(1):56–8.
[7] Carraro P, Plebani M. Errors in a stat laboratory: types and frequencies 10 years later. Clin Chem 2007;53(7):1338–42.
[8] US DHHS, FDA, CDER and CVM. Guidance for industry: bioanalytical method validation. US Department of Health and Human Services, Food and Drug Administration, Center for Drug Evaluation and Research and Center for Veterinary Medicine: Rockville, MD; 2013.
[9] Briscoe CJ, Stiles M, Hage DS. System suitability in bioanalytical LC/MS/MS. J Pharma Biomed Anal 2007;44(2):484–91.
[10] Kushnir MM, et al. Analysis of catecholamines in urine by positive-ion electrospray tandem mass spectrometry. Clin Chem 2002;48(2):323–31.
[11] Peeters RP. Thyroid hormones and aging. Hormones 2008;7(1):28.
[12] ISO 17511. Metrological traceability of values assigned to calibrators and control materials. In vitro diagnostic medical devices—Measurement of quantities in biological samples, ISO; Geneva, Switzerland; 2003.
[13] CLSI, Metrological traceability and its implementation. CLSI Document EP32-R2. Wayne, PA, Clinical and Laboratory Standards Institute; 2006.
[14] McGregor K, Makkai T. Self-reported drug use: how prevalent is under-reporting? Canberra, Australia: Australian Institute of Criminology; 2003.
[15] CLSI, Method comparison and bias estimation using patient samples; approved guideline—second ed. CLSI document EP9-A2. Wayne, PA, USA: CLSI; 2002.
[16] Erez A, Nagamani SC, Lee B. Argininosuccinate lyase deficiency—Argininosuccinic aciduria and beyond. Am J Med Genet C Semin Med Genet 2011;157C(1):45–53.
[17] Kushnir MM, et al. Liquid chromatography–tandem mass spectrometry assay for androstenedione, dehydroepiandrosterone, and testosterone with pediatric and adult reference intervals. Clin Chem 2010;56(7):1138–47.

[18] Zahir H, et al. Factors affecting variability in distribution of tacrolimus in liver transplant recipients. Br J Clin Pharmacol 2004;57(3):298–309.

[19] Mahalati K, Kahan BD. Clinical pharmacokinetics of sirolimus. Clin Pharmacokinet 2001;40(8):573–85.

[20] Elmlinger MW, Kuhnel W, Ranke MB. Reference ranges for serum concentrations of lutropin (LH), follitropin (FSH), estradiol (E2), prolactin, progesterone, sex hormone binding globulin (SHBG), dehydroepiandrosterone sulfate (DHEA-S), cortisol and ferritin in neonates, children, and young adults. Clin Chem Lab Med 2002;40(11):1151–60.

[21] CLSI, Evaluation of the linearity of quantitative measurement procedures: a statistical approach; approved guideline. CLSI Document EP6-A. Wayne, PA, Clinical and Laboratory Standards Institute; 2003.

[22] CLSI, Interference testing in clinical chemistry; approved guideline—second ed. CLSI Document EP7-A2. Wayne, PA, Clinical and Laboratory Standards Institute; 2003.

[23] Zabell APR, et al. Diagnostic application of the exponentially modified Gaussian model for peak quality and quantitation in high-throughput liquid chromatography–tandem mass spectrometry. J Chromatogr A 2014;1369:92–7.

[24] Dickerson JA, et al. Design and implementation of software for automated quality control and data analysis for a complex LC/MS/MS assay for urine opiates and metabolites. Clinica Chimica Acta 2013;415:290–4.

[25] Bennett P, Liang H. Overcoming matrix effects resulting from biological phospholipids through selective extractions in quantitative LC/MS/MS. 52nd ASMS Conference on Mass Spectrometry, Nashville, TN; 2004.

[26] Stokvis E, Rosing H, Beijnen JH. Stable isotopically labeled internal standards in quantitative bioanalysis using liquid chromatography/mass spectrometry: necessity or not? Rapid Commun Mass Spectrom 2005;19:401–7.

[27] Wright Y, Shuford C, Crawford M, Holland P, Dee S, Green M, Grant R. Challenges and solutions for high throughput TFC-LC-MS/MS in clinical toxicology. American Society of Mass Spectrometry. Baltimore; 2014.

[28] Kushnir MM, et al. Assessing analytical specificity in quantitative analysis using tandem mass spectrometry. Clin Biochem 2005;38(4):319–27.

[29] Dolan JW. Peak fronting, column life, and column conditioning. LC GC N Am 2005;23(6):566–74.

[30] Dolan JW. Column Triage. LC GC N Am 2011;29(10).

[31] CLSI, Liquid chromatography-mass spectrometry—approved guideline. CLSI Document C62-A. Wayne, PA, Clinical and Laboratory Standards Institute; 2014.

[32] Chernushevich IV, Loboda AV, Thomson BA. An introduction to quadrupole–time-of-flight mass spectrometry. J Mass Spectrom 2001;36(8):849–65.

[33] Eckers C, Wolff JC, Haskins NJ, Sage AB, Giles K, Bateman R. Accurate mass liquid chromatography/mass spectrometry on orthogonal acceleration time-of-flight mass analyzers using switching between separate sample and reference sprays. 1. Proof of concept. Anal Chem 2000;72:3683–8.

[34] Croxatto A, Prod'hom G, Greub G. Applications of MALDI-TOF mass spectrometry in clinical diagnostic microbiology. FEMS Microbiol Rev 2012;36(2):380–407.

[35] Klee GG. Cobalamin and folate evaluation: measurement of methylmalonic acid and homocysteine vs vitamin B12 and folate. Clin Chem 2000;46(8):1277–83.

[36] Hörster F, Hoffmann GF. Pathophysiology, diagnosis, and treatment of methylmalonic aciduria—recent advances and new challenges. Pediatr Nephrol 2004;19(10):1071–4.

TOXICOLOGY: LIQUID CHROMATOGRAPHY MASS SPECTROMETRY

6

K.L. Lynch

Department of Laboratory Medicine, University of California, San Francisco, CA, United States

1 TOXICOLOGY TESTING IN THE CLINICAL LABORATORY

Clinical toxicology is defined as the analysis of drugs and chemicals in body fluids for the purpose of patient care. In practice, it is difficult to test for all clinical toxins that may be encountered. The clinical toxicology testing provided by any given laboratory depends on the pattern of local illicit and prescription drug use and the available resources of the institution. Routine clinical toxicology testing usually includes drug screening for common drugs of abuse in urine. This testing includes a panel of tests that is usually referred to as the urine toxicology (utox) screen or the drug of abuse screen. For utox screens, immunoassays for classes of abused drugs are routinely used followed by targeted confirmations for select drug classes (i.e., opiates or amphetamines) using chromatographic techniques coupled to mass spectrometry.

While immunoassay screens are rapid, automated, and readily available, methods do not exist for all drugs and metabolites of clinical interest. Most immunoassays have limited specificity for individual drugs or metabolites, but rather are designed to detect a drug class. Also, some immunoassays have limited sensitivity, and identification is based on detection about a given cutoff. Immunoassays also vary significantly in terms of cross-reactivity with other unrelated compounds. For these reasons, other analytical methods are used for confirmatory testing and detection of drugs and metabolites not detected by the currently available immunoassays. Chromatographic techniques coupled to mass spectrometry are the most common methods used for confirmatory testing following immunoassay screens.

Other assays are offered in addition to the utox screen. These range from specific tests for common intoxicants (i.e., acetaminophen, salicylate, ethanol, digoxin) in serum to additional targeted panels of drugs (i.e., stimulant panel or benzodiazepine panel) in urine or to a broad spectrum drug screen in urine or serum capable of detecting over 100 drugs and metabolites. A method or combination of methods designed to detect a large number of drugs and metabolites, including prescription medications is often referred to as a broad spectrum drug screen. There are other terms that are also used to describe this type of testing including comprehensive drug screen, general unknown screening or systematic toxicological analysis. Broad spectrum drug screening is capable of detecting classes of drugs for which immunoassays have not been developed. These include, but are not limited to, antidepressants, nonbenzodiazepine hypnotics, neuropathic pain medications, muscle relaxants, antipsychotics, β-blockers, calcium-channel blockers, oral antidiabetics, synthetic opioids, and novel psychoactive substances.

Mass Spectrometry for the Clinical Laboratory. http://dx.doi.org/10.1016/B978-0-12-800871-3.00006-7

Some of the common indications for ordering a utox screen or a broad spectrum drug screen include: (1) workplace drugs testing, (2) to monitor drug use during pregnancy, (3) to evaluate drug exposure or withdraw in newborns, (4) to monitor compliance in patients on prescription medications for chronic pain, (5) to monitor patients in drug treatment programs, and (6) to aid in the diagnosis of toxicity or drug overdose. The later indication is somewhat controversial. In cases of toxicity, prompt diagnosis and treatment is essential. Diagnosis is generally made by history and clinical symptoms and treatment is usually supportive. The results of broad spectrum drug screening are usually not provided in real time and thus may not have an immediate impact on patient care. In addition, they are routinely done in urine and the drugs and metabolites detected may not be reflective of the patient's current clinical state. The usefulness of the information acquired from a broad spectrum drug screen is dependent on the time of drug exposure and the window of detection for the drug and/or metabolites of interest. However, the clinical history in cases of drug toxicity and overdose is not always available or reliable and the symptoms may not be enough to diagnosis drug toxicity. This is especially true in cases of polysubstance use or overdose. If provided in a timely manner, broad spectrum drug screening results can aid in the rule-in or rule-out of drug toxicity compared to other suspected causes on the diagnosis differential. The results may also be valuable for clinical follow-up. For example, the presence of a drug that is known to induce seizures may negate the need for a complete epilepsy workup in a case of a patient presenting to the emergency room with a seizure of unknown origin. In the future, broad spectrum drug screening in serum provided in real time could prove invaluable in cases of drug toxicity or overdose. For the other indications of drug screening listed earlier, urine will remain the specimen of choice due to the longer detection windows of drugs and metabolites in urine compared to serum or plasma. This chapter will focus on the use of liquid chromatography mass spectrometry (LC-MS) for toxicology testing in the clinical laboratory, specifically for urine drug confirmation testing and broad spectrum drug screening.

2 CHROMATOGRAPHIC METHODS FOR TOXICOLOGY TESTING IN THE CLINICAL LABORATORY

There are a number of combinations of chromatographic and detection techniques that can be used for drug confirmation testing and broad spectrum drug screening. The three main types of chromatographic methods used for clinical toxicology testing include thin-layer chromatography (TLC), gas chromatography mass spectrometry (GC-MS), and LC-MS.

2.1 THIN LAYER CHROMATOGRAPHY

Chromatography made its way into clinical toxicology laboratories via TLC. For many years, TLC was the method of choice for drug confirmations and broad spectrum drug testing. TLC may be used to detect a large number of drugs and metabolites. It is a versatile methodology that requires no instrumentation and is therefore inexpensive and operationally simple. In TLC, a thin layer of sorbent is spread uniformly on a solid planar support which serves as the stationary phase. Each sample is applied to a small spot near one edge of the plate. The edge of the plate is placed in the mobile phase that migrates up the plate via capillary action. The sample components retention is dependent on the solutes equilibrium between the stationary phase and mobile phase. Sample components are identified by color-generating reagents or ultraviolet fluorescence and comparison to standards on the same plate. Prepared TLC plates for drug confirmation testing and broad spectrum drug screening are available commercially for TLC applications in the clinical laboratory, but given limitations in sensitivity and specificity, GC and LC based methods have largely supplanted TLC.

2.2 GAS CHROMATOGRAPHY MASS SPECTROMETRY

GC-MS is the focus of the Chapter 7 in this book. Briefly, GC is widely available and used for qualitative and quantitative drug analysis. Various detectors can be used in combination with GC for drug screening; however, GC-MS and GC-MS/MS offer the greatest sensitivity and specificity. GC-MS is still often considered the gold standard for broad spectrum drug screening. Compounds that are nonpolar, volatile, and have a small molecular weight are suitable for analysis by GC. Electron ionization (EI) with full-scan mass detection or selected ion monitoring (SIM) are the most widely used approaches for drug confirmation testing and broad spectrum drug screening using GC-MS. Full-scan mass detection is an untargeted approach and is capable of potentially detecting all compounds in the sample which is advantageous for broad spectrum drug screening. Identification of unknown compounds can be achieved by comparison of their full mass spectrum with a mass spectral library or database. SIM is a targeted approach and will only identify the compounds for which specific acquisition parameters are entered into the analytical method. SIM is considered to be more sensitive compared to full-scan methods for the drugs and metabolites that the SIM method is designed to detect.

The major disadvantage of using GC-MS for drug confirmation testing or broad spectrum drug screening is that GC-MS methods are not capable of directly analyzing drugs that are nonvolatile, polar, or thermally labile. Derivatization is required to increase the volatility and thermal stability of these compounds. This involves derivatizing one or more polar groups on a compound to a less polar group. Derivatization can also be used to achieve increased sensitivity, selectivity, or specificity for a given chromatographic separation. For drug confirmation testing and broad spectrum drug screening, lengthy sample preparations, which include hydrolysis and derivatization, are required prior to GC-MS analysis. This significantly lengthens the time of sample analysis compared to most LC-based methods.

One of the major advantages of GC-MS compared to LC-MS is the high reproducibility of generated mass spectra using EI. The electron impact ionization process, used in GC-MS, is a hard ionization that results in the production of very reproducible mass spectra from one instrument to another. Large transferable EI-mass spectral libraries exist and are available for library searching of acquired spectra from any instrument [1]. Due to the availability of these libraries, individual laboratories do not have to develop their own in-house library, which can be very time consuming and costly.

2.3 LIQUID CHROMATOGRAPHY MASS SPECTROMETRY

The focus of this chapter is the use of LC-MS for toxicology testing in the clinical laboratory. The use of LC-MS has become increasingly more prominent in clinical toxicology laboratories for drug confirmation testing and broad spectrum drug screening.

2.3.1 Advantages and Disadvantages of LC-MS for Toxicology Testing

Unlike GC-MS, LC-MS using electrospray ionization (ESI) is capable of detecting nonvolatile, polar, and thermally labile compounds and provides a means of detecting a broad menu of drugs and metabolites without the need for lengthy sample preparations. The sample preparation time is usually significantly reduced for LC-MS methods compared to GC-MS methods since hydrolysis and derivatization are not required. In addition, sample run times can be shorter for LC-MS drug confirmation tests and broad spectrum drug screens compared to some GC-MS methods. These features of LC-MS methods are beneficial if results are needed rapidly for patient care. The ESI process, results in a soft ionization and thus fragmentation is produced by in-source collision-induced dissociation (CID) for single MS systems or CID in the collision cell, for tandem MS systems.

One current disadvantage of LC-MS for broad spectrum drug screening is that product ion spectra produced by CID differ markedly between instruments [2]. This poses a problem for the creation of large product ion spectral libraries utilized by chromatographic deconvolution software for broad spectrum drug screening. Despite attempts to standardize the production of the product ion spectra, ion relative intensities are highly variable requiring the need for the development of search algorithms that do not put a large amount of weight on relative intensities but rather on the absence or presence of ions. Instrument specific product ion spectral libraries have been created and vendors are beginning to market predeveloped broad spectrum drug screening methods, with chromatographic deconvolution software for library searching and compound identification. Some laboratories are capable of developing their own in-house; however, most laboratories do not have dedicated research and development staff for such undertakings.

Another disadvantage of ESI LC-MS techniques is ion suppression and matrix effects. Ion suppression results from the presence of less volatile compounds in a sample matrix that can change the efficiency of droplet formation or evaporation. This affects the amount of charged ions in the gas phase that ultimately reach the detector. Ion suppression should be extensively evaluated during method development and sample clean-up and separation should be optimized to eliminate significant ion suppression in clinical toxicology methods. Despite these limitations, LC-MS is the technique of choice for most drug analysis laboratories.

2.3.2 Types of Mass Analyzers for Toxicology Testing

LC-MS encompasses a variety of different LC and MS combinations, including LC-MS, LC-MS/MS, and LC-high resolution mass spectrometry (HRMS), to name a few [3]. Mass spectrometry analyzers are classified into two broad groups, beam-type and trapping-type instruments. In beam-type instruments, the ions travel through the mass analyzer in a single beam before they strike the detector. The two most common beam-type instruments used for toxicology include quadrupole and time-of-flight (TOF). Trapping-type instruments hold the ions in a spatially confined area for a period of time before they travel to the detector. The ion trap is the most common trapping-type instrument used in clinical laboratories. Selection of the best mass analyzer for each laboratory's specific testing needs is very crucial. The following section will discuss the specific mass analyzers commonly used and their advantages and disadvantages for drug confirmation testing and broad spectrum drug screening.

2.3.3 Single Quadrupole

Single quadrupole analyzers are more commonly coupled to GC systems, but can be coupled to LC systems as well. They are often referred to as mass filters because of their ability to select ions of a single m/z value for the analysis, while deflecting all other ions. Quadrupole analyzers consist of four parallel electrically conductive rods arranged in a square formation, forming a channel through which the ions pass. The rods are charged by direct current (DC) and radio frequency (RF) voltages with the opposite rods having likes charges. The ions interact with the imposed electrical fields; however, an ion will follow a stable trajectory through the channel only under a specific set of voltages. The rods can be scanned from low to high mass to allow ions of increasing mass to travel through the channel. This scanning mode is referred to as "full scan" and the DC and RF voltages are continuously varied to scan a range of m/z values. Alternatively, the DC and RF voltages can be fixed to select individual m/z values. This is referred to as SIM which allows for monitoring of specific ions from target analytes. The amount of time the analyzer remains at a given set of voltages is referred to as the dwell time. During the dwell time only one m/z is detected. SIM allows for longer dwell times and therefore higher sensitivity, however, full scan provides more information since ions not specifically selected in SIM are filtered out and not detected.

Despite being the most affordable, LC-MS single quadrupole instruments are not common in clinical toxicology laboratories due to their limited degree of sensitivity and specificity compared to other available mass analyzers, such as tandem mass spectrometry (MS/MS) instruments. In a single quadrupole instrument, there is no physical place for collision-induced dissociation of a selected mass to occur. In-source fragmentation is possible with some instruments; however, fragmentation patters using in-source fragmentation are highly variable. Compound identification is based on retention time, nominal mass, and in-source fragmentation pattern (if available).

2.3.4 Time-of-flight

TOF mass analyzers use an electrical field to pulse gas phase ions through a flight tube toward the detector. The time required for the ion to travel from the pulser to the detector is determined by the m/z ratio. Ions with a smaller m/z will move faster than larger ions. The ion velocity is inversely related to the square root of the m/z value times the voltage. There are different types of TOF analyzers, some contain linear flight tubes and others make use of reflectrons and ion mirrors to increase the length of the flight path and thus increase the resolution. Unlike quadruple analyzers, TOF analyzers are not capable of filtering out or retaining specific ions. TOF analyzers detect a large mass range and have the advantage of being able to determine the accurate mass and molecular formula of the sample components being analyzed. While TOF analyzers are very sensitive and have a high mass accuracy and resolution, their dynamic range is limited and thus they are not commonly used for quantitation.

TOF mass analyzers are more expensive than single quadruple MS analyzers, but typically are less expensive than sensitive MS/MS mass analyzers. Compound identification is based on retention time, accurate mass/molecular formula, and isotope pattern. In-source fragmentation is capable with some TOF analyzers; however, this is not common. TOF analyzers operate in full scan mode and therefore are capable of potentially seeing any drug or metabolite in a clinical sample, however, accurate mass, isotope pattern, and retention time is often not adequate for definitive identification of a drug or metabolite. One molecular formula for a small molecule can correspond to a number of different compounds. For example, morphine and hydromorphone have the same molecular formula and thus the same exact mass. Unless, all possible isomers are separated chromatographically definitive identification is not possible. TOF mass analyzers can be powerful tools for toxicology screening, but confirmation by a different analytical approach is preferred.

2.3.5 Tandem Mass Spectrometry

The detection of an analyte by the m/z value alone does not always provide the needed specificity for definitive identification for drug confirmatory testing or broad spectrum drug screening. In tandem mass spectrometry (MS/MS), two mass analyses occur sequentially to improve selectivity and analytical detection limits. These two mass analysis steps can occur tandem in space or tandem in time. In a tandem in space instrument, three quadrupoles are linked sequentially. The first quadrupole (Q1) is used to scan a preset m/z range and select the ions of interest. These ions are called "precursor ions." The second quadrupole (Q2), functions as a collision cell. The selected ions are accelerated by an electrical potential to high kinetic energy and then collide with neutral gas molecules to fragment the ions. This is called CID. The ions formed during fragmentation are called "product ions." The third quadrupole (Q3) analyzes the product ions generated in Q2.

In tandem in time instruments, the two mass analysis steps and CID are performed sequentially within the same spectrometer, which is typically an ion trap. Ion trap analyzers are the most common trapping-type instrument used in clinical toxicology laboratories. They are similar to quadrupole analyzers in that they are able to select and filter specific m/z values for analysis. Ion traps function by using RF fields to retain and trap ions within a three-dimensional space for a period of time. After the period of ion accumulation, the electric field adjusts, destabilizing the trapped ions and selectively sending them to the detector. All ions initially enter the trap, however only a specific RF frequency will cause ions with a particular m/z value to leave. There are two main types of ion traps, quadruple ion traps (QIT) and linear ion traps (LIT). In a QIT, the four rods are arranged to form a three-dimensional sphere rather than existing in parallel. A LIT is structural designed in the same manner as the quadrupole analyzer with the addition of electrostatic fields applied to the ends to prevent ions from exiting the mass analyzer. The trapping function can be turned on and off to allow the trapped ions to be sent to the detector.

There are a number of different scanning modes that can be used for compound identification in MS/MS (Fig. 6.1). The two most common scan modes for targeted drug confirmation testing and broad spectrum drug screening are product ion (PI) scan and selected reaction monitoring (SRM), both of which are targeted detection methods. LC-MS/MS instruments are also capable of performing a full scan, however, this scan mode in isolation is less common for broad spectrum drug screening compared to PI scans and SRM. In a PI scan, Q1 is set to select defined precursor ions. The ions are then

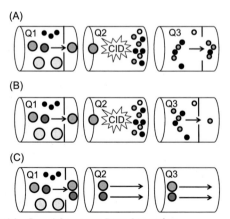

FIGURE 6.1 MS Scan Modes used for Broad Spectrum Drug Screening

The three most common MS scan modes used in broad spectrum drug screening include (A) product ion scan, (B) selected reaction monitoring, and (C) full scan.

fragmented in Q2 and all of the resulting PIs in a specified mass range are detected in Q3, which functions in full scan mode. A full mass spectrum of the fragmented analyte is acquired (Fig. 6.1). The acquired mass spectrum can be compared to mass spectral libraries for compound identification or used for structural elucidation of unknowns. SRM is similar to PI scan except that specific PIs are monitored in Q3 rather than acquiring a mass spectrum of the entire mass range (Fig. 6.1). The precursor and PI pair is referred to as a transition. In many clinical applications, two ion transitions are monitored per analyte. The most abundant PI is referred to as the quantifier ion and is used for quantification purposes and the second monitored PI is referred to as the qualifier ion and is used as a second form of identification for the analyte. The ratio of the chromatographic peak area of the more abundant PI to the less abundant PI can be compared to what is expected for the analyte of interest to further enhance specificity.

SRM and PI scans can be used in the same method with the instrument alternating between the two different scan modes. A common approach for broad spectrum drug screening is to perform an SRM scan followed by data dependent acquisition (DDA) of PI spectra. DDA allows for specific experimental conditions to be set to allow for the collection of PI spectra of the compounds of interest. Not all MS/MS analyzers are capable of performing these scan modes simultaneously in the same method. If this functionality is preferred for a clinical toxicology laboratory they must insure that the instrument they purchase is not designed solely for SRM based methods.

Tandem mass spectrometers are the instrument of choice for most clinical toxicology laboratories. For drug confirmation testing, many laboratories utilized SRM based methods with two transitions monitored per analyte and calculation of the ion ratio between the two most abundant or most selective PIs. However, for broad spectrum drug screening for over 100 drugs and metabolites monitoring two transitions per analyte can be cumbersome. Because of this, many broad spectrum drug screens utilize SRM-DDA-PI based methods. The PIs collected are searched against mass spectral libraries. The library match score is used in addition to retention time and SRM monitoring for compound identification.

The main disadvantage of MS/MS method for drug analysis is that most methods are designed to be targeted methods. They will only detect compounds for which specific compound dependent parameters are developed and entered into the method for identification, unlike full scan single quadruple methods or TOF analyzers that acquire data in an untargeted fashion and do not select specific masses for analysis. This is a disadvantage for the detection of all possible intoxicants in broad spectrum drug screening. However, due to the targeted nature of MS/MS methods, they are usually more sensitive and specific for the analytes of interest making them ideal for drug confirmation testing. In addition, the majority of drugs and metabolites of interest for broad spectrum screening can be incorporated into one MS/MS based method.

2.3.6 High Resolution Mass Spectrometry

In recent years there has been an emergence of the use of HRMS methods for drug screening [3,4]. These technologies can achieve high mass accuracy due to their high resolving power, making them ideal technologies for detecting hundreds of compounds simultaneously in complex biological matrices. HRMS technologies can determine the accurate mass of the sample components being analyzed. Some HRMS technologies that can be used for broad spectrum drug screening include TOF (described earlier), Orbitrap, and QTOF mass analyzers. LC-Orbitrap and LC-QTOF instruments merge the LC-TOF advantage of high mass resolution and LC-MS/MS advantage of mass filtering and fragmentation in one instrument. These instruments operate in the same manner as MS/MS instruments, but provide high resolution measurements for both precursor and PIs present in the acquired mass spectra. The resolution of a mass spectrometer is defined as the measured mass of an analyte divided by the full width of the corresponding peak at half maximum (FWHM). TOF analyzers typically achieve resolutions of 10,000–100,000 FWHM, whereas Orbitrap analyzers can achieve resolutions of 100,000–250,000 FWHM. Unlike TOF analyzers that calculate the time it takes for ions to traverse a flight tube, Orbitrap analyzers measure the stable orbit of an ion. Ions are injected tangentially to an electric field between an outer barrel and inner spindle like electrode. The stable orbit achieved by an ion is proportional to the *m/z* ratio. Compound identification using an LC-Orbitrap or LC-QTOF instrument is based on retention time, accurate mass, molecular formula, isotope pattern, and mass spectral matching.

The major disadvantage to the use of these technologies in the clinical laboratory is the expense of the instrumentation and the technical expertise required for method development, implementation, and routine use.

3 TOXICOLOGY LC-MS APPLICATIONS

This next part of this chapter will focus on two specific toxicology LC-MS applications that are common in toxicology laboratories including opiate confirmations and broad spectrum drug screening in urine. For each application, key clinical and technical decisions involved in the implementation of a method will be discussed. Each section will describe consideration for the assay development, validation, and ongoing method evaluation for each application. Each section will end with an example of the method developed in use in the author's laboratory.

3.1 OPIOID CONFIRMATION TESTING

Opioids are routinely prescribed for anesthesia, pain management, and to treat opioid addiction. They also have significant potential for abuse and dependence. Clinicians who prescribe opioids routinely order urine drug testing for their patients to monitor compliance, deter diversion, and insure nonillicit use of nonprescribed opioids, such as heroin. In most laboratories, this testing involves a preliminary immunoassay screen for opiates followed by targeted confirmation using mass spectrometry. A major limitation of this approach is that opiate immunoassays are sensitive for the naturally occurring opiates which include morphine and codeine; however, most have low cross-reactivity with the semi-synthetic and synthetic opioids commonly prescribed. There are other opioid immunoassays available (i.e., oxycodone, methadone, fentanyl, buprenorphine), however, these assays are still only used for screening purposes. False-negative and false-positive immunoassay results can have significant implications for patients; therefore, confirmation of screening results is necessary. Also, confirmation testing can determine which specific opioid is present in the sample which is necessary for compliance monitoring.

3.1.1 Opioid Metabolism

The metabolism of opiates and opioids is complex which has implications for method development and results interpretation. After administered, heroin is rapidly deacetylated in the blood to 6-monoacetylmorphine (6-MAM). 6-MAM is subsequently deacetylated to morphine. While morphine is a metabolite of heroin, it is also a prescribed medication. The primary route of metabolism of morphine is via conjugation with glucuronide to form, morphine-3 and 6-glucuronide. Codeine is metabolized by O-demethylation to morphine and undergoes glucuronidation to form codeine-6-glucuronide. Both morphine and codeine can be N-demethylated to form normorphine and norcodeine, respectively. Minor metabolic pathways convert morphine to hydromorphone and codeine to hydrocodone. Hydrocodone and hydromorphone are not just metabolites, but are also commonly prescribed for pain management. Hydrocodone is converted to hydromorphone via O-demethylation and norhydrocodone via N-demethylation. Hydromorphone is glucuronidated to hydromorphone-3-glucuronide.

Oxycodone is metabolized to oxymorphone and noroxycodone. Both oxymorphone and noroxycodone can be further metabolized to noroxymorphone. Oxymorphone can also be conjugated with a glucuronide to form oxymorphone-3-glucuronide. Some opioid confirmation assays include other synthetic opioids, such as methadone, fentanyl, and buprenorphine. Fentanyl undergoes N-dealkylation to norfentanyl. Methadone is N-demethylated to form EDDP. Buprenorphine is N-dealkylated to norbuprenorphine and both are glucuronidated.

3.1.2 Considerations for Method Development

One of the first challenges faced in the development of an opioid confirmation assay is determining which opioids and metabolites to monitor. Most assays are designed to detect morphine, codeine, hydromorphone, hydrocodone, oxycodone, oxymorphone, and their respective metabolites. Given that a significant portion of opioids are primarily detected as glucuronide metabolites in the urine, either the conjugated drugs need to be hydrolyzed to the free drug prior to LC-MS testing or the glucuronide metabolites need to be monitored directly in addition to the free drug. Also, the half-life of many opioids is short and therefore it is essential to test for their metabolites which have a longer window of detection in urine. Another challenge for method development is the structural similarity of these opioids and their isomeric nature. For example, morphine and hydromorphone are isomers and have similar, but not identical, fragmentation patterns. Care should be taken in method development to ensure that these two compounds are chromatographically separated and that unique PIs are monitored. The same is true of codeine and hydrocodone.

There are a number of published methods in the literature that differ in all aspects of testing, including which drugs and metabolites are monitored, choice of sample preparation, liquid chromatography columns and conditions, and mass spectrometry settings. Table 6.1 highlights a number of opioid confirmation methods published for comparison [5–14]. Traditionally GC-MS was the method of choice for opiate confirmations; however, LC-MS/MS has largely supplanted GC-MS for this application due to the relative ease of sample preparation. Most LC-MS/MS assays employ acid or enzyme (β-glucuronidase) hydrolysis to cleave the glucuronide linkage and measure total drug, however, there are a few methods that directly monitor the conjugated drug. Following hydrolysis, there are three main sample preparation techniques used for the analysis of opioids: protein precipitation (PPT), dilution, and solid phase extraction (SPE). All three methods have shown to be effective strategies for sample clean-up. The advantage of using PPT or dilution for sample preparation is the time and cost savings compared to SPE. However, the sample may be diluted such that some low concentration analytes are below the limit of detection. Also, the sample preparation is not removing other matrix components that could coelute with an analyte of interest and effect ionization efficiency. Therefore, matrix effects and ion suppression could reduce the sensitivity and precision of the method and must be evaluated during method validation no matter which sample preparation is selected in any laboratory.

Table 6.1 LC-MS/MS Opioid Confirmation Methods

Opioids Detected	Sample Preparation	Stationary Phase	Mobile Phases	Mass Analyzer	Acquisition Mode	Analysis Time (min)	Year [Reference]
M, C, HM, HC, OC, OM, 6AM, M3G, M6G, NC	No hydrolysis, Dilution	Phenyl	A: 10 mM AF, 0.05% FA in 95:5 H_2O:ACN B: 10 mM AF, 0.05% FA in 95:5 ACN:H_2O	ESI-MS/MS	SRM—1 transition	6	2005 [5]
M, C, M3G, M6G, CG	No hydrolysis, PPT	Fusion RP	A: 10 mM AA in H_2O B: ACN C: MeOH	ESI-MS/MS	SRM—2 transitions with ion ratio	15	2005 [6]
M, C, HM, HC, OC, OM, M3G, M6G, F, NF, BUP, TRAM, METH	No hydrolysis, SPE	C12 Max RP	A: 5 mM AF in 90:10 H_2O:ACN pH3 B: 5 mM AF in 90:10 ACN:H_2O pH3	ESI-MS/MS	SRM—2 transitions with ion ratio	35	2006 [7]
M, C, 6AM, M3G, M6G, CG, EM, EMG	No hydrolysis, Dilution	C18	A: 25mM FA in 99:1 H_2O:ACN B: 25 mM, FA in 90:10 ACN:H_2O	ESI-MS/MS	SRM—2 transitions with ion ratio	13	2007 [8]
M, C, DHC, 6AM, M3G, EDDP, BUP	No hydrolysis, Dilution	C8	A: 25 mM AA in 95:5 H_2O:MeOH B: 0.05 mM FA in 98:2 MeOH:IPA	ESI-MS/MS	SRM-IDA-EPI	12.5	2010 [9]
M, C, M3G, M6G, CG	No hydrolysis, SPE	HILIC	A: 10 mM AF, pH6.4 B: 10 mM AF in 90:10 ACN:H_2O, pH6.4	ESI-TOF, in source CID	Full Scan	11	2010 [10]
M, C, HM, HC, OC, OM, MG, CG, HMG, OMG	No hydrolysis, SPE	Amide HILIC	A: 10 mM AF, 0.125% FA in 50:50 ACN:H_2O B: 10 mM AF, 0.125% FA in 90:10 ACN:H_2O	ESI-MS/MS	SRM—2 transitions with ion ratio	17	2011 [11]
M, C, HM, HC	Hydrolysis, Dilution	PFP Propyl	A: 10 mM AF, 0.0005% FA in H_2O B: 10 mM AF, 0.0005% FA in MeOH	ESI-MS/MS	SRM-IDA-EPI	11	2012 [12]
M, C, HM, HC, OC, OM, 6AM, M3G, M6G, CG, HMG, OMG, BUP, NBG, F, NF, MEP, NMEP, METH, PP	No hydrolysis, Dilution	HSS T3	A: 2 mM AA, 0.1% FA in H_2O B: 0.1% FA in ACN	ESI-MS/MS	SRM—2 transitions with ion ratio	9	2012 [13]
M, C, HM, HC, OC, OM, DHC, NHC	No hydrolysis, SPE	PFP Propyl	A: 20 mM AF in H_2O, pH3 B: MeOH	ESI-MS/MS	SRM—3 transitions with ion ratio	12.5	2013 [14]

AA, ammonium acetate; ACN, acetonitrile; AF, ammonium formate; 6AM, 6-acetylmorphine; BUP, buprenorphine; C, codeine; CG, codeine-glucuronide; CID, collision induced dissociation; DHC, dihydrocodeine; EDDP, methadone metabolite; EM, ethylmorphine; EMG, ethylmorphine-glucuronide; EPI, enhanced product ion; ESI, electrospray ionization; F, fentanyl; FA, formic acid; HC, hydrocodone; HILIC, hydrophilic interaction liquid chromatography; HM, hydromorphone; HMG, hydromorphone-glucuronide; H_2O, water; IDA, information dependent acquisition; IPA, propan-2-ol; M, morphine; MeOH, methanol; MEP, meperidine; METH, methadone; M3G, morphine-3-glucuronide; M6G, morphine-6-glucurinde; MS/MS, tandem mass spectrometry; NBG, norbuprenorphine-glucuronide; NC, norcodeine; NF, norfentanyl; NHC, norhydrocodone; NMEP, normeperidine; OC, oxycodone; OM, oxymorphone; OMG, oxymorphone-glucuronide; PFP, pentafluorophenyl; PP, propoxyphene; PPT, protein precipitation; RP, reverse phase; SPE, solid phase extraction; SRM, selective reaction monitoring; TOF, time-of-flight.

The chromatography conditions for opioid confirmation methods are dependent upon which analytes are being monitored. Chromatography can be challenging due to the varying degrees of polarity between the drugs and their metabolites, which are highly polar. Reverse-phase chromatography (RPC) with gradient elution is the most common type of chromatography for opioid analysis. In RPC, molecules are bound to the hydrophobic matrix (stationary phase) in an aqueous buffer (polar) and eluted from the matrix using a gradient of organic solvent (nonpolar). The more polar analytes, such as glucuronide metabolites, will be poorly retained on the column and will elute prior to the less polar parent drugs. Various stationary phases have been used for opioid analysis. The more common RPC columns used allow for the analysis of both polar and nonpolar compounds. Different columns that have been used and reported in the literature are listed in Table 6.1. For detection by mass spectrometry, most published methods utilize selective reaction monitoring. Two ion transitions are monitored for each drug or metabolite of interest and the ion ratio between the two PIs is calculated and compared to the standards and controls to confirm positive identification. Most moderate sensitivity tandem mass spectrometers are adequate for opioid confirmation testing given that when taken as prescribed most opioids are present in high concentrations in urine.

3.2 OPIOID CONFIRMATION TESTING: EXAMPLE METHOD

The following section will describe the method that is currently used in the clinical laboratory at San Francisco General Hospital and Trauma Center (SFGH) [14a].

3.2.1 Rational for Method Development

All utox screens that test positive for opiates by immunoassay are automatically reflexed to opioid confirmation testing at SFGH. Prior to implementation of the LC-MS/MS method, opioid confirmations were done by GC-MS. The sample preparation included acid hydrolysis, derivatization, and SPE. The batch size was also limited to the equipment available in our laboratory for sample preparation. Other limitations included a 15 min instrument run time per sample, a difficult dilution scheme to limit sample carryover, interferences from large peaks, and a lengthy data analysis process. Over the last 5 years, opioid confirmation testing has increased by 40% and it has become essential to transition to a method that resulted in labor and cost savings for the laboratory. The GC-MS confirmation method tested for total morphine, codeine, hydrocodone, hydromorphone, oxycodone, and oxymorphone. Our laboratory had seen an increase in the number of requests for confirmatory testing for 6-MAM, fentanyl, methadone, and buprenorphine. Therefore, we decided to include these drugs and their metabolites in the method development. We also sought to eliminate the use of hazardous chemicals, such as derivatizers, acid hydrolysis reagents, and methylene chloride, which required special training by our staff and special hazardous waste disposal processes.

3.2.2 Method Development

We decided to include enzyme hydrolysis in our sample preparation in order to decrease the number of peaks that are needed to be reviewed by the technologists and additional standards and controls that are required to be tested and evaluated for the glucuronide metabolites. In addition, the interpretation of the results is more straightforward for the clinicians ordering the testing. We evaluated the hydrolysis efficiency of β-glucuronidase from Red Abalone, *E. coli*, *Patella vulgata*, and a recombinant β-glucuronidase. The recombinant β-glucuronidase (IMCSzyme, Integrated Micro-Chromatography

System, Columbia, SC) exhibited the highest hydrolysis efficiency for the seven opioid glucuronide conjugates. Initially a simple sample dilution after enzyme hydrolysis was evaluated but resulted in issues with column lifespan and build-up of matrix components in the source of the mass spectrometer on the curtain plate. Therefore, compound recovery, matrix effects, and precision of three PPT 96-well plates were evaluated for sample clean-up following hydrolysis. The plate that had the highest recovery and lowest ion suppression was selected (Supelco PPT filter plate, Sigma-Aldrich, St. Louis, MO).

A new LC-MS/MS system was not purchased for this application. We chose to develop the assay on one of our preexisting LC-MS/MS systems based on instrument availability and functionality. Any LC-MS/MS system with moderate sensitivity should be adequate for opioid confirmation testing. We chose to use an LC system that was equipped with a 96-well plate autosampler so that the sample preparation could be done in 96-well plates to cut-down on the number of manual pipetting steps for our technologists. Our laboratory does not have automated sample preparation equipment. Equipment does exist to automate sample preparation for opioid confirmations if the testing volume in a laboratory justifies the expense of the additional equipment. Separation of the 13 opioids included in the method was performed using an Agilent 1260 Infinity liquid chromatography system (Agilent, Santa Clara, CA) using a Phenomenex Kinetex 2.6 μm Phenyl-Hexyl column (50 × 4.6 mm) thermostatted at 30°C. A binary mobile phase consisting of (1) 10 mM ammonium formate and (2) 0.1% formic acid in methanol was ramped linearly from 20% to 100% B over 4.5 min. The column was washed at 100% for 1.5 min, followed by a 2 min reequilibration at 20% B at a flow rate of 0.7 mL/min. Data were collected using a 5500 QTRAP equipped with a TurboIon Spray source controlled by Analyst 1.5.1 software (ABSciex). Positive ionization experiments were performed using the following settings: ion spray voltage, 5000 V; curtain gas, 20 psi; ion source gas 1, 40 psi; ion source gas 2, 40 psi; CAD gas, medium; and ion source temperature, 700°C. Selective reaction monitoring was used with two transitions monitored for each compound and one transition for the internal standards (Table 6.2). The raw signal for each compound was normalized to the internal standard and concentration was calculated from the calibration curve.

3.2.3 Method Validation

Validation of the final method included determination of the lower limit of detection (LOD), linearity, precision, matrix effects, recovery, carryover, and interference for each analyte. The method was validated as a quantitative method, however, in routine practice cutoffs are used and qualitative results, positive or negative, are reported. For this reason, LOD was determined rather than the lower limit of quantitation (LOQ). The LOD was defined as the lowest concentration at which the signal-to-noise ratio was greater than 20:1. Linearity was determined using a 10-point curve tested in triplicate. Within-run precision was estimated by analyzing five independent aliquot of each of the two quality control (QC) samples within one run. Between-run precision was estimated by analyzing five specimens of each level per day over 5 days. Matrix effects were determined by spiking drug standards into water and five drug-free urine matrices at 500 ng/mL (50 ng/mL for fentanyl, norfentanyl, 6-MAM). It was calculated using the following equation: (B−A)/A × 100%, where B is the mean signal intensity in the urine matrix and A is the mean signal intensity in water. To calculate the percentage of extraction recovery, 500 ng/mL of opioids were spiked in three different aliquots of drug-free urine before extraction. Three additional aliquots of drug-free urine were taken through the extraction, and spiked with 500 ng/mL of opioids after extraction. The mean peak area of the samples spiked before extraction was divided by the mean peak area of the samples spiked after extraction. Interference studies were carried out by analyzing 140 compounds spiked in 10 drug free urine samples.

Table 6.2 Validation of the Opioid LC-MS/MS Assay

	Retention time (min)	Transitions	Linear Range (ng/mL)	Matrix Effect (%)	Recovery (%)	Within-Run Precision (%)		Between-Run Precision (%)		Cutoff (ng/mL)
						80 ng/mL	120 ng/mL	80 ng/mL	120 ng/mL	
Morphine	1.55	286.1 → 165.1/152.2	25–2000	90	98	4	7	11	10	300
Codeine	2.40	300.2 → 165.2/215.1	25–2000	90	101	5	5	14	13	100
Hydromorphone	1.99	286.1 → 185.2/157.2	25–2000	90	103	4	4	10	12	100
Hydrocodone	2.62	300.2 → 199.0/171.0	25–2000	83	103	7	5	11	13	100
Oxymorphone	2.52	302.1 → 227.2/198.2	25–2000	94	96	8	5	12	14	100
Oxycodone	1.66	316.2 → 298.2/241.2	25–2000	113	98	6	3	7	7	100
6-MAM	2.50	328.1 → 165.1/211.1	2.5–200	97	102	7	5	11	11	10
Methadone	4.58	310.0 → 105.2/77.1	25–2000	122	97	10	4	11	16	100
EDDP	4.12	278.2 → 234.2/186.2	25–2000	113	96	6	3	6	10	100
Buprenorphine	4.06	468.3 → 396.2/414.3	10–200	135	88	9	6	8	12	10
Norbuprenorphine	3.01	414.3 → 83.1/101.3	25–2000	115	100	12	7	16	15	100
Fentanyl	4.50	337.3 → 188.2/105.1	2.5–200	107	98	5	3	5		10
Norfentanyl	3.75	233.2 → 150.1/84.1	2.5–200	85	100	5	3	14	13	10

In order to determine an appropriate dilution scheme for patient samples that contained very high level of opioids, samples were tested straight and at various dilutions based on the immunoassay value. It was determined that samples with an opiate immunoassay value over 18,000 would undergo a 1:20 dilution, and samples that had an immunoassay value between 15,000 and 18,000 would undergo a 1:10 dilution. Samples that had an oxycodone immunoassay value over 300 were subjected to a 1:10 dilution. In the validation experiments, three blank samples were inserted after all diluted samples to ensure that no carryover was detected. Urine samples of 62 patients that screened positive by the opioid immunoassay and were analyzed by GC-MS, were reanalyzed by the LC-MS/MS method for patient comparisons.

The validation data for all opioids is shown in Table 6.2. Calibration curves of each analyte exhibited consistent linearity and reproducibility in the range of 25–2000 ng/mL (2.5–200 ng/mL, fentanyl, norfentanyl, 6-MAM, and 10–200 ng/mL, buprenorphine), with regression coefficient (r) $\geq$ 0.990 for all analytes. The LOD of each analyte was 10 ng/mL (1 ng/mL, buprenorphine, fentanyl, norfentanyl, 6-MAM). The cutoffs of the clinical reporting range were selected to be 300 ng/mL for morphine, 10 ng/mL for buprenorphine, fentanyl, norfentanyl and 6-MAM, and 100 ng/mL for all other analytes. No significant ion suppression was detected for any of the 13 analytes, with a few showing ion enhancement (83–135%). Extraction recovery ranged from 88% to 103%, with an average of 98%. Within-run coefficient of variation (CV) ranged from 3% to 12%. Between-run CV ranged from 5% to 16%. No interferences were found in the samples spiked with 1 μg/mL of 140 drugs that are common in our patient population. Samples were injected straight or diluted (1:10 or 1:20) based on the opiate and oxycodone immunoassay values. Carryover in the blank samples, following all samples with high opioid levels, was less than the LOD for all analytes. Authentic urine samples ($N = 62$) were analyzed by GC-MS and the LC-MS/MS method described. The two methods were 97% concordant. The LC-MS/MS method detected additional opioids compared to the GC-MS method. These additional opioids were all confirmed by patient prescription information or were supported by the presence of metabolites and/or parent drugs in the same sample. For example, one sample was positive for only hydrocodone by GC-MS, but both hydrocodone and hydromorphone were detected using LC-MS/MS.

3.2.4 Postimplementation Monitoring

Testing is done in batch using 96-well plates. Each batch includes a double blank, blank, 3 calibrators, and 2 sets of QC samples, each containing one low QC and one high QC, injected before and after the patient samples. The level of opioids in the QC samples bracket the cutoff for each analyte with the low QC containing 80% of the cutoff value and the high QC containing 120%. For example, the cutoff for oxycodone is 100 ng/mL, therefore the target value for the low QC is 80 ng/mL and 120 ng/mL for the high QC. QC ranges were initially established during validation using the between-day precision samples and were set as the mean plus or minus two standard deviations. The QC ranges were reevaluated after 2 months of testing and adjusted if needed. Levy-Jennings charts are routinely monitored by the laboratory supervisor to look for shifts in QC values. The internal standard (IS) area is plotted for all samples in a batch and all samples with an IS peak area that is outside two standard deviations of the mean for that plate are inspected by a supervisor to determine if they need to be retested on the next batch. For proficiency testing, our laboratory subscribes to the College of American Pathologists (CAP) Urine Toxicology Survey. Since implementation of this method, all CAP results have been acceptable for all results reported from the opioid confirmation method.

3.3 **BROAD SPECTRUM DRUG SCREENING**

As mentioned above, broad spectrum drug screening allows for the simultaneous identification of a large panel of illicit and pharmaceutical drugs and metabolites. Methods exist in the literature for various sample types including urine, serum, plasma, and/or whole blood. Urine is the most common sample type in clinical toxicology laboratories; however, blood is more reflective of what a patient has been exposed to acutely.

Methods for broad spectrum drug screening in clinical laboratories have been established for a variety of LC-MS/MS systems [15–19]. Recently more methods using LC-HRMS have been described [20–24]. Table 6.3 outlines a selection of published broad spectrum drug screening methods, using LC-MS/MS or LC-HRMS, which are capable of detecting greater than 100 drugs/metabolites. Sample clean-up prior to MS analysis for broad spectrum drug screening primarily consists of SPE or liquid–liquid extraction, however some methods using a simple dilution for urine have been described. The primary stationary phase that has been used for broad spectrum drug screening is a C18 column. Some of the newer methods have utilized other phases, such as pentafluorophenyl and biphenyl. ESI operating in positive mode is the primary ionization technique used since it is capable of detecting a wider range of drugs and metabolites compared to atmospheric pressure chemical ionization (APCI), however broad spectrum drug screening methods using APCI have been described.

3.3.1 *Considerations for Method Development*

One of the first challenges faced in the development of a broad spectrum drug screen is determining whether to develop a targeted or untargeted data acquisition method. Targeted methods are designed to detect a predetermined list of drugs and metabolites; however, it is often difficult to decide which drugs and metabolites to include in the analysis. Laboratorians should consult with ordering clinicians, emergency room doctors, the local poison control center, and pharmacists to ensure that the selected drugs and metabolites for the method meet the specific needs of the patient population. In most targeted approaches, triple-quadrupole or triple quadrupole linear ion-trap instruments perform an SRM survey scan to detect a single transition for each compound of interest. If the transition is detected and data dependent acquisition (DDA) criteria are met, the precursor ion is selected, and a product ion spectrum is obtained. Product ion spectra are then compared to a spectral library for compound identification. An alternative approach is to monitor two SRM transitions per analyte and calculate the ion ratio. The major limitation of SRM-based methods is that they are targeted in nature, meaning that information will only be collected on the presence or absence of the compounds included in the acquisition method. This precludes the identification of new or unexpected drugs in a sample, which is a major disadvantage in terms of broad spectrum drug screening. However, these methods are usually very sensitive and specific for the compounds they are designed to detect. Another disadvantage is the time consuming and difficult process of method development. For targeted SRM methods, compound specific parameters have to be determined for all analytes in the method and tested to ensure sensitivity and specificity.

The primary alternative to this testing approach is using an untargeted data acquisition method. In these methods, the initial MS scan is a full scan that monitors all masses within a specified range. Following the full scan, PI spectra can be obtained in either a data dependent (DDA) or data independent manner (DIA). Using DDA, spectra are only acquired for masses that are identified of meeting specified criteria. One disadvantage of this approach is that product ion spectra are only acquired for the "n" (as defined by the user) most abundant ions detected in the initial full scan at any moment in time. The presence of abundant small molecules in the sample that are not of toxicological interest may

Table 6.3 Broad Spectrum Drug Screens Detecting Greater Than 150 Drugs/Metabolites

Compounds	Matrix	Sample Preparation	Stationary Phase	Ion Source	Mass Analyzer	Targeted/ Untargeted	Acquisition Mode	Identification Criteria	Run Time	Year [Reference]
301	Serum	LLE or SPE	C18 (150 × 2 mm, 4 µm)	ESI +	MS/MS (QqLIT)	Targeted	SRM-IDA-EPI	RT, SRM, library search	33 min	2005 [15]
700	Urine, Serum	Dilution or LLE	PFP (50 × 2.1 mm, 5 µm)	ESI +	MS/MS (QqLIT)	Targeted	SRM-IDA-EPI	RT, SRM, library search	17.5 min	2010 [16]
~1000	Serum, Urine	SPE	C18 (100 × 2.1 mm, 3.5 µm)	ESI +/−	MS/MS (QqLIT)	Untargeted	full scan-IDA-EPI	RT, nominal mass, library search	25.5 min	2006 [17]
365	Serum, Urine	On-line SPE	C18 (125 × 4 mm, 3 µm)	APCI +/−	MS/MS (LIT)	Untargeted	full scan-DDA-PI	RT, nominal mass, library search	23 min	2010 [18]
202	Blood	SPE	C18 (150 × 2.1 mm, 1.8 µm)	ESI +	MS/MS	Targeted	SRM, 2 transitions, ion ratio	RT, SRM, ion ratio	15 min	2011 [19]
735	Urine	Hydrolysis, SPE	C18 (100 × 2 mm, 3 µm)	ESI +	TOF	Untargeted	HR full scan	RT, accurate mass, isotope pattern	21 min + post time	2006 [20]
~300	Urine	LLE	HSS T3 (100 × 2.1 mm, 1.8 µm)	ESI +	TOF	Untargeted	HR full scan, in-source CID	RT, accurate mass, isotope pattern, library search	17 min	2009 [21]
78 validated	Blood	SPE	C18 (150 × 2.1 mm, 1.8 µm)	ESI +	MS^E-TOF	Untargeted	HR full scan—DIA—all ion fragmentation	RT, accurate mass of precursor and 2 product ions	15 min	2013 [22]
169 validated	Urine	dilution	C18 (50 × 3 mm, 2.6 µm)	ESI +	QTOF	Untargeted	HR full scan—DDA—EPI	RT, accurate mass, isotope pattern, library search	14 min	2015 [23]
Not specified	Urine, Blood	SPE	C18 (100 × 1 mm, 3 µm)	ESI +	QTOF	Untargeted	HR full scan—DIA—EPI (SWATH)	RT, accurate mass of precursor and 5 product ions, library search	10 min	2015 [24]

APCI, atmospheric pressure chemical ionization; CID, collision-induced dissociation; ESI, electrospray ionization; HR, high resolution; HSS, high strength silica; IDA, information dependent acquisition; LIT, linear ion trap; LLE, liquid–liquid extraction; MS/MS, tandem mass spectrometry; PFP, pentafluorophenyl; QqLIT, hybrid triple quadrupole linear ion trap; QqQ, triple quadrupole; RT, retention time; SPE, solid phase extraction; SRM, selected reaction monitoring; TOF, time-of-flight.

result in the acquisition of product ion spectra for those compounds at the exclusion of drugs and metabolites that are less abundant. To circumvent this, data independent approaches can be used and are primarily used with LC-HRMS instruments. These approaches allow for the acquisition of product ion spectra for all theoretical masses in the sample. However, most instruments do not select a single mass prior to fragmentation. They either fragment all ions that coelute together or fragment a small mass range of ions. The resulting acquired product ion spectra often need to be deconvoluted by software prior to comparison with a spectral library for compound identification. The decision to use a targeted or untargeted data acquisition method, with or without DDA criteria, could depend on the available instruments in a given laboratory and the breadth of drugs and metabolites a laboratory needs to monitor. The development of untargeted full scan methods is simplified compared to targeted methods since compound dependent parameters are not established. Instead general parameters that are sufficient for all analytes of interest are determined and retention time is established for all validated analytes.

Depending on the mass spectrometry systems available in a laboratory for broad spectrum drug screening, different analytical approaches may be preferred. If your laboratory only has a tandem mass spectrometer without any trapping functionality to allow for the sensitive collection of product ion spectra, an SRM approach with two transitions per analyte will suffice. If a tandem mass spectrometer with trapping functionality is available, a method that utilizes SRM with DDA of product ion spectra is usually preferred. TOF mass spectrometers can also be used for broad spectrum drug screening, however, a second complementary and confirmatory method is required due to the isomeric nature of many of the compounds of interest and the decreased specificity of accurate mass alone compared to match a product ion spectra to a library spectra for compound identification. HRMS systems, such as QTOF or Orbitrap, offer the highest level of specificity due to the ability to determine the accurate mass of the precursor and PIs and provide isotope pattern information for all compounds in the sample.

3.4 BROAD SPECTRUM DRUG SCREENING: EXAMPLE METHODS

The following section will describe two LC-MS methods (LC-MS/MS and LC-HRMS) that are currently used in the clinical laboratory at SFGH for broad spectrum drug screening [23].

3.4.1 Rational for Method Development

Numerous new approaches to broad spectrum drug screening have made their way into clinical toxicology laboratories in the past 10 years. Prior to 2007, our laboratory used two complementary methods for comprehensive drug screening. These included the Bio-Rad REMEDi drug profiling system, which was a liquid chromatography ultraviolet detection (LC-UV) platform designed specifically for toxicology testing in the clinical laboratory, and a GC-MS full scan method. The REMEDi system served as a black-box turn-key solution for toxicology testing, however, Bio-Rad discontinued support of this system in 2008. Prior to this our laboratory purchased a hybrid triple quadrupole linear ion-trap instrument for future toxicology applications and testing. We developed and validated a new LC-MS/MS method to replace the REMEDi method in 2007 and have since made updates and improvements to this method in 2014. Our laboratory acquired an LC-QTOF in 2011 and subsequently developed and validated a broad spectrum drug screening method on this instrument. The following sections will highlight the method development, validation, and monitoring of the LC-MS/MS method. The end of this section will provide a brief summary of the LC-QTOF method and how it compares to the LC-MS/MS method.

3.4.2 Method Development

For sample preparation, a simple 1:5 dilution of urine in the starting mobile phase conditions containing internal standard was tested and provided adequate limits of detection for our testing needs. Chromatographic separation was performed on an Agilent 1200 series liquid chromatography system using a Phenomenex Kinetex 2.6-µm C18 column (3 × 50 mm) thermostated at 30°C. A binary mobile phase consisting of (1) 0.05% formic acid in 5 mM ammonium formate and (2) 0.05% formic acid in 50:50 methanol/acetonitrile was ramped linearly from 2% to 98% B over 10 min. The column was washed at 100% B for 2 min, followed by a 2 min reequilibration at 2% B. The flow rate was 400 µL/min. Data were collected using a 3200 QTRAP equipped with a TurboIon Spray source controlled by Analyst 1.5.1 software (ABSciex). Positive ionization was performed using the following settings: ion spray voltage, 5500 V; curtain gas, 20 psi; ion source gas 1, 40 psi; ion source gas 2, 40 psi; CAD gas, high; interface heater temperature, 40°C; declustering potential, 40 V; and entrance potential, 10 V.

Data were collected using a scheduled SRM scan followed by collection of product ion spectra when a compound met DDA criteria. The scheduled SRM window was 1 min, which means that the mass spectrometer would only look for a specific SRM 30 s before and after the specified retention time of that analyte. The method contained 169 transitions, which were independently established from infusion of each drug or metabolite of interest and ramping of all compounds dependent parameters to determine the optimal conditions for each. Dynamic background subtraction of the survey scan was performed to increase sensitivity. Fragmentation was performed in the collision cell with a collision energy spread of 35 ± 15 V. DDA criteria were set to select for the three most intense peaks that exceeded 400 cps. Former target ions were excluded for 15 s after three occurrences. The mass tolerance was set to 250 mDa. The EPI scan rate was 4000 Da/s. Q0 trapping was enabled, and the LIT fill time was set to 50 ms.

The product ion spectra produced by CID can differ markedly between LC-MS/MS instruments. This poses a problem for the creation of large product ion spectral libraries. Despite attempts to standardize the production of the product ion spectra, ion relative intensities are highly variable between instruments. Correct identification and ease of identification of compounds in biological samples using LC-MS/MS is highly dependent upon the quality of the mass spectral library used, therefore, we developed our own mass spectral library that contains spectra acquired for all drugs and metabolites of interest that were acquired on our instrument. MS vendors often provide libraries for initial testing; however, we have found in our experience that library match scores are higher when using a library developed in-house.

3.4.3 Method Validation

Lower LOD was determined by spiking drug standards into drug-free urine at various concentrations (5, 10, 25, 50, 100, 125, 250, and 500 ng/mL). Samples were injected in duplicate on two different occasions. The LOD was defined as the lowest concentration at which the drug was called positive by our scoring criteria and had a signal to noise ratio greater than 20:1. For the 169 drugs and metabolites in the method, 51% had an LOD of 5 ng/mL, 14% had an LOD of 10 ng/mL, and 11% had an LOD of 25 ng/mL. The remaining 24% had LODs ranging from 50–500 ng/mL.

Matrix effects were determined by spiking drug standards into three different drug-free urine matrices at 100 or 500 ng/mL. Samples were injected in triplicate producing a total of 9 data points for each drug. Peak intensities were recorded and compared to that of the drug spiked in water. Matrix effects were calculated using the following equation: (B–A)/A × 100%, where B is the average signal intensity in urine and A is the signal intensity in water. The matrix effects were 3% $\pm$ 76%. A few compounds were enhanced greater than 200% by matrix compared to water.

With-in run and between-run precision were evaluated using the positive control which contains 12 representative analytes. The positive control was tested in replicates of 5 on 5 different batches run by three different technologists. The coefficient of variance was determined for the combined score and peak intensity. The within-run CV for the combined score was 1–6% and 3–15% for peak intensity. The between-run CV for the combined score was 3–8% and 5–20% for peak intensity. The software program used to analyze the data does not integrate the peaks, but rather only provides peak intensity which can explain the higher CV since it is a less reproducible feature of an analytical peak than peak area. Also, the intensity was not normalized to the internal standard intensity since the method does not contain a unique internal standard for all drugs and metabolites in the method. This is not necessary for a qualitative assay.

100 patient urine samples were evaluated using the LC-MS/MS broad spectrum drug screen. All samples were also tested using our standard immunoassay drug screen, GC-MS and LC-MS/MS drug confirmation methods, and LC-Orbitrap and LC-QTOF broad spectrum drugs screens. In addition, patient prescription information was available for all samples. This information was used to determine which drugs and metabolites were expected in each sample. The method identified a total of 596 drugs in the 100 urines samples, of which, 531 (89%) were confirmed by another method. In general, a major limitation of any evaluation of a broad spectrum drug screen is that there is no gold standard method for comparison.

For data analysis, positive identification of a compound was based on retention time score and library score. Different weighting parameters were evaluated for retention time score and library match score to produce a combined score. After systematic analysis of known samples (LOD validation samples), it was determined that a combined score of 70 or greater using the following equation would be used: combined score = 50% (retention time score) + 50% (library score).

3.4.4 Postimplementation Monitoring

Since this is a qualitative test each batch only contains a negative and positive control in addition to the patient urine samples. The negative control is drug free urine. The positive control contains 12 representative analytes. The negative and positive controls are monitored in the same manner as the patient urine samples. In order for the controls to pass, there must be no compounds identified in the negative control except for the internal standard. All expected compounds and the internal standard should be identified in the positive control and meet reporting criteria (combined score >70). When reviewing patient results, the technologist has to pay special attention to isomers in the method. For example, morphine and hydromorphone have the same mass, similar fragmentation patterns, and close retention times. In most instances, one peak for morphine will be identified as both morphine and hydromorphone using the data analysis software. The technologist then has to select which has the highest combined score and not report the other unless there are two distinct peaks present for both morphine and hydromorphone.

For proficiency testing (PT), our laboratory subscribes to the CAP Urine Toxicology Survey. Since implementation of this method, all CAP results have been acceptable. The only compounds that we have missed are compounds spiked into the urine PT sample that are not included in our targeted acquisition method. This highlights one of the limitations of a targeted method; a targeted method is not capable of detecting all possible drugs and metabolites, only the ones included in the acquisition method. There was one instance in which our method detected clomipramine, norclomipramine, and imipramine in a sample that was only supposed to contain clomipramine and norclomipramine. The identification of imipramine was graded as unacceptable; however, the participant summary noted that several laboratories reported the presence of imipramine in the PT sample and that analysis by a reference laboratory confirmed that the challenge sample did contain trace amounts of imipramine below the minimum reporting concentration of 250 ng/mL set by CAP. Since our method is qualitative we do not use cutoff values, but rather report a drug if it meets our reporting criteria. In this instance, imipramine met all of our reporting criteria and thus should be reported as positive even if only small amounts exist in the sample. There is no consensus on whether cutoffs should be used for broad spectrum drug screening and if yes, what the cutoff should be for each drug and metabolite that might be included in the method. In order for cut-offs to be used the method must be at the very least semi-quantitative and would require that calibrators be included for all drugs and metabolites in the method with each batch. This significantly increases the labor time associated with method development and utilization while providing little additional clinical value. Urine drug concentrations are highly variable due to hydration status and other factors and are generally not reflective of blood levels, drug compliance, or acute exposure.

3.5 LC-QTOF FOR BROAD SPECTRUM DRUG SCREENING

Our laboratory has also developed and validated an LC-QTOF broad spectrum drug screen. The sample preparation, LC conditions, and data analysis software are identical to the LC-MS/MS method described earlier. This method is also used for serum, plasma, and blood samples. For these samples, a simple protein precipitation is performed for sample preparation.

For this method, chromatographic separation was performed on a Shimadzu Prominence LC-20ADXR system using the same column, mobile phases, gradient, and injection volume as described for the LC-MS/MS method. Data were collected using a 5600 QTOF system equipped with an ESI source controlled by Analyst 1.5.1 (ABSciex). Ionization was performed using the following settings: ion source gas 1, 30 psi; ion source gas 2, 30 psi; curtain gas, 25 psi; temperature 500°C; ion spray voltage floating, 5500 V; declustering potential, 100 V. The method consisted of a TOF MS full scan from 50–700 Da and DDA triggered acquisition of product ion spectra (DDA threshold = 100 cps) for up to 20 candidate ions per cycle. The instrument was calibrated using the automatic calibrant delivery system which injected manufacturer-supplied calibration solution every 5 injections. The same validation experiments were conducted for the LC-QTOF method as described earlier for the LC-MS/MS method. A considerable amount of time was invested in determining the optimal data analysis parameters and criteria for positive compound identification using the LOD samples with known compounds spiked into drug free urine at known concentrations. It was determined that a drug would be called positive if it had a combined score greater than 70 using the following equation: combined score = 10% (mass score) + 10% (retention time score) + 10% (isotope score) + 70% (library score).

In comparisons with the LC-MS/MS method, the LC-QTOF method had the same LOD for 76 compounds; the LC-QTOF LODs were lower for 33 drugs, and the LC-MS/MS LODs were lower for 60 drugs. Overall, the LC-MS/MS method was more sensitive than the LC-QTOF method, but in many cases, this was only a difference of 5–15 ng/mL. The matrix effects were much more significant for the LC-QTOF method compared to the LC-MS/MS method, which is to be expected considering the method is an untargeted acquisition method. In the 100 urine samples tested for patient comparison data, the LC-QTOF method identified 515 drugs, of which, 500 (97%) were confirmed.

The major advantage of the LC-QTOF method is the ability to identify unexpected compounds by suspect or untargeted data analysis. The comparison of the LC-MS/MS and LC-QTOF methods above only compares their ability to detect the same subset of drugs and metabolites in patient urine samples. However, the LC-QTOF method is acquiring data in an untargeted manner and can be interrogated for additional compounds using various data analysis software tools. Identified compound can be confirmed by a PI match to a reference library or testing an analytical standard for comparison.

4 CONCLUSIONS

In summary, LC-MS is a powerful tool for toxicology applications in the clinical laboratory. The primary use of an LC-MS in a toxicology laboratory is for drug confirmation testing following an immunoassay screen and for broad spectrum drug screening. Multiple different LC-MS platforms including, LC-MS, LC-MS/MS, LC-TOF, LC-QTOF, and LC-Orbitrap have been used for toxicology testing. In addition, multiple different data acquisition modes have been described; including, targeted and untargeted data acquisition methods with and without data dependent acquisition of product ion spectra (DDA or DIA). This chapter highlighted two LC-MS applications; including opioid confirmation testing and broad spectrum drug screening and one laboratories experience with method development, validation, and implementation. Several different approaches to both of these applications have been described in the literature, are referenced in this chapter and can be used as resources for toxicology laboratories considering the implementation of LC-MS.

REFERENCES

[1] Maurer HH, Pfleger K, Weber A. Mass spectral and GC-data of drugs, poisons, pesticides, pollutants and their metabolites. 3rd ed. Weinheim, Chichester: Wiley-VHC; 2007.

[2] Jansen R, Lachatre G, Marquet P. LC-MS/MS systematic toxicological analysis: comparison of MS/MS spectra obtained with different instruments and settings. Clin Biochem 2005;38:362–72.

[3] Ferrer I, Thurman EM. Liquid chromatography/mass spectrometry, MS/MS and time of flight MS. Cary: Oxford University Press; 2003.

[4] Jiwan JL, Wallemacq P, Hérent MF. HPLC-high resolution mass spectrometry in clinical laboratory? Clin Biochem 2011;44(1):136–47.

[5] Edinboro LE, Backer RC, Poklis A. Direct analysis of opiates in urine by liquid chromatography-tandem mass spectrometry. J Anal Toxicol 2005;29:704–10.

[6] Murphy CM, Huestis MA. LC-ESI-MS/MS analysis for the quatification of morphine, codeine, morphine-3-β-D-glucuronide, morphine-6-β-D-glucuronide, and codeine-6-β-D-glucuronide in human urine. J Mass Spectrom 2005;40:1412–6.

[7] Musshoff F, Trafkowski J, Kuepper U, Madea B. An automated and fully validated LC-MS/MS procedure for the simultaneous determination of 11 opioids used in palliative care, with 5 of their metabolites. J Mass Spectrom 2006;41(5):633–40.

[8] Gustavsson E, Andersson M, Stephanson N, Beck O. Validation of direct injection electrospray LC-Ms/MS for confirmation of opiates in urine drug testing. J Mass Spectrom 2007;42(7):881–9.

[9] Dowling G, Regan L, Tierney J, Nangle M. A hybrid liquid chromatography-mass spectrometry strategy in a forensic laboratory for opioid, cocaine and amphetamine classes in human urine using a hybrid linear ion trap-triple quadrupole mass spectrometer. J Chromatogr A 2010;1217(44):6857–66.

[10] Kolmonen M, Leinonen A, Kuuranne T, Pelander A, Ojanpera I. Hydrophilic interaction liquid chromatography and accurate mass measurement for quantification and confirmation of morphine, codeine and their glucuronide conjugates in human urine. J Chromatogr B 2011;878:2959–66.

[11] French D, Wu AH, Lynch KL. Hydrophilic interaction LC-MS/MS analysis of opioids in urine: significance of glucuronide metabolites. Bioanalysis 2011;3(23):2603–12.

[12] Fitzgerald RL, Griffin TL, Yun Y-M, Godfrey RA, West R, Pesce AJ, Herold DA. Dilute and shoot: analysis of drugs of abuse using selected reaction monitoring for quantification and full scan product ion spectra for identification. J Anal Toxicol 2012;36:106–11.

[13] Dickerson JA, Laha TJ, Pagano MB, O'Donnell BR, Hoofnagle AN. Improved detection of opioid use in chronic pain patients through monitoring of opioid glucuronides in urine. J Anal Toxicol 2012;36:541–7.

[14] Langman LJ, Korman E, Stauble ME, Boswell MV, Baumgartner RN, Jortani SA. Therapeutic monitoring of opioids: a sensitive LC-MS/MS method for quantitation of seven opioids including hydrocodone and its metabolites. Ther Drug Monit 2013;35(3):352–9.

[14a] Yang HS, Wu AHB, Lynch KL. Development and validation of a novel LC-MS/MS opioid confirmation assay: evaluation of β-glucuronidase enzymes and sample cleanup methods. J Anal Toxicol 2016;40:323–9.

[15] Mueller CA, Weinmann W, Dresen S, Schreiber A, Gergov M. Development of a multi-target screening analysis for 301 drugs using a QTrap liquid chromatography/tandem mass spectrometry system and automated library searching. Rapid Commun Mass Spectrom 2005;19:1332–8.

[16] Dresen S, Ferreiros N, Gnann H, Zimmermann R, Weinmann W. Detection and Identification of 700 drugs by multi-target screening with a 3200 QTRAP LC-MS/MS system and library searching. Anal Bioanal Chem 2010;395:2521–6.

[17] Sauvage FL, Saint-Marcoux F, Duretz B, Deporte D, Lachatre G, Marquet P. Screening of drugs and toxic compounds with liquid chromatography-linear ion trap tandem mass spectrometry. Clin Chem 2006;52:1735–42.

[18] Sturm S, Hammann F, Drewe J, Maurer HH, Scholer A. An automated screening method for drugs and toxic compounds in human serum and urine using liquid chromatography-tandem mass spectrometry. J Chromatogr B 2010;878:2726–32.

[19] Rosano TG, Wood M, Swift TA. Postmortem drug screening by non-targeted and targeted ultra-performance liquid chromatography-mass spectrometry technology. J Anal Toxicol 2011;35:411–23.

[20] Ojanperä S, Pelander A, Pelzing M, Krebs I, Vuori E, Ojanperä I. Isotopic pattern and accurate mass determination in urine drug screening by liquid chromatography/time-of-flight mass spectrometry. Rapid Commun Mass Spectrom 2006;20:1161–7.

[21] Lee HK, Ho CS, Iu YP, Lai PS, Shek CC, Lo YC, Klinke HB, Wood M. Development of a broad toxicological screening technique for urine using ultra-performance liquid chromatography and time-of-flight mass spectrometry. Anal Chim Acta 2009;649(1):80–90.

[22] Rosano TG, Wood M, Ihenetu K, Swift TA. Drug screening in medical examiner casework by high-resolution mass spectrometry (UPLC-MS^E-TOF). J Anal Toxicol 2013;37:580–93.

[23] Thoren KL, Colby JM, Shugarts SB, Wu AHB, Lynch KL. Comparison of information-dependent acquisition using QqTOF and QqLIT mass spectrometers for broad-spectrum drug screening in urine. Clin Chem 2016;62(1):170–8.

[24] Arnhard K, Gottschall A, Pitterl F, Oberacher H. Applying 'sequential window acquisition of all theoretical fragment ion mass spectra' (SWATH) for systematic toxicological analysis with liquid chromatography-high-resolution tandem mass spectrometry. Anal Bioanal Chem 2015;407:405–14.

TOXICOLOGY: GCMS

P.B. Kyle

University of Mississippi Medical Center, Jackson, MS, United States

1 USE AND STRENGTHS OF GC-MS IN CLINICAL TOXICOLOGY

The advent of inexpensive computerization allowed GC-MS to migrate into clinical laboratories over 30 years ago. The consumer cost of GC-MS instruments has not increased over the years to the extent that one would expect, but has remained nearly constant as has the cost of personal computers. Today, single quadrupole and ion trap GC-MS instruments are commonly available with an autosampler and spectral drug library for approximately $80,000 USD, approximately the same price that they were 10 years ago [1]. Therefore, for a laboratory seeking to bring in a "gold standard" methodology, a GC-MS is relatively inexpensive compared to other platforms.

Early on, the emission voltage during electron impact ionization (EI) was standardized to 70 eV in GC-MS, which promoted the development of extensive spectral libraries. The application of EI promotes reliable ionization with extensive fragmentation that results in robust spectra. In this regard, GC-MS is still far ahead of LC-MS. One reason is that the collision energies required for reliable fragmentation in LC-MS/MS platforms vary widely depending upon the analyte/s being targeted, solvent flow rate, auxiliary gas flow, and interface temperature. Individual toxicology laboratories began developing spectral libraries for LC-MS/MS years ago [2], whereas commercially available LC-MS/MS libraries have only recently become available. These include the iMethod Forensic Toxicology Library available from AB Sciex and the *Maurer/Wissenbach/Weber LC-MSn Library of Drugs, Poisons, and Their Metabolites* [3]. However, these contain far fewer spectra than those for GC-MS. In contrast, comprehensive libraries have been available to GC-MS users for 30 years. Reputable examples include the NIST/EPA/NIH, Wiley Registry of Mass Spectral Data, and the Maurer/Pfleger/Weber libraries. The Maurer/Pfleger/Weber library, now published by Wiley, was developed for clinical toxicology applications using EI in positive ionization mode. The earliest version (Pfleger/Maurer/Weber) became commercially available in 1985, and included EI mass spectra for 1500 drugs, poisons, and metabolites. The latest 2011 version contains EI spectra for over 8500 substances and more than 5000 metabolites [4]. The NIST/EPA/NIH and Wiley Registry of Mass Spectral Data libraries are very broad with each containing spectra for over 100,000 compounds.

As with other computerized platforms, GC-MS instrumentation continues to evolve into smaller and more sensitive units with impressive capabilities. Battery-operated instruments are now available in portable briefcase sizes, which may be used by HAZMAT and military teams to evaluate chemicals present in urban, rural, and battlefield settings. These units are dwarfed by the instruments of the 1980s era and exemplify evolution of this technology. Given recent acts of terrorism, the use of GC-MS has been implemented in many airports allowing the detection of chemicals on traveler's hands or luggage as well as in the air space surrounding an individual [5].

The clinical applications of GC-MS are many and varied, which is due to the flexibility of the instruments. For example, an ion trap instrument may be configured to perform chemical ionization or EI, and may be used in positive ionization or negative ionization modes. Chemical ionization uses a gas, often nitrogen or argon, to soften the impact of electrons and promote a softer fragmentation than with EI. The resulting spectra almost always contain the parent ion, which can be helpful for structural identification. Detectors may be set to acquire full-scan data for the purposes of general unknown screening or selective ion monitoring (SIM) for targeted analysis. Use of SIM greatly increases analytical sensitivity because the instrument is set to acquire only a few specified ions per second, which significantly increases the dwell time. By comparison, instruments may acquire ions from 0 to 500 m/z in 1 s during full scan acquisition mode. Instruments may be obtained with ion trap, single quadrupole, triple quadrupole, time of flight (TOF), quadrupole-trap, quadrupole-TOF, as well as Fourier transform capabilities. While GC-MS instruments are not perfect, these instruments allow laboratories to utilize "gold standard" technology for the detection of drugs, metabolites, pesticides, and other compounds at a relatively low entry cost.

2 LIMITATIONS OF GC-MS IN CLINICAL TOXICOLOGY

Although GC-MS has been the mainstay of clinical and forensic toxicology laboratories for the past 30 years, the use of LC-MS in toxicology has recently become more common in order to bridge the analytical gaps associated with GC-MS. One of the requirements and therefore limitations of GC-MS is that analytes must be volatilized into the gas phase during injection into the instrument. It is for this reason that most if not all GC-MS analyses require extraction of the analytes from the biological matrices that are often encountered in toxicology. This usually involves some form of liquid–liquid extraction (LLE) or solid phase extraction (SPE) [6]. The purpose of the extraction procedure is to transfer the analytes of interest from the biological matrix into a solvent that is easily volatilized when injected into the instrument. The benefits of extraction often include (1) concentration of the analytes, and (2) removal of biological contaminants, such as proteins and phospholipids. Sample/analyte extraction typically requires 10–60 min depending on the length of the extraction protocol and the type and amount of the final solvent that must be evaporated prior to injection. This is in stark contrast to LC-MS analyses in which sample volatilization is not required. Most LC-MS protocols perform best when analytes are in an aqueous matrix as this prevents band spreading upon injection onto the analytical column. In fact, some laboratory protocols involve the injection of pure urine or diluted urine into the LC-MS for analysis [7].

Another limitation of GC-MS involves the chemical and physical characteristics of the analytes-required for gas chromatographic separation. These include heat stability, low polarity, and relatively low boiling point. Currently, most clinical toxicology laboratories utilize wall-coated open tubular (WCOT) type GC columns for drug separation. These are currently made of fused silica with internal diameters ranging from 0.1 to 0.5 mm and lengths from 10 to 150 m. A stationary phase coats the interior of the wall but does not occlude the entire bore of the column. The stationary phases used for drug separation are composed of polysiloxanes with various user-selected functional groups that affect column polarity. The separation of analytes is based on their interaction with the stationary phase and their partitioning with the carrier gas. Drugs with lower boiling points often, but not always elute from the column faster than those of higher boiling points. Factors that influence boiling point include the compound's molecular weight, cyclic ring saturation, and associated functional groups. Compounds greater than approximately 450 amu often do not chromatograph well and often result in poor or nonexistent peaks. In addition, polar drugs whose side chains include carboxylic acid, such as salicylic acid, ibuprofen, and 11-nor-Δ9-carboxy-THC may not chromatograph well unless they are either chemically modified (derivatized) or injected onto specialized polar columns for analysis. Drug metabolites are usually more polar than their parent compounds, which can impede their analysis by GC-MS. For these reasons, some compounds are either not amenable to GC-MS analysis, or may require derivatization and extraction procedures prior to analysis.

3 INSTRUMENT SELECTION

Prior to selecting an instrument, one should evaluate each analytical method that will be performed by GC-MS in the laboratory. This is critical because GC-MS analyzers can vary widely in their sensitivity, scanning capability, linear response, and cost. The acquisition of multiple instruments should be considered because a secondary or backup instrument can be invaluable during instances when the primary instrument is down for maintenance. Users often find that specific analytical columns or derivatization procedures are required for specific applications. Therefore, it is often helpful if not imperative for the laboratory to employ at least two or three GC-MS instruments in order to accommodate multiple applications. Laboratories seeking to perform drug confirmations should seek a robust platform with linear response if quantitative analyses are to be included. This most often involves a quadrupole or triple quadrupole analyzer as discussed later. Clinical laboratories seeking to support the emergency department during instances of unknown drug intoxication should consider robust instruments with fast scan rates and comprehensive drug libraries that include prescription and nonprescription drugs, drug metabolites, pesticides, and other compounds. In contrast, laboratories performing pain management testing have a much narrower focus that may only involve about 50 drugs and metabolites. Some clinical laboratories perform forensic analyses in addition to clinical testing. These laboratories will require instruments with linear response for quantitative analyses. Finally, those in university medical centers and research institutions should keep in mind the potential for future research and should consider acquiring instruments with higher capabilities than those necessary for clinical testing.

The concentrations of analytes one anticipates should be considered prior to instrument acquisition. Analyte concentrations involved during medical applications will include physiological concentrations as detailed in the medical literature. However, laboratories seeking to include forensic drug confirmation should determine the concentrations designated by the applicable regulatory agencies for preemployment, workplace, or forensic testing. The analyte concentrations involved in unknown toxic ingestions can vary and will depend on the potency and bioavailability of the exact compounds. Very sensitive instruments are desired in situations of suspected drug ingestion.

The target analytes have a large influence on the type of autosampler and injector, an instrument should be outfitted with. The wide majority of clinical analyses are performed using direct injection. However, specific applications for volatile analytes, such as gamma hydroxybutyric acid or hydrogen sulfide may require headspace techniques as detailed in sections later. Many laboratories employ instruments with the flexibility to accommodate split and splitless injection modes as well as EI and CI ionization modes. Instruments with the capability of direct probe analysis can also be useful for the analysis of tablets or liquids that are received with patient specimens.

3.1 INSTRUMENT MODEL/MANUFACTURER CONSIDERATIONS

There are several reputable instrument manufacturers currently producing GC-MS platforms. After evaluating the laboratory's needs, one should determine type of analyzer/s required. The performance of new instrument models or instrument types should always be carefully evaluated because months may be required to correct software bugs or other issues after market introduction. Similarly, less downtime may be expected with products from long-term GC-MS manufacturers.

Prior to purchase, individuals should evaluate the performance specifications of several instruments. Specifications, such as sensitivity, signal to noise ratio, mass range, scan speed, MRM scan speed, mass stability, and polarity switching are usually listed in the manufacturers' instrument brochures. Other sources of information include published instrument evaluations found in the medical and research literature. Regional and national colleagues are often the best sources of information for specific instruments or they may at least explain nuances of specific equipment and software. It is also wise to ask your salesperson for references that are willing to discuss their experience with the particular equipment of interest, and how they chose a specific instrument over others. Those unfamiliar with GC-MS analysis may consider negotiating a site visit in order to evaluate instrumentation during actual laboratory use. Most salespeople will be happy to contact their clients and arrange for travel in order to secure a sale.

Other considerations in choosing an instrument involve the service and support provided by the vendor. Given that GC-MS is not used in every hospital, laboratories experience varying response times for onsite service, which often depends on geographic location. One should inquire into the location of the nearest service technicians and expected response times. Same day response may not be available for laboratories located away from hubs of research or industry.

3.2 ANALYZER TYPES

3.2.1 Ion Trap

Of the GC-MS platforms used in clinical toxicology, the quadrupole ion trap (ion trap) is the least expensive due to its simple design and construction. The trap itself is composed of a ring and two end-cap electrodes (Fig. 7.1). Ion trap instruments utilize electrical potential and radio frequency (RF) to retain and concentrate ions within the trap until they are selectively ejected to the detector. Sensitivity is increased with the addition of a low pressure gas, which serves to dampen molecular vibrations and reduces the number of random collisions with the electrodes. Ion trap instruments offer 0.1 m/z resolution, low cost, and high sensitivity. With all things being equal, ion traps are often the most sensitive detectors because target ions are effectively concentrated within the trap before being ejected to the electron multiplier. However, nuances in electronics, scan speed, lens design, and detector configuration have profound effects on detector sensitivity.

Early ion trap instruments exhibited limited dynamic range due to space charge effects that distorted resolution when the trap was filled with too many ions. Users were required to set the ion accumulation time (in milliseconds), which would be determined by the analyte concentrations anticipated in a given experiment. A significant amount of experience was required to achieve high sensitivity without loss of

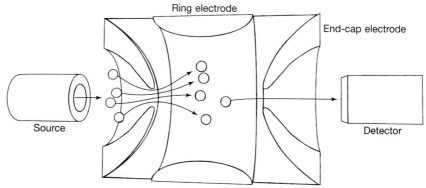

FIGURE 7.1 Ion Trap Illustrating the Flow of Ions From the Ion Source Into the Trap

Several ion species are being retained by the trap and one species is being ejected to the detector.

resolution. Contemporary instruments have an automatic gain control setting in which the instrument determines the ion acquisition time based on prescan signal intensities.

The sensitivity of any mass spectrometer is a function of its duty cycle. The duty cycle reflects the percentage of time the instrument is actually scanning ions during a scan cycle. Higher sensitivity is directly proportional to the time spent detecting a specific ion (m/z). Therefore, sensitivity can be increased by increasing scan speed and/or the dwell time. An ion trap instrument captures and concentrates ions of all masses in the trap then ejects them to the detector via mass-selective instability scanning. This way, ionization and mass analysis occur consecutively in an ion trap, and the duty cycle is determined by the scan rate relative to the ion accumulation time. For example, an ion trap using a typical 25 ms accumulation time during a 200 ms scan cycle exhibits a 12.5% duty cycle. The reduced duty cycle of a quadrupole is due to the use of mass-selective stability scanning, which transmits one m/z at a time through the quadrupole filter to the detector. Ionization and mass analysis occur simultaneously in quadrupole instruments, and the duty cycle is determined by the scan range. Therefore, a quadrupole instrument scanning 0–500 m/z each second spends 1/500 s (2 ms) detecting each m/z during a scan resulting in a duty cycle of 0.2%. For this reason, ion trap instruments exhibit great sensitivity, and lend well to general unknown toxicology screening. However, due to their ion concentrating characteristics, ion trap detectors usually exhibit nonlinear analyte response. Therefore, ion trap analyzers are limited in their quantitative capabilities unless the ion accumulation times have been set manually and assay validation has been performed.

3.2.2 Quadrupole Mass Filter
3.2.2.1 Single Quadrupole Analyzers
Single quadrupole analyzers are the work horses of the laboratory and may be found in nearly every clinical and forensic toxicology laboratory. Quadrupole analyzers offer adequate sensitivity, 0.1 m/z resolution, linear response, and rugged performance. Currently, the base cost of a fully configured GC-MS with quadrupole analyzer is approximately $90,000 USD, which is slightly higher than a comparable ion trap, but significantly less than other designs.

Quadrupole mass spectrometers function by employing four parallel rods in a square array connected to an electrical circuit (Fig. 7.2). Voltage is applied to diagonally opposite rods (electrical poles)

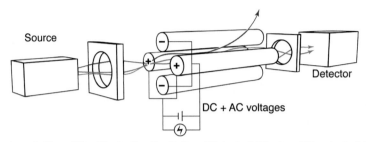

FIGURE 7.2 A Quadrupole Mass Filter Illustrating the Paths of Selected (Blue) and Unselected (Red) Ions

in an alternating fashion, which also generates an RF field. As ion fragments pass from the ion source into the quadrupole, the voltage and RF are automatically adjusted so that ions of specific m/z ratios are transmitted to the electron multiplier. This process is known as mass-selective stability scanning. The combination of alternating current and RF in the quadrupole causes ions to be transmitted in a swirling fashion through the length of the quadrupole to the electron multiplier. Ions not specified by the analysis swirl widely and, either collide with the electrical poles, or swirl outside the quadrupole field and are taken away in vacuum. In this way, the quadrupole acts as a mass filter. The voltages and RF are ramped through the mass range while scanning, but are held constant during SIM mode.

Quadrupole mass spectrometers may exhibit decreased sensitivity with larger ions because increased molecular mass causes increased radial oscillation amplitudes within the molecule. As a result, a higher percentage of larger ions will collide with the electrical poles and be lost. Recent experimentation showed that a currently-produced single quadrupole instrument exhibited over 2 times the sensitivity to methanolic standards of butalbital at 224 m/z than to trifluoperazine at 407 m/z [8].

Significantly increased sensitivity is obtained when using quadrupole analyzers in SIM mode, which is often used for quantitative analysis. During SIM, the instrument is set to record only the m/z of the specific analytes of interest. Compared to a full-scan experiment involving a 500 m/z range, the use of SIM to monitor a few ions can result in a 100-fold increase in sensitivity. During SIM, the user defines the ion dwell times so that each ion is scanned once or several times per second. This limited but rapid data acquisition even allows users to collect quantitative data from coeluting compounds. For this reason, many laboratories incorporate deuterium- or ^{13}C-labeled internal standards into their analyses.

3.2.2.2 Triple Quadrupole Analyzers

Triple quadrupole analyzers are often used when higher sensitivity and specificity is required. They may also be used to generate additional fragmentation data from ions of interest. The instruments are rugged, offer 0.1 m/z resolution, exhibit linear response, and resulting MS/MS spectra can be helpful in structural identification. The base price for a fully outfitted triple quadrupole GC-MS/MS is approximately $120,000 USD depending upon configuration. A triple quadrupole analyzer is configured so that the ions of interest pass through a sequence of three quadrupole mass filters. The first quadrupole (Q1) selects ions of interest that were generated in the ion source. The second quadrupole (Q2) is typically filled with nitrogen and is used as a collision chamber to generate fragment or daughter ions. Some use argon as the collision gas because it is more dense than nitrogen and promotes more robust fragmentation. The third quadrupole (Q3) is used to select and conduct specified fragment ions to the electron multiplier. The use of Q2 for ion fragmentation is optional and allows the user to collect simple MS data or MS/MS data depending upon specific needs. Triple quadrupole instruments are quite flexible and allow the user to monitor and collect: (1) simple MS ion data, (2) full-scan fragmentation data, (3) specific ion fragmentation data (selected reaction monitoring), (4) neutral loss data, or (5) precursor ion data.

Triple quadrupole analyzers are often applied to quantitative analyses due to their linear detector response. Their flexibility also allows toxicology laboratories to utilize chromatographic and MS data for drug screening or use MS/MS data to perform drug confirmations. Recent software advances allow data-dependent acquisition in which, instruments operated in full-scan mode can be set to acquire MS/MS data whenever a peak intensity exceeds the predefined threshold. This allows the user to screen for drugs and to confirm drug identity in a single run. Triple quadrupole analyzers are also used when higher sensitivity and specificity is required. Use of Q1 as a mass filter allows the user to focus on ions of interest. Applications include the quantitation of critical or low abundance compounds, such as THC metabolites found in oral fluid [9].

3.2.3 Time of Flight Analyzers

TOF detectors offer high resolution (2–5 ppm), fast scan rates, and the capability to measure ions of high m/z. These detectors have been comparably slow to make their way into the clinical setting, likely due to their high cost and limited number of applications that cannot be performed by instruments already in place. Until recently, TOF detectors have exhibited relatively poor resolution at low m/z ratios. This is due to the mechanisms involved in the actual determination of the ion TOF required for m/z determination.

TOF analysis is akin to a race in which all competitors start at the same time and place. Each competitor's race time is recorded at the finish line. Given the same kinetic energy, the smaller (less massive) ions accelerate faster and arrive at the detector sooner than larger ions. In this regard, a great deal of effort has been focused on getting ions of all m/z assembled at the starting point at the same time so that they can begin their transit to the detector at the same time. The continuous ionization of EI and CI coupled to TOF, a pulsed technique, creates an inherent mismatch. This appears to be a key factor that limits resolution at lower m/z ranges, which can drop off significantly below 600 m/z. This problem has been largely overcome with orthogonal acceleration TOF. During orthogonal acceleration, ions are injected perpendicularly into the flight tube relative to the flight axis. The injection region is allowed to fill with ions while the acceleration voltage is off. The acceleration voltage is then turned on and the race to the detector begins. Mass resolution at lower m/z ranges will improve as this process becomes more refined.

TOF instruments boast extremely fast scan rates, and typically exhibit higher sensitivities than quadrupole instruments used in scan mode [10]. However, TOF instruments may exhibit lower sensitivity than other analyzer types because they take a snapshot in time of all ions being transmitted. Higher sensitivity has been demonstrated with a quadrupole analyzer in SIM mode than current TOF instruments [11]. TOF analyzers also boast accurate mass measurement capability, which allows the user to confirm the molecular formula of a compound via its isotope abundance. The fast scanning capability of TOF instruments also allows differentiation of coeluting compounds [12], and allows users to exploit faster chromatography to achieve high sample throughput [13]. Some instruments utilize a reflectron at the end of the flight tube to reflect ions back through the tube to a detector positioned beside the starting point. This arrangement essentially doubles the length of the flight tube and results in higher m/z accuracy.

3.2.4 Orbitrap Analyzers

Orbitrap analyzer is an ion trap variant that has been available for several years in LC-MS platforms, but has only recently become available in GC-MS configuration. Orbitrap analyzers exhibit high mass resolution ($> 150,000$) with 1–5 ppm mass accuracy [14]. However, this performance comes with a high price tag that is comparable to some TOF analyzers. Orbitrap analyzers utilize a central rod-like electrode surrounded by a barrel-shaped outer electrode. Ions are injected into the barrel of the trap and swirl around the central electrode. Ion trapping occurs axially and m/z is determined using the harmonic oscillations in combination with Fourier transforms. These analyzers exhibit comparatively slow scan speeds, but high sensitivity due to their trapping capability. Like TOF analyzers, the accurate mass capability of orbitrap analyzers utilizes isotope abundance data to confirm the molecular formula of a compound. Although new to the market, these instruments should lend well to general unknown toxicology screening and drug confirmation protocols.

3.2.5 Hybrid Analyzers

Hybrid analyzers include a combination of components from multiple types of mass analyzers. The Q-TOF and the Q-Orbitrap are two examples in which a single quadrupole serves to filter and transfer ions to the TOF or orbitrap analyzer. These arrangements offer greater filtering capabilities and MS/MS capability that lend well to qualitative and quantitative analyses. In addition, the accurate mass capabilities of both TOF and orbitrap analyzers offer greater structural identification capabilities due to higher mass accuracy. These instruments are not as common as others found in clinical toxicology laboratories and often serve in specialized applications or perform multiple roles in well-funded clinical research laboratories.

It should be noted that these instruments collect greater amounts of data which requires greater storage space and greater computing power. Whereas a single quadrupole or triple quadrupole analyzer in scanning mode might acquire 1–10 MB in a single 20-min analysis, a Q-TOF or Q-Orbitrap might acquire 1000 MB data during the same time. Therefore, a second computer system dedicated to postacquisition data analysis might be advised so that the primary data acquisition system is not overwhelmed.

3.3 METHODS OF SAMPLE INTRODUCTION

3.3.1 Solvent Injection

The most common form of sample injection involves direct injection of a solvent containing the analytes extracted from biological samples. During injection, heat within the injection port vaporizes the injection mixture which then enters the chromatographic column. Many protocols specify a starting column temperature that is 100 degrees cooler than the injection port, which promotes concentration of the analytes onto the head of the column. A temperature gradient can then be used to promote the elution of analytes from the column. As discussed elsewhere, the cleanliness of the injection solvent greatly affects the accumulation of contaminants, chromatography, detector sensitivity, and column life. Therefore, a great deal of attention should be given to the extraction procedures employed. In some procedures the extraction solvent is injected into the GC-MS. However, extraction solvents often contain acidic, basic, and/or polar solvents, such as ammonium hydroxide, acetic acid, and isopropanol that may negatively affect the chromatographic system. Cleaner injections often involve evaporation of the extraction solvent and subsequent reconstitution of the analytes in ethyl acetate, dichloromethane, or other solvent that is better suited for injection. Ideally, solvents that exhibit low boiling points and low expansion volumes, such as ethyl acetate and dichloromethane should be used for injection.

Given the high sensitivity of current instrumentation, many laboratories employ split instead of splitless injection modes. During a split injection, only a portion of the aliquot injected into the instrument is directed onto the analytical column. In this way, most of the injection solvent vapor and some nonvolatile analytes are swept away to waste. In addition to reducing column contamination, split injection modes improve chromatography by reducing peak tailing. Split ratios ranging from 1:5 to 1:50 are commonly employed.

3.3.2 Headspace Injection

Headspace injection is a clean injection technique often used for the analysis of volatile analytes. Prior to injection, an aliquot of the patient sample is transferred to a glass headspace vial with any associated internal standard/s and matrix modifiers. The vial is then sealed, and heated to 40–60°C for a prescribed amount of time. The volatile analytes become dispersed equally between the liquid and the gaseous headspace during this heated incubation period. A syringe pierces the septum cap of the headspace vial and withdraws an aliquot of the headspace gas, which is injected into the instrument. In this way, the analyst eliminates the requirement for cumbersome extraction techniques. Headspace injection is particularly effective when complex viscous biological samples, such as whole blood must be analyzed. It has long been used for the analysis of ethylene glycol, ethanol, and other volatiles by gas chromatography. It has also been applied to the analysis of gamma-hydroxybutyric acid in various biological matrices [15], and to propofol in combination with solid phase microextraction (SPME) as described later in this chapter [16].

3.3.3 Solid Phase Microextraction (SPME)

SPME is a relatively new extraction technique that involves a sorbent-coated rod or fiber that is placed into the sample vial. Compounds from liquid or gaseous samples are adsorbed onto the fiber. Once the extraction is complete, the SPME fiber is inserted into the heated injection port of the GC-MS, where analytes are thermally desorbed from the fiber and enter the analytical column. The fibers are typically composed of a 1–2 cm length of fused silica coated with polar and nonpolar sorbents similar to those available in WCOT GC columns. SPME is most often applied for the extraction and analysis of volatile analytes and pesticides during food and environmental monitoring. However, SPME has been used in assays for selecting drugs in various biological matrices [16–18].

3.3.4 Direct Probe Analysis

Direct probe analysis involves the insertion of a solid or liquid sample into the ion source of the analyzer via a probe. The end of the probe houses a cup or tube used to contain the sample. This technique is often used for the analysis of powders and solid dose medications in forensic laboratories. The ability to completely bypass the chromatographic process is the primary advantage of direct probe analysis. Therefore, it can be useful for the analysis of compounds that are thermally labile as well as those that are difficult or impossible to elute chromatographically. While direct probe analysis is not frequently used in clinical toxicology laboratories, it could be useful for the analysis of tablets or powders found with obtunded patients.

4 DEFINING AN ASSAY METHOD

4.1 CHROMATOGRAPHY

4.1.1 Column and Temperature Gradient

After an instrument has been installed and outfitted, one must define the assays that will be performed. One typically begins with the chromatography necessary to separate the target analytes, which often centers around column selection. A 30 m, 0.25 mm ID, capillary column with a 25 µm film thickness and DB-5, Rtx-5, or other G27-equivalent phase is a good starting point for general unknown toxicology screening. This setup will result in good separation for a variety of drugs, metabolites, and pesticides. The newer DB-5ms, Rtx-5ms, and equivalent phases are even more inert and were engineered for decreased column bleed and increased temperature stability. Characteristics of these columns include: reduced tailing of polar compounds, lower baseline noise, cleaner mass spectra, and improved spectral matching compared to less inert columns.

Oven and column temperature is very important given that analyte elution is largely determined by boiling point. Temperature gradients are useful, if not required, for adequate analyte separation. For general unknown screening, a relatively cool starting temperature of 50–60°C will cause most drugs to condense onto the head of the column. This prevents band spreading and improves peak shape. In order to promote elution, the oven temperature should be ramped up at a reasonable rate, such as 20°C per minute to a high temperature that is 20–40°C below the maximum column temperature. The maximum column temperature is usually around 280–300°C depending upon the exact column used. Finally, many will program the instrument to maintain this high temperature for 10 min or longer in order to clean the column and boil off many of the contaminants carried over from the sample extraction process.

Targeted drug confirmations often employ shorter 15 m analytical columns with the stationary phases listed above. The use of these shorter analytical columns results in shorter analytical run times and increased throughput. Other methods used to decrease analytical times include the use of decreased column film thickness to limit the depth of drug diffusion into the stationary phase, and higher starting column temperatures. For example, methods for opiate confirmation might specify a starting oven temperature of approximately 100°C, whereas methods for less volatile analytes, such as the cannabinoids might specify higher starting oven temperatures [19].

4.1.2 Filament Start Delay

Regardless of analysis type, most agree that the filament should not be powered on until 1–5 min after injection. This is known a solvent delay or filament start delay, and allows the majority of the injected solvent to pass through the system before the filament is powered on. The filament start delay effectively increases the life of the filament by avoiding rapid cooling from the solvent, which is known to cause rapid filament degradation. The solvent delay also prolongs the life of the electron multiplier by preventing large boluses of solvent ions during each injection. An appropriate solvent delay time can be determined for a new analytical method by injecting a small volume of solvent, such as 0.1 µL, in split mode with a 1:10–1:20 split ratio.

4.1.3 Split/Splitless Injection

Given the highly sensitive instruments currently produced, the split injection mode is recommended for general toxicology work as it will assist in protecting the filament from solvent, and will confer superior chromatography than splitless mode. Split injections also help prevent column contamination because most of the nonvolatile contaminants are taken to waste. This can decrease the frequency of column trimming and instrument downtime.

4.1.4 Carrier Gas Flow Rate

The flow rate of the carrier gas is primarily determined by the diameter and length of the analytical column. Column manufacturers usually recommend specific flow rates for their products in order to achieve a prescribed number of theoretical plates. One should know that backpressure will increase with temperature. Therefore, a constant pressure will not ensure a constant flow when a temperature gradient is used. Modern instruments utilize electronic flow controllers that allow user to select either constant pressure or constant flow. Typically, more consistent relative retention times (RRT) are obtained using constant flow rates, which can be helpful for analyte identification and quantitation. In some instances, a user may choose a column with a thick film surface in order to achieve the desired analyte selectivity. The increased analyte retention times associated with high film thickness may be overcome to some degree by increasing the carrier gas flow rate.

4.2 INTERNAL STANDARDS

An internal standard is a substance added in a constant amount to calibration standards and patient samples that allows one to calibrate a system by determining the ratio of the analyte to the internal standard. Analytes in patient samples are then quantified by comparing the analyte:internal standard ratio to the calibration curve. The internal standard method of quantification is more accurate than the external standard method because analyte losses during extraction and transfer are accounted for. One should choose an internal standard that would not be expected to be present in any patient sample. Chemicals, such as SKF-525A, pentafluorobenzene, aminochlorobenzophenone, and others are commonly used as internal standards for GC-MS analyses. An ideal internal standard should be similar to but distinct from the analyte of interest. The internal standard should also be amenable to the same preanalytical manipulations as the analyte, including: hydrolysis, derivatization, and extraction.

Whenever possible, isotope-labeled internal standards should be included in quantitative analyses. Most internal standards are labeled with deuterium (^{2}H) or carbon-13 (^{13}C) because these are stable isotopes. Deuterium is an isotope of hydrogen that contains a proton in its nucleus causing it to have a mass of 2.014 amu instead of 1.007 amu. Labeling a compound consists of replacing single or multiple hydrogen molecules with deuterium, which increases the mass of the compound but does not affect the compound's tertiary structure nor its hydrophobicity. Isotope labeling of a drug results in an internal standard that exhibits the same extraction properties and chromatographic properties as the drug analyte. Any issue that affects extraction efficiency during sample extraction or ion suppression during analysis will affect the analyte and the internal standard to the same degree. Therefore, best practice would dictate including an isotope-labeled internal standard for each analyte being quantified. This may not be feasible in analyses involving many analytes, but the user should choose internal standards for the most critical or most prevalent analytes. Some high resolution mass spectrometers are accurate to 0.0001 m/z. Methods involving high resolution instruments should utilize internal standards labeled with a minimum of 5 ^{2}H or ^{13}C atoms so that neither quantitation nor identification is affected.

4.3 SAMPLE PREPARATION

4.3.1 Analyte Extraction

Sample preparation is one of the most important steps in the analytical process because it largely determines the analytical precision and accuracy of the analytical method. Sample/analyte extraction is required to move drug analytes from an aqueous matrix into a solvent that may be injected into the GC-MS. Aqueous solutions should never be injected onto GC-MS due to the problems that include: (1) backflash due to high vapor expansion volumes, (2) poor chromatography due to condensation of water on the column, and (3) severe column contamination from salts and solutes in the sample [20]. The importance of clean extraction techniques cannot be overemphasized because the accumulation of sample contaminants can affect nearly every component of the instrument and ultimately patient results. Clean sample extraction techniques should be employed whenever possible. Differences in sample cleanup can be readily observed in the analysis of drug-free urine specimens. For example, a higher number of extraneous peaks are observed in Fig. 7.3 after a nonbuffered LLE technique compared to a reputable SPE technique.

4.3.2 Liquid–Liquid Extraction

LLE is one of the oldest and widely used extraction techniques. The principle of LLE involves transferring an analyte from an aqueous matrix into an extraction solvent that may be analyzed by GC-MS. LLE is often applied to a variety of matrices including blood, serum, urine, and gastric contents. Forensic laboratories have long relied on LLE for the extraction of additional matrices, such as bile and liver or kidney homogenates. Disadvantages of LLE include use of large solvent volumes, multiple extractions, and long extraction times.

The extraction solvent used for LLE should be immiscible in the aqueous matrix so that the two liquids can easily be separated. Analytes should be soluble in the extraction solvent, and should ideally exhibit high partition coefficients in the solvent. Common extraction solvents include dichloromethane, chloroform, 1-chlorobutane (*n*-butyl chloride), toluene, and hexane. Some protocols include sodium chloride in the mixture, which dissolves into the aqueous phase and helps to force drug analytes into the organic phase, a process known as "salting out." The concepts of acid-base chemistry apply to LLE in that nonionized analytes will readily transfer from the polar aqueous matrix into a nonpolar solvent. Therefore, the pH of the aqueous matrix should be adjusted two pH units higher than the pKa of a basic analyte, or two pH units lower than the pKa of an acidic analyte in order for 99% of the analyte to be uncharged. The use of an acidic or basic pH is highly recommended so that impurities, such as phospholipids and cholesterol esters are not extracted. During the extraction process, the matrix-solvent mixture should be shaken or inverted many times over a period of minutes to ensure that the two liquid phases receive ample contact with each other. Vigorous shaking may form emulsions, which should be cleared with centrifugation. Extraction solvents that are less dense than water may be removed using a handheld pipette, whereas more dense solvents can be evacuated from the bottom of a separatory funnel. Two or even three extractions may be required in order to remove the majority of the analyte from the matrix. Some protocols require back extraction of the analyte into a small volume of acidified methanol or other polar solvent in order to eliminate many of the neutral compounds that may interfere with GC-MS analysis.

4.3.3 Solid Phase Extraction

SPE involves the adsorption of drug analytes from the biological matrix onto a solid sorbent, and subsequent elution of the analytes from the sorbent into an organic solvent that may be injected for

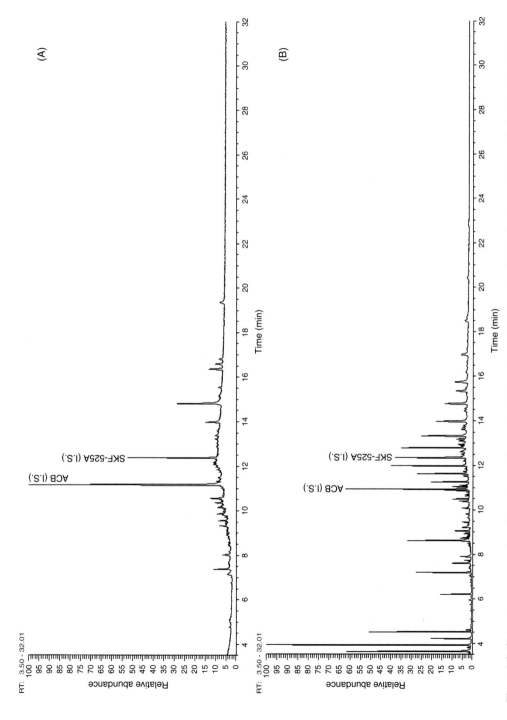

Figure 7.3 Chromatograms of a drug-free urine specimen extracted with a reputable SPE method (A) and a nonbuffered liquid–liquid method (B). Differences in extraction cleanliness are illustrated by significant differences in the numbers of extraneous peaks. The internal standards Aminochlorobenzophenone (ACB) and SKF-525A are indicated.

analysis. SPE techniques became vogue in the 1980s with the advent of several C18 cartridge-type columns. Many types of sorbents are produced for drug extraction including C2, C8, C18, ion exchange, and mixed mode sorbents composed of combinations of the previous [21]. Commercial SPE products are available as cartridges, tubes, and 96-well plates. Advantages of SPE include low solvent consumption, ease of use, efficient drug extraction, and automation capabilities. One disadvantage of SPE is the limitation to extraction of nonviscous aqueous samples. The extraction of whole blood, gastric contents, or tissue is not compatible with many SPE products.

For best results, many SPE sorbents are required to be conditioned with water or methanol prior to introduction of the biological matrix. After conditioning, the biological matrix is applied and, in the case of SPE columns, drawn through the column with a small amount of vacuum. Use of a vacuum manifold allows the extraction of one or more samples simultaneously, and several devices have been developed to automate this process. The sample application step is very important and requires the sample to move through the column slowly, so that drug analytes have ample time to bind to the sorbent. Extraction efficiency will be reduced if sample application is rushed. The next step involves washing the sorbent with water to remove unwanted buffers and contaminants. The sorbent may then be prepared for elution using an acidic or basic buffer. Finally, drugs are eluted from the column with an organic solvent, and captured in a glass vial or reaction tube. Many protocols specify that most of the solvent be evaporated in order to concentrate the analyte. However, cleaner extracts may be obtained after complete evaporation of the solvent and reconstitution of the extract with a pure solvent that is more appropriate for injection onto GC-MS.

Extraction procedures can vary greatly in their specificity, cleanliness, and extraction efficiency [21], so great care should be given to the choice of a sorbent and extraction procedure. Newer polymer-based sorbents composed of styrene-divinylbenzene cores tend to have increased surface areas, more consistent particle sizes, and the absence of amides. These characteristics reportedly result in reproducible packing and less ion suppression due to reduced protein and lipid retention. SPE methods that require low solvent volumes tend to require shorter amounts of time for extraction. There are several manufacturers that produce SPE sorbents for targeted and untargeted analyses. Users should pick several methods from reputable vendors and perform an in-laboratory comparison to determine which works best in their hands.

4.3.4 Supported Liquid Extraction (SLE)

SLE, also known as solid supported liquid extraction, is a highly efficient extraction technique. SLE is based on the principle of LLE, however the aqueous phase is supported by a solid media composed of synthetic materials or purified diatomaceous earth. Numerous products are commercially available as packed SPE-size cartridges or 96-well plates. The extraction process begins as an aqueous sample is poured onto the dry support media where it is adsorbed onto the surface and distributed via capillary action. Due to the porosity of the support media, the sample is dispersed into a very thin layer with a large surface area. An immiscible solvent is then poured on top and allowed to migrate through the media where it extracts the analytes from the aqueous phase. Advantages of SLE over LLE include low solvent consumption, no shaking or mixing, no emulsion formation, no centrifugation, and faster extraction times. Some products incorporate an acidic or basic buffer into the media that allows users to selectively target acidic or basic drugs.

4.3.5 Dispersive Liquid–Liquid Microextraction (DLLME)

DLLME is a miniaturized form of LLE that requires microliter volumes of solvent and short extraction times [22]. The technique involves a high-density extraction solvent, such as chloroform or dichloromethane and a dispersive solvent that is soluble with the extraction solvent and water. Rapid injection of the extraction solvent and dispersive solvent mixture into an aqueous sample results in a turbulent solution that forms a cloudy emulsion. Formation of the emulsion increases the surface area of the extraction solvent with the sample. As a result, the equilibrium state is quickly achieved and the analyte/s are extracted from the biological solution and concentrated into a small of volume organic solvent. DLLME, which is its primary advantage over other methods [23]. The emulsion is cleared with centrifugation and the extraction solvent is removed from the bottom of the vial.

The use of DLLME relies on the analyte distribution coefficient, which is the ratio of the analyte concentration in the extraction solvent versus the aqueous sample. The distribution coefficient should be above 500 in order to achieve adequate analyte extraction. Use of a buffer may be employed to ensure that analytes are not ionized in order to increase their distribution into the extraction solvent. The extraction solvent should be immiscible in water to ensure final phase separation during centrifugation. The purpose of the dispersive solvent is to reduce the interface tension between the extraction solvent and the aqueous sample. This ensures the formation of small droplet sizes and subsequent emulsion. Acetone, methanol, and acetonitrile are commonly used as disperser solvents. The advantages of this technique include short extraction times and low solvent consumption.

4.3.6 Solid Phase Microextraction
SPME is a technique in which analytes are adsorbed onto a sorbent-coated fiber placed into the sample vial. The fiber is then inserted into the injection port and the analytes are desorbed by heat. Additional details may be found in the previous section describing injection techniques.

4.3.7 Hydrolysis
During metabolism, many drugs are conjugated to glucuronic acid or sulfate respectively by UDP-glucuronosyltransferase and various sulfotransferases within liver hepatocytes. Conjugation increases the compound's water solubility resulting in the metabolites being easier to excrete through the bile and/or urine. Many laboratories that perform targeted drug confirmation choose to hydrolyze the glucuronide from such analytes in order to increase the concentration of unconjugated drug. This often results 10–20 times higher concentrations of free drug compared nonhydrolyzed samples as exhibited in (Fig. 7.4). The most commonly hydrolyzed analytes include the opiates, benzodiazepines, and cannabinoids. Protocols often involve either acid hydrolysis or enzymatic hydrolysis via bacterial, snail, or conch sources [24]. Enzymatic hydrolysis is often preferred over acid hydrolysis because the latter is known to decompose some benzodiazepines into common benzophenone forms, which makes unequivocal identification impossible. Interestingly, laboratories that perform testing for pain management clinics may not hydrolyze their samples, because detection of the conjugated metabolites ensures that the medication was consumed. It was determined that some patients may seek to adulterate their urine by scraping off a portion of a prescription tablet into the urine container during sample collection. This practice allows them to divert their medications while they appear to adhere to their prescription.

4.3.8 Derivatization
Derivatization is the process of chemically altering an analyte or analytes. Laboratorians often choose to derivatize particular analytes in order to improve their chromatography, thermal stability, or their identification. Derivatization for GC-MS typically involves silylation, alkylation, or acylation reactions. The reagents used for silylation react with active hydrogen, whereas the reagents for acylation

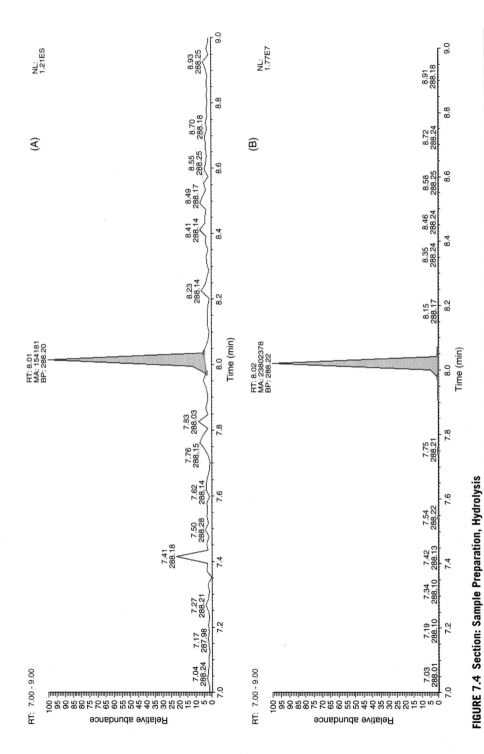

FIGURE 7.4 Section: Sample Preparation, Hydrolysis

Chromatograms of codeine illustrating a 150-fold increased peak area of free drug from a single specimen that was not hydrolyzed (A) and hydrolyzed (B). Peak areas of the nonhydrolyzed and hydrolyzed aliquots were 154,181 and 23,802,378 respectively.

react with polar functional groups. Reagents for alkylation react with active hydrogens on amines and acidic hydroxyl groups [25]. Derivatives exhibit differing rates of stability, and some may require prompt analysis due to degradation within 12–24 h. Therefore, one should not only evaluate the literature to determine the most appropriate derivatives to be used, but should also verify their stability during method validation protocols. Some reagents used for derivatization are highly reactive and can inactivate the column. The portion of the column nearest the injection port is affected first, and more frequent column trimming may be required to maintain consistent chromatography. It may be wise to designate specific instruments for the analysis of derivatized samples.

The cannabinoids are often derivatized in order to improve their chromatography and detection. The derivatization of the hydroxyl and carboxyl groups on THC, 11-OH-THC, and THC-COOH is required to increase their volatility is a prerequisite for the detection of THC-COOH by GC-MS. Silylation and methylation are the preferred means of derivatization for the cannabinoids [26], although acylation and methylation have also been used.

The thermal decomposition of diazepam, nordiazepam, oxazepam, and temazepam results in formation of the same aminobenzophenone making the identification of the original compound impossible [27]. Therefore, many choose to derivatize the benzodiazepines prior to GC-MS confirmation in order to increase the identification rates. Derivatization with trimethylsilane (TMS), tertiarybutyldimethylsilane (TDMS), or propionyl chloride forms adducts that result in specific fragments that allow identification of these compounds. Derivatization also improves the chromatography of benzodiazepines by reducing peak tailing.

The amphetamines are often derivatized prior to GC-MS confirmation in order to increase confidence during identification. EI of amphetamine results in only two ions 44 m/z and 91 m/z, whereas the EI spectra of methamphetamine only contains 58 m/z and 91 m/z. Many sympathomimetic amines and similar synthetic stimulants also exhibit these base peaks with few other ions. This can prevent confident identification especially when low concentrations are encountered. Derivatization with pentafluoropropionic anhydride (PFPA) or other agent results in at least three significant ions [28], which significantly increases confidence during their identification.

5 ANALYSIS

Prior to the advent of computerized mass spectral libraries, spectral matching was performed manually by printing the mass spectrum of a given peak and visually comparing it to those printed in reference books. While computerization has simplified the task of spectral matching, it has not eliminated the need for well-trained mass spectrometrists. Mass spectrometry-based tests are high-complexity assays that require specially trained individuals to operate and troubleshoot the instruments as well as to verify the accuracy of analyses.

5.1 QUALITATIVE SCREENING

Laboratories may perform qualitative screening and identification for a variety of reasons including: (1) the detection of drugs involved in toxic ingestions, (2) to ensure medication compliance, (3) identification of abnormal or unexpected drug metabolism, and (4) to identify patients that may be diverting their medications. Therefore, qualitative screening may involve methods for targeted as well as untargeted analyses. Instrument software has evolved to allow the automatic detection and identification of analytes [29]. However, these automated functions require users to establish peak area/height cutoffs in order to determine the peak spectra that will be compared to the onboard library. Criteria for matching retention times and mass spectra must also be established and critically validated to reduce the chance of misidentification [30]. While these automated functions can simplify analyses, it is still recommended that results from each analysis be evaluated by a trained user.

During emergency toxicology screening, medications suspected of being ingested or those known to be in the home may be related to the laboratorian. In addition to performing a general search, the user should also perform a targeted search for selected ions in the background milieu of full scan data. After data acquisition, the user should use the instrument software to display only ions consistent with the base peak of the suspected drug/s. This technique eliminates all ions except for those selected, and often aids in the identification of potent compounds that may be present in low concentrations.

Other techniques, such as background subtraction can aide in the identification of a drug when multiple peaks coelute. Background subtraction prior to and/or after the peak of interest can also increase the confidence in identification whenever extraneous ions are present, which often results from a bleeding column or septum.

5.2 SCREENING WITH SUBSEQUENT CONFIRMATION

Some chemists and toxicologists have elected to perform all drug screening with mass spectrometry instead of immunoassay. Those with GC-MS/MS instruments may screen for drugs using retention time indexes and m/z, and then confirm drug identity with fragmentation spectra acquired with the MS/MS feature. The primary advantage is decreased reagent cost, whereas disadvantages include increased preanalytical work and increased turn-around times. Screening and confirmation for all drug classes can be performed, but multiple injections may be required.

5.3 DRUG CONFIRMATION

GC-MS is often used to confirm the results of positive or negative immunoassay drug screens. Laboratories may perform qualitative and/or quantitative confirmation, depending upon the client's needs. The two are very similar in concept except that quantitative confirmation involves additional steps during validation in order to prove linearity and precision. Quantitative drug confirmation is most often preferred, even for urine, because concentration data can be used to help determine incidental use of OTC medications, discriminate between opioid and poppy seed ingestion, help identify new drug use, and indicate variances in drug metabolism that may put the patient at risk.

Both qualitative and quantitative methods require that the laboratory selectively determine the analytes to be targeted. It is not necessary to identify every drug and metabolite in each drug class. Rather, most seek to confirm only those parent drugs and metabolites that will provide the information required to complete the clinical picture. As an example in the class of opiates, the laboratory may seek to target 6-monoacetyl morphine because its detection offers conclusive evidence of heroin use. The laboratory may not wish to target dihydrocodeine because it is a metabolite of two different parent compounds, is not a parent drug itself, and its detection will not provide meaningful information. These decisions should be discussed with relevant clinical personnel. It should be recognized that small amounts of drug impurities are allowed in current pharmaceutical production practices that may cause confusion during result interpretation. For example, the United States Pharmacopeia (USP) allows up to 1.0% hydrocodone as impurity in oxycodone preparations [31]. Therefore, the laboratory should provide clinical consultation as needed.

5.4 ASSAY PARAMETERS

Amphetamines Confirmation
Sample Matrix: Urine
Analytes: Amphetamine, Methamphetamine, Methylenedioxyamphetamine (MDA), Methylenedioxymethamphetamine (MDMA)
Hydrolysis: None
Derivatization: Incubate with PFP and PFPA at 95°C for 15 min
Extraction: Solid phase extraction using UCT Clean Screen column

Instrument Parameters
Column: Restek Rtx-5MS, 15 m × 0.25 mm, 25 μm df
Detection Mode: Positive ionization, full scan 40–500 m/z
Electron Energy: 70 eV
Filament Start Delay: 4.3 min
Helium Velocity: 1 mL/min (constant flow)
Injection Mode: Direct injection with 1:10 Split
Injection Volume: 1 μL
Injector Flow Rate: 40 mL/min
Split Flow: 10 mL/min
Sweep Gas: 5 mL/min (constant septum purge)
Injection Port: 200°C
Transfer Line: 280°C
Ion Source: 200°C
Oven Temperature: Start at 50°C, hold for 1 min
Ramp from 50 to 170°C @ 10°C/min
Ramp from 170 to 280°C at 30°C/min
Hold at 280°C for 5 min
Total Analysis Time: 22 min

Cannabinoid Confirmation
Sample Matrix: Urine
Analytes: Carboxytetrahydroxannabinol (THC-COOH)
Hydrolysis: Incubate with 10 M potassium hydroxide at 60°C for 20 min
Extraction: Solid phase extraction using UCT Clean Screen THC column
Derivatization: Incubate with BSTFA with 1% TMCS at 70°C for 20 min

Instrument Parameters
Column: Restek Rtx-5MS, 15 m × 0.25 mm, 25 μm df
Detection Mode: Positive ionization, full scan 40–500 m/z

Electron Energy: 70 eV
Filament Start Delay: 8 min
Helium Velocity: 1 mL/min (constant flow)
Injection Mode: Direct injection with 1:10 Split
Injection Volume: 1 μL
Injector Flow Rate: 50 mL/min
Split Flow: 10 mL/min
Sweep Gas: 5 mL/min (constant septum purge)
Injection Port: 260°C
Transfer Line: 280°C
Ion Source: 200°C
Oven Temperature: Start at 190°C, hold for 1 min
                   Ramp from 190 to 270°C @ 40°C/min
                   Hold at 280°C for 5 min
Total Analysis Time: 8 min

Cocaine Metabolite Confirmation
Sample Matrix:
Analyte: Benzoylecgonine
Hydrolysis: Incubate with 1 M glacial acetic acid
Extraction: Solid phase extraction using Agilent Bond Elut Certify column
Derivatization: Incubate with PFP and PFPA at 75°C for 20 min

Instrument Parameters
Column: Restek Rtx-5MS, 15 m × 0.25 mm, 25 μm df
Detection Mode: Positive ionization, full scan 75–435 m/z
Electron Energy: 70 eV
Filament Start Delay: 4 min
Helium Velocity: 1 mL/min (constant flow)
Injection Mode: Direct injection with 1:10 Split
Injection Volume: 1 μL
Injector Flow Rate: 40 mL/min
Split Flow: 10 mL/min
Sweep Gas: 5 0.2 mL/min (constant septum purge)
Injection Port: 230°C
Transfer Line: 280°C
Ion Source: 200°C
Oven Temperature: Start at 90°C, hold for 1 min
                   Ramp from 90 to 280°C @ 30°/min
                   Hold at 280°C for 5 min
Total Analysis Time: 11 min

Opiates Confirmation
Sample Matrix: Urine
Analytes: Codeine, Morphine, Heroin, Hydrocodone, Hydromorphone, Oxycodone
Hydrolysis: Incubate with concentrated HCl at 120°C for 30 min

Extraction: Solid phase extraction using UCT Clean Screen column
Derivatization: Incubate with PFP and PFPA at 75°C for 25 min

Instrument Parameters
    Column: Restek Rtx-5MS, 30 m × 0.25 mm, 25 μm df
    Detection Mode: Positive ionization, full scan 50–650 m/z
    Electron Energy: 70 eV
    Filament Start Delay: 4 min
    Helium Velocity: 1 mL/min (constant flow)
    Injection Mode: Direct injection with 1:10 Split
    Injection Volume: 1 μL
    Injector Flow Rate: 40 mL/min
    Split Flow: 50 mL/min
    Sweep Gas: 5 mL/min (constant septum purge)
    Injection Port: 270°C
    Transfer Line: 280°C
    Ion Source: 200°C
    Oven Temperature: Start at 100°C, hold for 1 min
                        Ramp from 100 to 280°C @ 20°C/min
                        Hold at 280°C for 6 min
    Total Analysis Time: 16 min

General Toxicology Screening
    Sample Matrix: Urine
    Analytes: Hundreds
    Hydrolysis: None
    Derivatization: None
    Extraction: Solid phase extraction using UCT Clean Screen column

Instrument Parameters
    Column: Restek Rtx-5MS, 30 m × 0.25 mm, 25 μm df
    Detection Mode: Positive ionization, full scan 40–500 m/z
    Electron Energy: 70 eV
    Filament Start Delay: 3.3 min
    Helium Velocity: 1 mL/min (constant flow)
    Injection Mode: Direct injection with 1:10 Split
    Injection Volume: 1 μL
    Injector Flow Rate: 40 mL/min
    Split Flow: 10 mL/min
    Sweep Gas: 5 mL/min (constant septum purge)
    Injection Port: 250°C
    Transfer Line: 280°C
    Ion Source: 200°C
    Oven Temperature: Start at 60°C, hold for 1 min
                        Ramp from 60 to 280°C @ 20°C/min
                        Hold at 280°C for 20 min
    Total Analysis Time: 32 min

6 ASSAY VALIDATION

Assay validation is the process of testing and verifying the performance of an analytical method through a set of experiments designed to estimate the reliability and reproducibility of the method. The goal of the validation is to demonstrate that the method is capable of successful analytical performance for its intended purpose and to identify the method's limitations under normal operating conditions. One should recognize that GC-MS assays are considered laboratory-developed tests. As such, they require stringent validation protocols in order to ensure satisfactory results.

6.1 RECOMMENDATIONS AND GUIDELINES

Laboratories should ensure that their validation procedures comply with local and federal regulations as well as relevant accreditation agencies. At the time of this writing, the laboratory director holds the final approval for methods validated in each respective laboratory. Ultimately, the laboratory director is held responsible for all testing performed in his or her laboratory. Published guidelines addressing the validation of clinical and forensic tests are available and should be consulted. There are numerous publications by the Clinical and Laboratory Standards Institute (CLSI) with specific recommendations for evaluation of precision, linearity, bias, limits of detection, and other parameters for clinical assays that may be useful [32–35]. The CLSI has also published recommendations for clinical toxicology testing [36], and specifically for the GC-MS confirmation of drugs [37]. Methods for qualitative analyses may not require the same protocols as those for quantitative analyses, but should include experiments to prove analytical sensitivity, specificity, method comparison, and analyte carryover.

6.2 VALIDATION PARAMETERS

6.2.1 Sensitivity and Lower Level of Quantitation

The laboratory should evaluate the sensitivity of the assay for each analyte being targeted. Sensitivity is often determined as a detectable peak that is 3x greater than baseline noise. The sensitivity of an assay should always be determined using the same sample matrices used for analysis so that matrix effects are accounted for. The lower level of quantitation (LLOQ) is the lowest concentration the assay can quantify with reliability. Laboratories often set the LLOQ of an assay as the concentration at which the assay exhibits less than 20% coefficient of variability.

6.2.2 Specificity

The laboratory should test the method's ability to correctly differentiate and quantify targeted analytes in the presence of common matrix components. Specificity should be tested using at least 5–10 patient or donor specimens with each matrix. Common biological interferences, such as hemolysis, icterus, and lipemia should be evaluated as well as the presence of common medications in order to prove specificity and the absence of cross reactivity. Common medications include acetaminophen, ibuprofen, caffeine, nicotine, and others. Laboratories may consult several references including the *Red Book: Pharmacy's Fundamental Reference* [38] for lists of the top drugs prescribed in the United States.

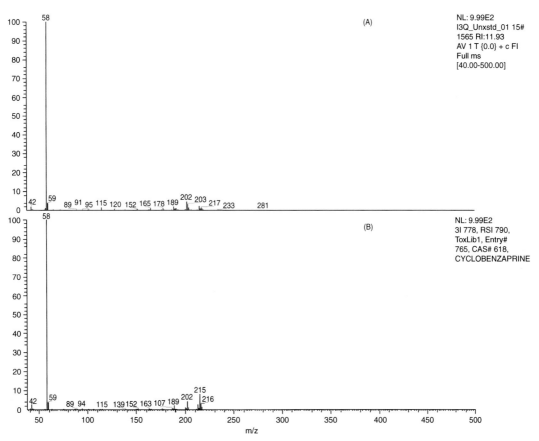

FIGURE 7.5 EI Spectra of Two Compounds That are Easily Misidentified by Their Mass Spectra

Amitriptyline (A) and cyclobenzaprine (B) both exhibit a 58 *m/z* base peak, and ions 189 *m/z*, 202 *m/z*, and 215 *m/z* in lesser abundance. Retention times and/or relative retention times should always be factored into drug identification because some compounds exhibit similar spectra.

Medical and research literature should also be evaluated for citations of interfering substances. For example, it is known that cyclobenzaprine may be misidentified as amitriptyline and vice versa due to their similar mass spectra [39] (Fig. 7.5). Given this, the importance of retention time matching cannot be overemphasized. Laboratories should also recognize the ability or inability to differentiate chiral compounds based on the equipment and methods employed. Common medications, such as pseudoephedrine cannot be differentiated from ephedrine unless chiral derivatization techniques or chiral chromatography is employed. Another chiral interference includes that of L-methamphetamine, an ingredient in some OTC nasal inhalers, and the illicit drug D-methamphetamine.

6.2.3 Precision
The precision of an assay should be demonstrated using a minimum of two concentrations of known material. Less precision is typically observed at low analyte concentrations. Intraday as well as interday

precision should be determined so that the laboratory can anticipate how the assay will perform. Intra-day precision is often determined using 5–10 replicates at multiple concentrations. Interday precision is similar, but should be determined over a minimum of 3 days.

6.2.4 Accuracy

The accuracy of a new method can be determined by analyzing known control materials, proficiency test materials, and by comparing patient samples to a current reference method. Some inaccuracy or bias should be anticipated in new quantitative methods, and this should be documented. Data obtained from a method comparison may be used to determine constant and proportional bias, respectively, by determining the slope and y-intercept during linear regression.

6.2.5 Linear Response

The linear response of a qualitative assay should be validated in order to determine the analytical measurement range. Analyte concentrations outside this range may be concentrated or diluted into the analytical measurement range so long as the procedure can be demonstrated with accuracy and repeatability. An assay's linear response is often demonstrated by measuring 4–6 concentrations of analyte in triplicate as illustrated by the calibration curve of benzoylecgonine in Fig. 7.6.

6.2.6 Method Comparison

Each new method should be established by comparing its performance with a previously established method. Qualitative analyses should involve a minimum of 20 patient specimens. However, the laboratory should make every effort to prove that the new method performs as well as or better than

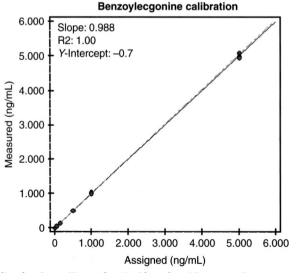

FIGURE 7.6 Standard Calibration Curve Illustrating the Linearity of Response Demonstrated with Triplicate Measurements of Benzoyecgonine from 15 to 5,000 ng/mL

a previously validated method. For example, the validation of an instrument for general unknown screening might involve the analysis of purchased controls, proficiency test specimens, spiked biological specimens, and 50 or more previously-analyzed patient samples that contain a variety of drugs and metabolites. One recent method validation included over 100 patient specimens with 387 drug "hits" and 98% concordance, in which the new instrument outperformed the previously-established method [8].

For quantitative methods, a minimum of 20 patient samples is considered acceptable as long as their concentrations span the analytical measurement range of the assay. Nineteen of the 20 values must be acceptable in order to achieve 95% agreement. The correlation of the two methods should be demonstrated by calculating the correlation coefficient (r) using least squares linear regression, or Deming regression.

6.2.7 Analyte Recovery
Analyte recovery is defined as the proportion of analyte that is recovered after sample extraction and handling. Recovery may be determined by the analysis of a neat sample compared to a patient sample spiked to the same concentration. Analyte recovery can be influenced by hydrolysis, extraction, derivatization, as well as matrix effects.

6.2.8 Analyte Carryover
The laboratory should determine the concentrations of analytes that cause analyte carryover to the next patient sample. It is not uncommon for high analyte concentrations to hang around in and "contaminate" the injection syringe and injection port. Patient samples with concentrations greater than the carryover cutoff should be followed by solvent blanks until the carryover has been eliminated. The software loaded on modern instruments often allows the user to set a concentration cutoff that will trigger a stop run signal or a blank solvent injection.

6.2.9 Analyte Stability
The stability of each analyte should be evaluated in every specimen matrix and specimen container specified by the laboratory. The stability of an analyte in a particular matrix or container should not be extrapolated to other matrices or containers. Validation experiments should evaluate the stability of analytes during sample collection and handling, short-term storage, and during long-term storage as defined by the laboratory. The validation conditions should reflect those likely to be encountered during normal use, such as benchtop storage at room temperature as well as short- and long-term refrigerated storage. The effects of freeze-thaw cycles should be evaluated if the laboratory intends to store frozen specimens. In addition, the laboratory should not neglect to test the stability of its working solutions made up in methanol or other solvents.

---

7 INSTRUMENT MAINTENANCE AND TROUBLESHOOTING
7.1 DAILY MAINTENANCE
Daily maintenance should include checks for proper system vacuum, ion source temperature, transfer line temperature, appropriate ion time, and a check for air leaks. Air leaks are easily checked by

powering on the filament and reviewing currently acquired mass spectra for m/z 18 (water). The instrument should also be tuned on a specific mass every day to ensure that its calibration has not drifted. This is often performed by introducing gaseous perfluorotributylamine (PFTBA) into the ion source. Many instruments include a small compressed vial of liquid PFTBA, which is released into the system electronically. PFTBA has a molecular weight of 671, but fragments significantly upon ionization with EI. Common tuning points include the predominant 69 m/z as well as other less common ions. Ideally several ions spanning the window of detection should be chosen as tuning points for the detector. Clinical toxicology laboratories performing common drug detection might include 69, 131, 219, 264, 414, and 501 when using PFTBA tune gas. Lastly, daily maintenance should include evaluation of the chromatographic and detection system. This is often performed by injecting a mixture of known analytes in a methanolic solution. This allows the user to ensure that chromatographic retention times and ion abundances are appropriate for the system.

7.2 WEEKLY MAINTENANCE

A full instrument tune should be performed every week or as recommended by the manufacturer. The full tune includes a mass tune, and other critical components of the analyzer. Depending upon the exact instrument, this can include automatic adjustment of the emission current, ion lens voltages, ion guide RF and voltage, quadrupole RF and voltage, gain control, and electron multiplier voltage. Weekly maintenance should also include a check for carrier gas leaks, which is easily performed with a handheld electronic leak detector. Currently-produced leak detectors are outfitted with an internal monitor, air pump, and a sipper probe. All connections and gas fittings should be checked for leaks.

7.3 MONTHLY MAINTENANCE

The injection port liner should be changed on a routine basis. Monthly or more frequent changes may be required depending upon the laboratory's sample volume. Injection port liners are usually made of borosilicate glass or other material. Samples are injected into the injection port liner where the components are volatilized prior to enter the column. Over time, the inner surface of the liner becomes coated with sample contaminants resulting in an active surface that impairs chromatography. The injection port liner should be replaced daily to monthly depending upon extraction cleanliness, use of derivatizing agents, and the number of samples analyzed. Prior to replace the injection port liner, the injection port should be cooled to room temperature in order to prevent thermal burns to the hands and fingers. The septum and associated O-ring seal should also be replaced any time the injection port liner is replaced. Most commercially available inlet liners have been silanized in order to deactivate the glass and minimize the adsorption of polar compounds onto the surface. However, untreated inlet liners as well as those treated with phosphoric or other acid may also be obtained, which can improve the analysis of acidic compounds. Inlet liners are engineered with a wide internal diameter so that the sample components will be quickly volatilized. This allows the sample to be loaded onto the head of the column in a narrow band for best chromatography. Most inlet liners range from 1.0 to 5.0 mm ID, and may be packed with 2–3 cm of glass or quartz wool, which prevents nonvolatile material from entering the column, and can aid in vaporization of the sample.

7.4 OTHER PERIODIC MAINTENANCE

7.4.1 Column Maintenance

With normal use, the stationary phase will eventually become saturated or inactivated with contaminates present after analyte extraction. Most reagents used for analyte derivatization are very reactive and can also inactivate the column. Obviously the portion of the column nearest the injection port is affected first. One way to remove the contamination is to trim a section of the column in order to restore the chromatography. Typically trimming between 12 and 36 in. from the column is all that is required and only minimum decreases in retention time are noted when using 30 m columns.

Increasing the temperature of the column is another method used to remove contaminants. Increasing the column temperature causes contaminates to move more quickly through the stationary phase. This is often performed by making multiple solvent injections using the programmed temperature gradient cycle. Alternately, the temperature of the column can be raised and held to within 20–30°C of the column's maximum stated temperature for several hours in a technique referred to as baking out. It should be recognized however, that these methods may only move the contaminants toward the detector and may not remove them entirely. For this reason, trimming the head of the column is recommended before increasing the column temperature.

Eventually, each column will require replacement due to the contamination associated with normal use. Currently produced columns should last for 6–12 months and should accommodate up to 500–1000 injections of biological samples. However, column life varies from laboratory to laboratory and is directly related to the use of derivatizing agents, extraction cleanliness, and the laboratory's annual sample volume. A 30 m column can often be trimmed several times to 25 m before retention times and relative retention times are significantly affected. Immediately after installation, a new column should be conditioned with heat for several hours while flushing with carrier gas to remove any water and volatiles from the manufacturing process. All column connections should also be checked for leaks after column replacement.

7.4.2 Ion Source and Ion Volume

High operating temperatures and voltage emission eventually oxidizes the surfaces of the ion source during normal operation. Once an ion source has become dirty, a user will often note poor ion peak shapes and/or high electron multiplier gain during tuning and/or decreased analyte response that is not resolved by cleaning the injection port liner or column. Cleaning the ion source typically involves brushing or scrubbing with alumina grit and subsequent rinsing with deionized water. The use of nitrile gloves is generally recommended during cleaning procedures. Users should take care to avoid contacting the internal detector components with any waxes, plastics, or detergents because these will increase background noise which results in reduced sensitivity.

7.4.3 Pump Oil

The pump oil should be changed every 6–12 months or as indicated by the manufacturer's recommendations. The oil not only serves to lubricate the pump, but also traps volatile components and contaminants from the samples injected. Therefore, gloves should be worn whenever handling the oil, which should be regarded as hazardous. Some pumps utilize a charcoal filter to ensure that contaminants are not emitted into the room air. The charcoal filter should be replaced whenever the oil is changed.

7.4.4 Filament

The filament may burn out in 6 months to 1 year depending on the use of the instrument. Some instruments are outfitted with two filaments, so that one can be switched on when the first filament fails. This decreases the time required to replace filaments and decreases instrument downtime. Filament life can be prolonged by programming a delayed filament ON signal after each sample injection. This delay prevents filament burning during the solvent front and is often referred to as a solvent delay. Relatively large amounts of solvent are introduced into the instrument upon injection. Solvents, such as methanol and dichloromethane differ significantly from the carrier gas, and promote rapid cooling of the filament. This rapid cooling has been found to degrade the filament. In addition, large amounts of burned solvent tend to foul the ion source and ion volume.

7.5 TROUBLESHOOTING

Troubleshooting typically involves the determination of chromatographic issues from detector issues. Air leaks are often the highest offenders, and can affect both chromatography and detection. Small air leaks allow the system to continue to function, but can cause poor retention time reproducibility, increased background noise, tailing peaks, increased column bleed, and eventually column failure. Large air leaks may cause the system to shut down if pressure control limits are exceeded, and can reduce filament life or cause the filament to be consumed. Another frequent problem involves dirty instrument components, which can result in a myriad of problems. During normal use, sample contaminants build up on the injection port liner, head of the column, and the ion source resulting in poor chromatography, ghost peaks, spurious ions, and decreased sensitivity.

Occasionally, preanalytical factors may influence the analytical results. These may include alterations in extraction efficiency, reagent contamination, analyte carryover, and others. At least one internal standard should be added to biological specimens prior to extraction. Some laboratories will include another internal standard in the solvent used to reconstitute sample residues after extraction. Dual internal standards, such as these can be used to monitor extraction efficiencies and to help identify the source of any problems. Contaminants can arise from any point in the process of sample extraction and analysis. Common sources of contamination involve water, air, analytical reagents, and others. Therefore, high quality mass spectrometry-grade solvents should be used whenever available. Contaminants common to lesser grade solvents can include: phthalates, propylene glycol, polyethers, and others [40].

Similarly, the water used for all GC-MS procedures should be high quality and exhibit 18-MΩ resistivity. This is often referred to as mass spectrometry-grade, ultrapure, Type I, or nanopure water. Ideally, the water has been filtered via reverse osmosis, ion exchange, carbon, ultraviolet, and 0.2 μm filters to remove contaminants. The use of lower quality water for making buffers or even washing glassware can introduce contaminants whose origins may be difficult to trace and eliminate. For example, at one institution that utilized an in-ground well as its sole water supply, polyunsaturated hydrocarbons were often detected in mass spectral analyses. In an attempt to identify the source of contamination, laboratory personnel repoured each organic solvent and remade each aqueous reagent until it was determined that the laboratory's water was contaminated. It was finally determined that, in an effort to avoid pulling the well pump to replace worn pump seals, maintenance personnel had been pouring cooking oil into the well for lubrication. To prevent recurrence, an ultrapure reverse osmosis water filtration system was acquired by the laboratory.

A filter should be installed on the gas lines carrying carrier gas to the GC-MS. Several combination-type filters are available that guarantee removal of 99.999% moisture, oxygen, hydrocarbons, and fuel gases. Color indicators are often incorporated into the filter materials so that activity can be verified by quick visual inspection.

It is rumored that every good mass spectrometrist withdraws Kimwipes from the bottom of the box, so that the wipe does not contact the plastic sleeve and contaminate his sample with plasticizers. While the authenticity of this rumor has not been verified, it does highlight the issue of contamination with plasticizers. Plasticizers are commonly use as mold-release agents during plastic manufacturing processes. These include various phthalates, benzoates, adipates, and others. Unfortunately, sample contamination with plasticizers can be difficult to avoid given the widespread use of plastics products in the laboratory. These include protective gloves, transfer pipettes, pipette tips, SPE columns, vial caps and septa, parafilm, and others. Many plasticizers are soluble in organic solvents and increased solubility may be noted with increased heat. Therefore, users should attempt to eliminate all solvent contact with plastic products during the sample preparation process. The use of PTFE-lined septa in injection ports and vial caps are recommended. The manufacturers of disposable pipette tips commonly use polypropylene with or without release agents. Users may need to evaluate multiple brands before finding tips that do not leach plasticizers into solvents. A significant amount of effort may be required to eliminate all plasticizers from analyses.

Some system components must be changed or cleaned regularly in order to prevent analytical problems. These include the septum, inlet liner, column, and sample syringe. While the septum is an inexpensive silicone seal employed to prevent the escape of carrier gas, it should not be overlooked when troubleshooting. Septa must endure dozens of punctures at high temperatures and continue to seal without affecting chromatography or ion spectra. High injection port temperatures will increase septum bleed as polysiloxanes within the septum are boiled away. Septum bleed can result in the inclusion of extraneous ions into mass spectra, which often include 281 m/z, 355 m/z, 429 m/z, and 503 m/z. Eventually, coring of the septum can occur from the syringe needle, and particles of septum can fall into the injection port. These particles can cause extraneous chromatographic peaks as solvent is injected onto them.

A high number of injections can cause the septum to leak as a hole forms from repeated punctures. Air contamination can reduce the life of columns with polar stationary phases. Air leaks are typically detected at the beginning of the day when checking the air/water ion ratios during daily startup. The source of an air leak may be determined by lightly spraying a compressed gas duster product at critical points, such as column connections. Many of the newer environment-friendly compressed duster products contain tetrafluoroethane or other hydrocarbon that promotes a distinct detector signal. Therefore, injection port liner and septum changes should be implemented at regular intervals.

As described previously, the analytical column is one of the most important components of the instrument system. Any alteration of the column can therefore affect the analyses and analytical results. Column bleed is a common problem the user faces. Most manufacturers recommend a specific temperature range for the best operation of their GC columns. Temperatures below the recommended range increase the viscosity of the phase and can affect chromatography by reducing analyte diffusion into the stationary phase. However, high temperatures cause the stationary phase to bleed out of the column. Increased bleeding is often noted from columns with thick stationary phases, so a thin film thickness of 0.18 or 0.2 μm is often preferred if possible. Use of temperatures higher than those recommended will cause excessive column bleed as well as reduce the life of the column. Significant column bleed often results in the inclusion of extraneous ions that can interfere with spectral matching. Ions commonly observed from column bleed include 207, 281, and 355 m/z.

The syringe is used to inject every analytical standard and biological extract into the instrument and may be easily contaminated. Needle washes that involve the aspiration and ejection of polar and nonpolar solvents are recommended. Ethyl acetate and dichloromethane are often recommended, but other solvents may be used. Performing 3–5 needle washes with a polar solvent and a nonpolar solvent both prior to and after each injection should be sufficient to remove all traces of analytes.

At some time, issues with electronics will begin to affect the computer system. Digital patient and QC data should be routinely backed up on secondary storage devices in order to maintain an ample amount of free hard drive space. Systems can begin to malfunction when available storage decreases to 20% or less. This may occur as incomplete data acquisition often noted as incomplete chromatograms. Instruments greater than 10 years old may begin to have issues with motherboards, communication boards, and operating systems. All electronic components are known to degrade with age and heat. Therefore, instruments should always be housed in cool environments.

8 PERSONNEL/TRAINING
8.1 BACKGROUND AND EXPERIENCE

The personnel best suited for mass spectrometry analyses are most often intelligent, internally driven, and motivated individuals who are dedicated to their job. These individuals are often not content with routine clinical chemistry-type analyses, but are intrigued by the more complex aspects of laboratory testing. Toxicology laboratories often employ an equal mix of medical technologists, biologists, and chemists. However, relevant state and local regulations should be consulted because specific qualifications may be required for patient testing. Due to their expanded education, individuals with multiple degrees often lend themselves well to mass spectrometry. The concepts of analyte extraction, ionization, and chromatography must be understood by those employed for mass spectrometry. Significantly more time and attention is required to use mass spectrometry than for automated analyzers within the clinical laboratory.

As compared to software used for routine clinical chemistry analyzers, the software used to acquire and analyze GC-MS data is often very powerful in nature and therefore quite intricate and extensive. As such, several months may be required for one to become proficient with the software, whereas many months may be required for one to master the software. These software systems not only allow users to specify the types of analyses to perform and the type of data to acquire, but also allow one to manipulate the data after acquisition.

Given the complexity of GC-MS software systems, it may be wise for the laboratory to purchase multiple instruments from the same manufacturer so that technologists do not have to learn multiple software systems. Laboratory directors seeking new instruments should always evaluate products from multiple manufacturers, but the use of a single manufacturer's software can greatly simplify laboratory operations. A manufacturer typically loads its proprietary software onto all the GC-MS platforms it manufactures. Variations of the same software might be used to operate a manufacturer's ion trap, quadrupole, and TOF analyzers. Therefore, use of a single vendor's software facilitates the development of expertise that can be applied to multiple instruments, and prevents having a single expert user for each instrument. Use of a single vendor's software also decreases the time required for personnel to become technically proficient whenever a new instrument is purchased. Purchasing a single manufacturer's instruments may not always be feasible, but it can reduce a number of complexities for new members of the laboratory.

As mentioned earlier, many months are usually required in order to become proficient with a GC-MS and its software. Years are often required for individuals to become proficient in GC-MS maintenance and troubleshooting. These individuals know how to determine the source of air leaks, spurious chromatographic peaks, evaluate spectral matches, and perform all maintenance. Expert users are not common and are therefore very valuable. Great efforts should be made to retain these individuals whenever possible.

REFERENCES

[1] Kyle PB, Spencer JL, Purser CM, Eddleman KC, Hume AS. Suspected pediatric ingestions: effectiveness of immunoassay screens vs. gas chromatography/mass spectrometry in the detection of drugs and chemicals. J Toxicol Clin Toxicol 2003;41(7):919–25.

[2] Dresen S, Ferreiros N, Gnann H, Zimmermann R, Weinmann W. Detection and identification of 700 drugs by multi-target screening with a 3200 Q TRAP LC-MS/MS system and library searching. Anal Bioanal Chem 2010;396(7):2425–34.

[3] Maurer HH, Weber AA, Wissenbach DK. Maurer/Wissenbach/Weber LC-MSn library of drugs, poisons, and their metabolites. 1st ed. Weinheim, Berlin: Wiley-VCG; 2014.

[4] Maurer HH, Pfleger K, Weber AA. Mass spectral library of drugs, poisons, pesticides, pollutants, and their metabolites. 1st ed. Weinheim, Berlin: Wiley-VCG; 2011.

[5] Gowadia HA, Settles GS. The natural sampling of airborne trace signals from explosives concealed upon the human body. J Forensic Sci 2001;46(6):1324–31.

[6] Peters FT, Drvarov O, Lottner S, Spellmeier A, Rieger K, Haefeli WE, Maurer HH. A systematic comparison of four different workup procedures for systematic toxicological analysis of urine samples using gas chromatography-mass spectrometry. Anal Bioanal Chem 2009;393(2):735–45.

[7] Gustavsson E, Andersson M, Stephanson N, Beck O. Validation of direct injection electrospray LC-MS/MS for confirmation of opiates in urine drug testing. J Mass Spectrom 2007;42(7):881–9.

[8] Kyle PB, Bhaijee F, Magee L, Booth D. A highly sensitivity quadrupole GC-MS for quantitative and qualitative toxicology. Mass spectrometry: applications to the clinical lab, Sixth Annual Conference. San Diego CA; 2014. p. 106, Poster 38a.

[9] Barnes AJ, Scheidweiler KB, Huestis MA. Quantification of 11-nor-9-carboxy-Δ9-tetrahydrocannabinol in human oral fluid by gas chromatography-tandem mass spectrometry. Ther Drug Monit 2014;36(2):225–33.

[10] Aebi B, Sturny-Jungo R, Bernhard W, Blanke R, Hirsch R. Quantitation using GC-TOF-MS: example of bromazepam. Forensic Sci Int 2002;128:84–9.

[11] Rockwood AL, Annesley TM, Sherman NE. Mass spectrometry. In: Burtis CA, Ashwood ER, Bruns DE, editors. Tietz textbook of clinical chemistry and molecular diagnostics. 5th ed. St Louis, USA: Elsevier Saunders; 2012. p. 329–53.

[12] Revelsky A, Samokhin AS, Virus ED, Rodchenkov GM, Revelsky IA. High sensitive analysis of steroids in doping control using gas chromatography/time-of-flight mass-spectrometry. Drug Testing Anal 2011;3(4):263–7.

[13] Barden D. TOF-MS takes on complex GC analyses. Chromatography techniques. Available from: http://www.chromatographytechniques.com/articles/2012/09/tof-ms-takes-complex-gc-analyses; 2015.

[14] Perry RH, Cooks RG, Noll RJ. Orbitrap mass spectrometry: instrumentation, ion motion, and applications. Mass Spectrom Rev 2008;27:661–99.

[15] Ingels AS, Neels H, Lambert WE, Stove CP. Determination of gamma-hydroxybutyric acid in biofluids using a one-step procedure with "in-vial" derivatization and headspace-trap gas chromatography-mass spectrometry. J Chromatogr A 2013;1296:84–92.

[16] Miekisch W, Fuchs P, Kamysek S, Neumann C, Schubert JK. Assessment of propofol concentrations in human breath and blood v yeans of HS-SPME-GC-MS. Clin Chim Acta 2008;395(1–2):32–7.

[17] Anzillotti L, Castrignano E, Strano Rossi S, Chiarotti M. Cannabinoids determination in oral fluid by SPME-GC/MS and UHPLC-MS/MS and its application on suspected drivers. Sci Justice 2014;54(6):421–6.

[18] Brown SD, Rhoodes DJ, Pritchard BJ. A validated SPME-GC-MS method for simultaneous quantification of club drugs in human urine. Forensic Sci Int 2007;171(2–3):142–50.

[19] Bergamaschi MM, Barnes A, Queiroz RHC, Hurd YL, Huestis MA. Impact of enzyme and alkaline hydrolysis on CBD in urine. Anal Bioanal Chem 2013;405(14):4679–89.

[20] Kuhn ER. Water injections in GC—How wet can you get? LCGC Asia Pacific 2002;5(3):30–2.

[21] Magiera S, Hejniak J, Baranaowski J. Comparison of different sorbent materials for solid-phase extraction of selected drugs in human urine analyzed by UHPLC-UV. J Chromatogr B 2014;958:22–8.

[22] Rezaee M, Assadi Y, Milani Hosseini MR, Aghaee E, Ahmadi F, Berijani S. Determination of organic compounds in water using dispersive liquid–liquid microextraction. J Chromatogr A 2006;1116:1–9.

[23] Rezaee M, Yamini Y, Faraji M. Evolution of dispersive liquid–liquid microextraction method. J Chromatogr A 2010;1217(16):2342–57.

[24] Fu S, Lewis J, Wang H, Keegan J, Dawson M. A novel reductive transformation of oxazepam to nordiazepam observed during enzymatic hydrolysis. J Anal Toxicol 2010;34:243–51.

[25] Grob R, Barry E. Modern practice of gas chromatography. Wiley Intersci 2004;817–8.

[26] Chericoni S, Battistini I, Dugheri S, Pacenti M, Giusiani M. Novel method for simultaneous aqueous in situ derivatization of THC and THC-COOH in human urine samples: validation and application to real samples. J Anal Toxicol 2011;35:193–8.

[27] Uddin MN, Samanidou VF, Papadoyannis IN. Bio-sample preparation and gas chromatographic determination of benzodiazepines—a review. J Chromatogr Sci 2013;51(7):587–98.

[28] Dobos A, Hidvegi E, Somogyi GP. Comparison of five derivatizing agents for the determination of amphetamine-type stimulants in human urine by extractive acylation and gas chromatography-mass spectrometry. J Anal Toxicol 2012;36:340–4.

[29] Choe S, Kim S, Choi H, Choi H, Chung H, Hwang B. Automated toxicological screening reports of modified Agilent MSD Chemstation combined with Microsoft Visual Basic application programs. Forensic Sci Int 2010;199(1–3):50–7.

[30] Marin SJ, Sawyer JC, He X, Johnson-Davis KL. Comparison of drug detection by three quadrupole time-of-flight mass spectrometry platforms. J Anal Toxicol 2014;39(2):89–95.

[31] Pesce A, West C, City KE, Strickland J. Interpretation of urine drug testing in pain patients. Pain Med 2012;13:868–95.

[32] CLSI. Evaluation of the linearity of quantitative measurement procedures: a statistical approach; approved guideline. CLSI document EP06-A. Wayne, PA: Clinical and Laboratory Standards Institute; 2003.

[33] CLSI. Evaluation of detection capability for clinical laboratory measurement procedures; approved guideline. CLSI docuemtn EP17-A2. 2nd ed. Wayne, PA: Clinical and Laboratory Standards Institute; 2012.

[34] CLSI. User verification of precision and estimation of bias; approved guideline. CLSI document EP15-A3. 3rd ed. Wayne, PA: Clinical and Laboratory Standards Institute; 2014.

[35] CLSI. Evaluation of precision of quantitative measurement procedures; approved guideline. CLSI document EP05-A3. 3rd ed. Wayne, PA: Clinical and Laboratory Standards Institute; 2014.

[36] CLSI. Toxicology and drug testing in the clinical laboratory: approved guideline. CLSI document C52-A3. 3rd ed. Wayne, PA: Clinical and Laboratory Standards Institute; 2015.

[37] CLSI. Gas chromatography/mass spectrometry confirmation of drugs; approved guideline. CLSI document C43-A2. 2nd ed. Wayne, PA: Clinical and Laboratory Standards Institute; 2010.

[38] Red book: pharmacy's fundamental reference, 2010 Edition. Thomson Reuters, Physicians' Desk Reference Inc. Montvale, NJ.

[39] Wong ECC, Koenig J, Turk J. Potential interference of cyclobenzaprine and norcyclobenzaprine with HPLC measurement of amitriptyline and nortriptyline: resolution by GC-MS analysis. J Anal Toxicol 1995;19(4):218–24.

[40] Keller BO, et al. Interferences and contaminants encountered in modern mass spectrometry. Anal Chim Acta 2008;627:71–81.

THERAPEUTIC DRUG MONITORING USING MASS SPECTROMETRY

8

P.J. Jannetto

Mayo Clinic, Department of Laboratory Medicine and Pathology, Toxicology and Drug Monitoring Laboratory, Metals Laboratory, Rochester, MN, United States

1 BACKGROUND/INTRODUCTION

Therapeutic drug monitoring (TDM), the quantitative measurement of medications in whole blood, serum, or plasma, plays a vital part in the clinical management of a patient. Drug concentrations in combination with other physical signs, symptoms, or laboratory measurements are often used to assess and adjust medication dosages to optimize the therapeutic effectiveness of a drug while minimizing adverse side effects. TDM is typically beneficial for medications that have narrow, well-defined therapeutic ranges that correlate with efficacy or toxicity. Traditionally, immunoassays have been commonly used by laboratories to measure drug concentrations. However, mass spectrometry (MS) is slowly becoming a key technology for laboratories to use for TDM due to its superior sensitivity and specificity. While MS has several advantages for TDM, laboratories need to fully comprehend the unique challenges and limitations of implementing this technology. This chapter will review the clinical, technical, and strategic considerations involved in the implementation of MS to perform TDM with a variety of clinical examples (tacrolimus, levetiracetam, busulfan, and methotrexate) from the author's laboratory.

2 MS CONSIDERATIONS FOR TDM

2.1 CLINICAL NEED/UTILITY

The decision to bring up any test in the laboratory begins with understanding the clinical necessity and requirements for that analyte. With TDM tests, it is recommended that the laboratory consult with and involve the physician specialty groups and/or pharmacists that would typically order the new test/analyte. In addition, the laboratory should conduct a thorough literature search and review the published clinical trials to determine the role and importance of TDM for that drug. It is important that the laboratory only develop tests for medications that have clinical value. Additionally, the laboratory will have to find out the answers to a number of other key questions during its conversations with the ordering physicians including: what is the expected/required turn-around-time (TAT) for the new TDM test to provide timely and appropriate patient management? For example, does the physician need the test result STAT (within 1 h of collection/receipt in the laboratory) or can it be batched (analyzed daily, weekly, etc.)? It is also important to get an idea about the anticipated test volumes as

Mass Spectrometry for the Clinical Laboratory. http://dx.doi.org/10.1016/B978-0-12-800871-3.00008-0

this will ultimately affect the methodology/technology that the laboratory will use to get the required throughput. The laboratory should also investigate other published methods used to quantify the drug and check to see if they have the capability and capacity on existing technology. Fig. 8.1 lists some of the therapeutic medications that are commonly monitored using MS and if a Food and Drug Administration (FDA)-cleared immunoassay is available.

The first TDM tests brought up using MS in the author's laboratory was for the immunosuppressive agents [i.e., tacrolimus, cyclosporin A (CsA), and sirolimus] for which TDM was well established [1]. Back in 2000, the author's laboratory was already using commercially available, FDA-approved immunoassays for tacrolimus and CsA. While immunoassay's can be automated and have good throughput, a major limitation of these assays is their specificity for the intended drug. At the time, the local transplant physicians were already concerned about their higher rejection rate in renal transplant patients receiving CsA. The physicians attributed this problem to the limitations of the existing immunoassay which had been shown to have varying degrees of cross-reactivity with the active parent drug and many of its inactive metabolites leading to varying degrees of bias and overestimation of the parent CsA concentration [2]. In addition, the transplant physicians wanted to start using and monitoring Sirolimus, a newly approved immunosuppressive agent, for the prophylaxis of organ rejection in renal transplant patients. However, no FDA-approved sirolimus immunoassay was available at that time and the physicians required same day TAT for newly transplanted patients so the dosage could be adjusted to maximize efficacy (prevent organ rejection) and minimize toxicity. With no commercially available immunoassay, the laboratory looked at chromatography based methods including: high performance liquid chromatography (HPLC), and liquid chromatography tandem mass spectrometry (LC-MS/MS) methods which had been used in some of the clinical trials. LC-MS/MS offered several advantages over HPLC with superior sensitivity and specificity which is what the transplant physicians required and specifically requested [3].

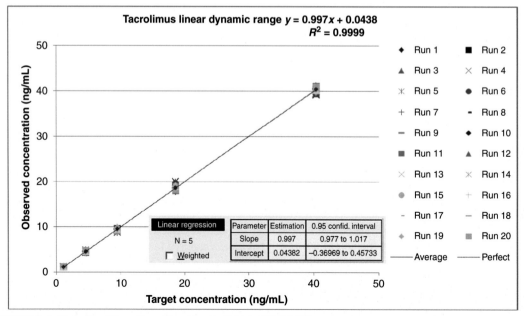

FIGURE 8.1 Linearity of Tacrolimus Over Twenty Runs by LC-MS/MS

Decisions to bring up other clinically useful TDM applications using LC-MS/MS were also examined. For example, the anticonvulsants (i.e., levetiracetam, lamotrigine, and lacosamide), and antineoplastic agents (i.e., busulfan and methotrexate) represented other applications with proven clinical utility that required same day TAT by the local physicians. The neurologists at our seizure clinic commonly treated complex seizure patients who were refractory to multiple medications. These patients came from around the world to the seizure clinic for specialty care. For this patient population, the neurologists wanted to get anticonvulsant medication levels drawn in the morning (trough values), so when the patient's arrived at their afternoon appointment the doctors already had their TDM results and could adjust the medication dosages if needed. In addition, the hematology oncologists demanded a quick TAT for patients undergoing bone marrow ablation with busulfan prior to hematopoietic stem cell transplant. With variable pharmacokinetics, dose adjustments of busulfan is recommended and guided by pharmacokinetic evaluation [4,5]. Since there were no commercially available immunoassays at the time, the physicians wanted to do these tests in-house. As a result, the laboratory had to look at other technologies (i.e., MS) to meet their clinical need. However, it should be noted that immunoassays for some of these anticonvulsant and antineoplastic medications are now available.

2.2 TECHNICAL CONSIDERATIONS

Other technical considerations also need to be examined when looking at various methodologies for TDM testing. As mentioned earlier, the laboratory first should rule out the use of existing technology since new instrumentation requires additional training, expertise, and has a steeper learning curve which is especially true for highly complex equipment like LC-MS/MS. These platforms require knowledgeable laboratory personnel who can develop, validate, operate, and perform basic trouble-shooting/repair for laboratory-developed tests (LDTs) on these analyzers. When considering MS for the clinical laboratory, it is also important to contemplate the different preventative maintenance and service contracts offered by MS vendors. Laboratories should ask the following questions: where is the location of the nearest service engineer? what is your maximum onsite response time (i.e., the time it takes a vendor engineer to arrive on site once a service call is placed)?; and what are the days/hours when remote technical assistance (telephone support) is available for troubleshooting? Unfortunately, the best service contracts offered by MS vendors typically allow up to 3 working days to respond to a service call and telephone support is usually only available on first or second shift depending on the location of their call center. As a result, laboratories need to use this service/support information along with the critical nature of the TDM tests they plan to analyze to determine what level of redundancy is required to support their institution. If you only have one MS system and it goes down, what other options do you have to get those TDM results? If you can send out those tests, what are the associated costs and TAT of the reference laboratory and will that meet your physician's needs?

Other technical considerations include what specific compound(s) and what concentrations need to be measured. For some medications, the parent compound is the active compound and those concentrations correlate with efficacy or toxicity (i.e., tacrolimus) [6]. In other cases, the parent drug is inactive and the metabolite is active so the metabolite needs to be measured (i.e., teriflunomide which is the metabolite of leflunomide) [7]. In addition, immunoassays may sometimes be contraindicated in certain clinical situations like the monitoring of methotrexate concentrations in the first 48 h following administration of Voraxaze (glucarpidase), which is indicated for the treatment of toxic plasma methotrexate concentrations [8]. If methotrexate measurements are required, a chromatographic method must be utilized to prevent the inactive metabolite of methotrexate that results from the treatment with glucarpidase causing an erroneous immunoassay measurement that overestimates the methotrexate concentration [9]. Furthermore, the laboratory needs to determine the lowest concentration they need to measure and if there are any endogenous compounds that could interfere with that measurement. For TDM tests, this is not usually a challenge like in other disciplines (i.e., endocrinology). The laboratory will also have to identify the target concentrations (i.e., reference ranges) they need to measure along with the full range of concentrations found in patient samples. Together this information can be used to help the laboratories decide how sensitive of a MS it will need. The MS vendors typically offer several tiers of MS with higher end versions having greater sensitivity, mass accuracy, dynamic measuring ranges, and price tags. Laboratories need to consider all the applications they plan to measure on the MS to appropriately determine which MS analyzer will best meet their needs.

2.3 PHYSICAL CONSIDERATIONS

Since laboratory's rarely have unused space, the laboratory needs to consider a number of physical requirements beginning with the area required for a MS analyzer. The laboratory not only has to determine the actual footprint (area around each side of the MS analyzer), but also the additional space needed for auxiliary equipment (i.e., HPLC, gas tanks or generators, etc.), and the countertop/bench space for technologists and/or computers that control the equipment and link to the laboratory information system (LIS). MS also have very specific electrical requirements (i.e., 220 V) and need a specific type and number of outlets. The MS and peripheral equipment will also emit a lot of heat (BTU/hour) so the laboratory's heating, ventilation, and air conditioning system has to be able to handle/dissipate that heat. Noise coming from the rough pumps for the MS can also be an issue, so noise dampening cabinets or measures also needed to be considered. In addition, if HPLC is used as the front end, the liquid waste has to either be collected or drains installed nearby (depending on the solvents used and local municipality regulations that dictate what can go down the drain). Internet jacks are also required not only for the LIS, but they can be used by the MS vendor for remote diagnostics which is highly recommended. Lastly, the laboratory shouldn't forget about supply space for both the MS and TDM assays. You will likely have consumable parts/supplies that need to be stored at room temperature, refrigerated, and/or frozen conditions. Common solvents used in TDM assays or extractions include methanol and acetonitrile, so a flammable cabinet is required. Depending on the type of sample extraction used, a ventilated hood/bench may also be required. In the end, you want to make sure to account for all these items along with the space to keep commonly replaced spare parts (i.e., cones, capillaries, etc.) on hand for the MS.

2.4 FINANCIAL CONSIDERATIONS

The fiduciary responsible of laboratories is only increasing as healthcare reimbursement continues to decline. Laboratories need to evaluate the financial aspects for any new equipment and look at the cost per test (CPT), net present value, and return on investment (ROI). For TDM tests, immunoassays typically range in cost from $5–15 per test depending on the laboratories volumes and institutional buying power/group. Back in 2000, my laboratory was paying $12/test for tacrolimus and $15/test for CsA immunoassays with yearly volumes of 20,000 and 13,000, respectively. Sirolimus was being sent out to a reference laboratory at a cost of $30/test and a yearly volume of 3,500. The combined yearly cost to perform testing for these three immunosuppressant's was $540 K. The cost to perform the simultaneous analysis of all three immunosuppressant's using LC-MS/MS was only $5/test for total yearly cost of \$182.5 K and could use existing personnel. The financial savings to switch to LC-MS/MS was \$357.5 K/year which lead to a favorable ROI. In the end, a detailed financial analysis should be done for every major equipment purchase.

2.5 STRATEGIC CONSIDERATIONS

When considering MS for TDM, it is also good to look at how it can place your laboratory for the future. Each year, the Food and Drug Administration (FDA) approves dozens of new medications. Some of these medications may require TDM, but are unlikely to have companion diagnostics or FDA-approved immunoassays at the time of approval. As a result, if a laboratory wants to measure a new drug, a LDT using MS or another chromatographic technique has to be considered. It is in this context that laboratories using MS have a strategic advantage since they have the ability to measure newer medications (i.e., third generation anticonvulsants like lacosamide and ezogabine) in a cost effective and timely manner (Table 8.1).

Table 8.1 Medications that have Clinical Value for Therapeutic Drug Monitoring and are Commonly Measured Using Mass Spectrometry

Drug Class	Drugs Measured	FDA Cleared Immunoassay Available
Antiarrhythmic agents	Amiodarone/Desethylamiodarone	No
	Flecainide	No
	Mexiletine	No
Anticonvulsants	Ezogabine/Retigabine	No
	Lacosamide	No
	Lamotrigine	Yes
	Levetiracetam	Yes
	Rufinamide	No
Antidepressants	Amitriptyline/Nortriptyline	No
	Imipramine/Desipramine	No
	Fluoxetine/Norfluoxetine	No
	Paroxetine	No
Antifungals	Posaconazole	No
	Voriconazole	No
Antineoplastics	Busulfan	No
Immunosuppressants	Cyclosporin A	Yes
	Everolimus	Yes
	Leflunomide metabolite (Teriflunomide)	No
	Mycophenolic acid	Yes
	Sirolimus	Yes
	Tacrolimus	Yes

3 ASSAY DESIGN/TEST LIFE CYCLE CONSIDERATIONS

3.1 PATENTS

Since MS tests are primarily LDTs, the laboratory at the very beginning of any new TDM assay should perform a right to use (RTU) patent search. The purpose of this search is to identify if anyone already has a patent on any of the steps in the proposed sample extraction and/or measurement. The laboratory doesn't want to waste time developing and validating an assay only to find out when it goes live with the test that another laboratory already holds the patent and either royalties/licensing fees may have to be paid or the testing stopped all together.

3.2 REGULATORY ASSESSMENT

At the beginning of each new TDM test developed using MS, our laboratory also does a regulatory assessment. As part of this assessment, the test is appropriately classified (i.e., LDT). In addition, our laboratory is New York State (NYS) certified, so the NYS test category is confirmed (i.e., therapeutic substance monitoring/quantitative toxicology). Both of these determinations are used to determine the appropriate documentation/paperwork that must be completed and submitted for approval to the institution and NYS.

3.3 BUSINESS MANAGEMENT

If your laboratory also has an outreach program/reference business, it is wise to do a quick survey of the market. How many other laboratories can offer the same TDM test? Would this new TDM test give you a unique competitive edge? If other laboratories do offer the test, what methodology is used? What is the price and TAT of the other commercial/reference laboratories? In the end, if the laboratory's primary intention is to commercialize a test it needs to determine how big the market is and who else offers the test. On the other hand, some laboratories solely develop TDM tests to control costs (minimize high send-out costs) and improve patient care (TAT) at their local institution. In the previous example, immunosuppressant's were originally switched to LC-MS/MS to improve patient care (improved specificity and TAT) and reduce send-out costs for the author's institution/physician's.

4 SAMPLE PREPARATION CONSIDERATIONS

4.1 SAMPLE TYPE

For TDM, one of the first key steps is to determine the appropriate test matrix (i.e., whole blood, serum, plasma) that should be used for measurement. In the case of the immunosuppressant's, the drug primarily resides inside the cellular components of blood, especially erythrocytes, so whole blood is the preferred sample type [10]. For other medications like the anticonvulsants, serum levels are the preferred sample type. However, the laboratory must also determine if serum gel separator tubes (gold top) or if plain serum tubes without gel (red top) must be used. Previous studies have already shown that some drugs can't be collected and/or stored in tubes with separator gels since the drug (i.e., phenytoin, antidepressants, or benzodiazepines) can bind to the gel and lead to artificially lower results [11,12].

Once the sample type is identified, the laboratory then needs to determine the appropriate sample preparation [i.e., protein crash, solid–phase extraction (SPE), liquid–liquid extraction (LLE)]. Excluding capital equipment costs and labor, the sample preparation time/supplies tends to be the next highest contributor to the cost/test. The most important goal of sample preparation is to clean up, potentially concentrate, and exchange the original sample matrix with a solvent/buffer that is going to be compatible with the MS assay. In general, laboratories generally attempt to minimize sample preparation steps which are appropriate to reduce labor and supply costs while enhancing the throughput of samples. However, laboratories must take care to evaluate the effects of the various sample preparation methods on the quality (accuracy) of their results, since the matrix effects can lead to ion suppression or enhancement of the signal. A rapid and simple approach that is often used for TDM assays is protein precipitation. In this approach, an organic solvent is typically added to the matrix to denature the proteins and samples are either filtered or centrifuged to remove the precipitation from the sample. This rapid and cost effective technique is used for levetiracetam in my laboratory where the serum is diluted with acetonitrile containing the internal standard. The protein precipitate is centrifuged and a portion of the supernatant diluted with mobile phase for detection by LC-MS/MS. LLE is another approach that can also be used with serum, plasma, and whole blood. In this case, an immiscible organic solvent is added to the sample and it forms a separate layer. With vortex mixing, the target analyte migrates from the original sample matrix to the organic layer. The organic layer is then further manipulated and used for the analysis. Another approach to use is SPE. This method uses a chromatographic media packed into a column or plate. The chemistry of the packing material is selected to target (bind) the analyte of interest. Briefly, the sample matrix is applied and allowed to flow through the sorbent. The retained analyte is then washed and eventually eluted with solvents and collected into a clean tube/plate for analysis by MS. This approach will clean up and concentrate the sample for analysis, but typically requires more time, labor, and expense. In the end, the laboratory should investigate the various sample preparation options to determine which meets their need for a particular analyte.

Lastly, it should be stated that the choice of internal standard is extremely important for TDM applications. The selection of an appropriate internal standard is required to compensate for sample to sample differences in extraction efficiency during sample preparation and ionization efficiency for LC-MS/MS [13]. For most TDM applications, a deuterated (^{2}H) or carbon (^{13}C) labeled internal standard is commercially available. However, for some assays another analog or drug may be used (i.e., ascomycin is the internal standard used for tacrolimus). The ideal internal standard behaves identical to the drug of interest during the sample preparation, chromatographic separation, and ionization/detection by MS. It is highly recommended that during method development, the laboratory look at the effects that different internal standards have on the results obtained by LC-MS/MS [14]. For most TDM applications in my laboratory, the final concentration of the internal standard used in the assay is around the 5–25% of the analytical measuring range (AMR) to minimize cost and effects due to incomplete labeling (i.e., the ascomycin concentration is 2.5 ng/mL in the tacrolimus LC-MS/MS assay where the AMR is 1–40 ng/mL).

4.2 MANUAL VERSUS AUTOMATION

When considering sample preparation methods, the volume of testing must also be taken into account along with the required TAT. The laboratory will have to determine if they are able to manually perform the protein precipitation, LLE, or SPE or if the volumes/labor time require more automation friendly format. As volumes increase, assay formats can be switched from individual tubes to 96- or

384-well plates. In addition, robotic liquid handlers (RLH) can be used to automate sample preparation for LC-MS/MS applications to improve efficiency, quality, and provide cost-savings [15]. In our laboratory, we use commercially available RLH to prepare 96-well plates for the analysis because of our high test volumes (i.e., immunosuppressant test volumes of > 500 samples/day). The automation allows the technologists to focus on MS review and the timely release of results while minimizing non-value added tasks and chances to make an error (i.e., mix-up pipetting a patient sample since the RLH reads the individual barcodes and tracks the sample identification to the plate well).

5 VALIDATION CONSIDERATIONS

Written guidelines, such as those published by the Clinical and Laboratory Standards Institute (CLSI), can provide guidance to laboratories. For example, CLSI released the C62-A guideline in October 2014 for LC-MS methods to provide direction for laboratories to reduce inter-laboratory variance and the evaluation of interferences, assay performance, another characteristics of clinical assays characterized during the verification of MS methods [16]. Some of the key studies required by the Clinical Laboratory Improvement Act (CLIA) are listed later.

5.1 IMPRECISION

It is important to determine the imprecision of a method at key concentrations (medical decision points). In the case of TDM assays, the laboratory will typically verify the imprecision at a subtherapeutic concentration, therapeutic concentration, and toxic concentration. The laboratory usually chooses these concentrations based on the proposed reference range from the literature and targets these as future quality control (QC) concentrations. Both within run and between run imprecision experiments are conducted following CLIA guidelines. Table 8.2 shows the imprecision data (within and between run)

Table 8.2 Imprecision Data (Within Run and Between Run) for Tacrolimus by LC-MS/MS			
Within Run Imprecision for Tacrolimus			
	$N = 20$	$N = 20$	$N = 20$
Expected (ng/mL)	5	15	30
Mean	4.20	14.95	30.60
SD	0.13	0.54	0.68
% CV	3.2%	3.6%	2.2%
Between Run (20 days) Imprecision for Tacrolimus			
	$N = 40$	$N = 40$	$N = 40$
Expected (ng/mL)	5	15	30
Mean	5.46	15.67	31.59
SD	0.28	0.77	1.81
% CV	3.1%	2.1%	2.7%

for tacrolimus by LC-MS/MS. The acceptance criteria in our laboratory is that the coefficient of variation must be $< 10\%$ for TDM assays.

5.2 LINEARITY OR REPORTABLE RANGE

Based on the information gathered when looking into the clinical utility of the analyte, the laboratory should have already identified the range of concentrations found in the human matrix being analyzed. The goal is to design the assay so you can measure patient samples in that concentration range while trying to minimize dilutions of high samples. In this experiment, the laboratory will define the upper and lower end of the reportable range. For TDM tests, the laboratory usually has to spike drug free patient pools at a minimum of five levels using commercially available standards, which are then calculated off an independently prepared calibration curve. It is a good idea to prepare multiple levels of samples (spiked drug free matrix) at levels below, within, and beyond the proposed linear dynamic range. Fig. 8.1 shows the linearity of tacrolimus over 20 runs by LC-MS/MS. The acceptance criteria for our laboratory is that the concentrations of the spiked standards must be within 10% of the expected value and the slope $= 1.0 \pm 0.1$, 95% confidence interval of the intercept should span 0, and the $r^2 > 0.98$.

5.3 ACCURACY

With any method validation, a comparison of methods experiment is required to estimate the inaccuracy or systematic error. TDM tests in our laboratory are typically verified using a combination of patient comparison samples with a reference laboratory and proficiency testing (PT) samples or commercial standards (certified reference material), if available. Fig. 8.2 shows the method comparison ($n = 69$) of the busulfan by LC-MS/MS to GC-MS (reference method) along with the difference plot (Bland-Altman plot). The results were compared using a standard linear regression. The acceptance criteria were that the mean difference between the results should be $< 10\%$ with no individual value $> 20\%$ and the slope $= 1.0 \pm 0.1$, 95% confidence interval of the intercept should span 0, and the $r^2 > 0.98$.

5.4 REFERENCE INTERVALS

For TDM tests, true reference range studies would be challenging for most laboratories to enroll patients taking the desired drug, collect samples at the appropriate time (peak/trough), and correlate it with clinical outcome (efficacy/toxicity). As a result, published reference ranges from clinical trials, clinical practice guidelines, or other sources are often cited. However, it is important to review how the published reference ranges were established and look at the methodology used, patient population, etc., to see how it compares to the laboratory's patient population. Where available, the laboratory should still verify the reference interval with local patient samples where clinical history can be obtained. Alternately, some laboratories have used the data collected from the reference laboratory in the method comparison study to calculate or adjust their reference range based on the mathematical relationship determined between the two analytical methods. This approach is dangerous since all MS assays aren't standardized and large variations can be seen. In the case of tacrolimus, studies have shown how independently calibrated LC-MS/MS assays without traceability to an accepted reference method or standard reference material could generate significantly different results [17].

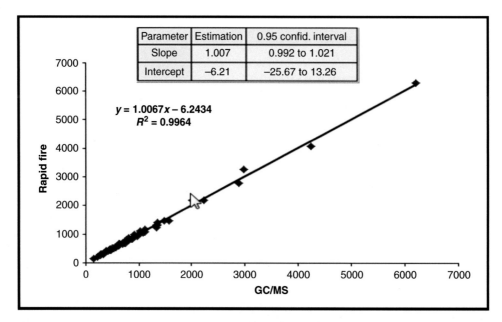

Parameter	Estimation	0.95 confid. interval
Slope	1.007	0.992 to 1.021
Intercept	−6.21	−25.67 to 13.26

$y = 1.0067x - 6.2434$
$R^2 = 0.9964$

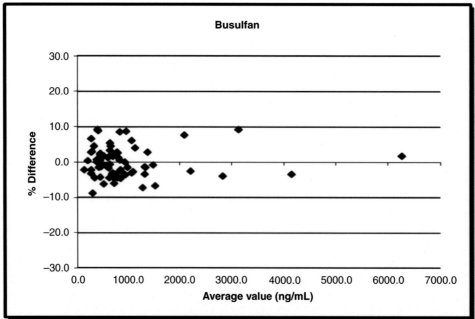

FIGURE 8.2 Method Comparison ($n = 69$) of Busulfan LC-MS/MS (RapidFire) Assay Versus GC-MS (Reference Assay) and the Difference Plot (Bland-Altman plot)

5.5 **SENSITIVITY**

The limit of detection (LOD) and limit of quantitation (LOQ) for each TDM assay must be defined. The LOD is the lowest analyte concentration that can be distinguished from the assay background, while the LOQ is the lowest concentration at which the analyte can be quantitated at defined levels for imprecision and accuracy (bias) [18]. In my laboratory, the LOD has been defined as the lowest concentration tested that has a peak height that is greater than or equal to the average of a blank sample (no analyte) plus three standard deviations (SD) of the blank. The acceptance criterion is that the LOD has to be less than 20% of the LOQ. The LOQ in our laboratory for TDM tests is defined as the lowest concentration where the between-day coefficient of variation is less than or equal to 10% with a signal-to-noise ratio for all MRM transitions greater than 10.

5.6 **SPECIFICITY (INTERFERENCES)**

Interference studies are conducted to determine if other commonly coprescribed, used, or abused medications might interfere with the TDM assay. In our laboratory, we run a standard list of the top 25 most commonly prescribed drugs [19], common drugs of abuse, and any other comedications and medications from the same drug class, which are all spiked at 10 μg/mL. Additionally, serum samples are tested for endogenous interferences from hemolysis, icterus, and lipemia. Serum samples are spiked at three different concentrations (subtherapeutic, therapeutic, and toxic) of drug and tested at five different concentrations of hemoglobin, bilirubin, or intralipid at concentrations up to 1,000, 60, and 2,000 mg/dL, respectively. The acceptance criterion is that all samples must match within 10–20% depending on the analyte.

5.7 **CARRY-OVER**

For TDM assays, the laboratory also assesses carryover. For this experiment, the laboratory spikes drug free matrix at concentrations at least two times the upper end of the AMR. The laboratory extracts and analyzes the spiked sample followed by five separate blank (drug-free) samples. The blank samples must be < 20% of the LOQ for the assay. In the end, the laboratory policy and standard operating procedure still requires a technologist to repeat any patient following a sample that is above the AMR, since the assay is not accurate in that concentration range and the true value may be 10 times or 20 times higher where carryover could occur.

5.8 **ION SUPPRESSION/ENHANCEMENT**

With MS, ion suppression/enhancement is a critical factor that can affect a patient's result. Sample matrix, coeluting compounds, and cross-talk can all contribute to ion suppression/enhancement so it is important that laboratories characterize MS assays [20]. Ion suppression studies for TDM assays are often done by comparing standards spiked into 10 different matrix pools and compared to the drug spiked into aqueous mobile phase. Three different concentrations that span the AMR are prepared and analyzed in triplicate. The acceptance criteria used by the laboratory is that the average ion suppression/enhancement should be < 20% for the analytes at all levels.

5.9 ADDITIONAL VALIDATION STUDIES

For TDM assays, the laboratory also performs matrix studies, prepared sample stability studies, and analyte stability studies. As mentioned previously, it is important to measure a drug in the appropriate matrix (i.e., serum, plasma, whole blood). As a result, the laboratory tests a variety of matrices to determine what can or should be accepted. Drug free matrices (serum-red top, sodium heparin plasma-green top, sodium fluoride/potassium oxalate plasma-gray top, potassium EDTA plasma-purple top, and serum separator tube-gold top) are spiked at concentrations level near the low, middle, and high end of the AMR. Pools are mixed and allowed to stand for at least 1 h at room temperature and then analyzed. The pools are then refrigerated and reanalyzed again the next day. The recoveries from the various matrices are then compared. For analyte stability, the results of freshly prepared pools are split and stored under ambient, refrigerated, and frozen (−20°C) conditions. At day 0, 1, 3, 7, 14, and 28, the samples are analyzed and compared back to day 0. Pools are spiked at concentrations near the low, middle and high end of the AMR. In addition, three freeze/thaw cycles are tested. The average difference for any analyte should be < 10% with no individual value > 20%. Based on the recovery results, one time-period for each storage condition is selected as the laboratory's specimen acceptance criteria along with a preferred storage/transport condition. Lastly, the laboratory examines the stability of the extracted samples by reinjecting the same plate (samples) at least 24 h after the original run on the same instrument while stored at the same temperature. Sample results are then calculated using the calibrators in the original run and recalculated using the calibrators from the reinjected run. This procedure is only done for internal documentation purposes to determine if samples can be reinjected if the MS has a problem during the analysis. The results must match within 10% otherwise the sample extraction process would have to be repeated on fresh patient samples if the MS had any issues.

6 QUALITY CONTROL/QUALITY ASSURANCE CONSIDERATIONS

Once a test is validated and goes live, the laboratory must continue to monitor the tests performance per regulatory guidelines, to minimize, identify, and correct analytical errors, ensure optimal MS and method performance, confirm method robustness, and confirm result quality. Three key postimplementation parameters to monitor are shown subsequently.

6.1 PROFICIENCY TESTING (PT)

It is a regulatory requirement that PT be performed for all tests. In the case of newer TDM tests, external PT may not be available from the College of American Pathologists (CAP) or other organizations so an alternate PT testing must be done. While each agency (i.e., CAP) has defined acceptance criteria (i.e., 80% pass rate), each laboratory should look for trends/shifts in the results compared to their peer group mean using the same methodology. For TDM tests, several providers are available, such as CAP and LGC Standards. Our laboratory prefers LGC for some of the new TDM analytes, since more LC-MS/MS users participate in that survey compared to CAP (i.e., back in 2014, levetiracetam PT for CAP only had ~16 laboratories using LC-MS/MS compared to ~35 laboratories for LGC). Table 8.3 shows the results of the PT for 2014 for levetiracetam in the LGC Standards PT program.

Table 8.3 2014 LGC Standards Levetiracetam Results for the Therapeutic Drugs Proficiency Testing Scheme

Month	Sample ID	Number of Laboratories Participating	Assigned Value	Method Mean	Laboratory Result	z Score
January	TM 123	37	18.70	18.7	18.3	−0.24
February	TM 124	30	38.75	39.4	35.6	−0.99
March	TM 125	32	75.80	74.2	87.0	1.55
April	TM 126	34	4.00	3.9	3.9	−0.11
May	TM 127	37	23.50	23.3	23.4	−0.04
June	TM 128	36	41.36	41.4	45.3	1.15
July	TM 129	36	7.50	7.6	6.8	−0.67
August	TM 130	30	55.82	56.0	57.0	0.24
September	TM 131	37	9.36	9.5	9.3	−0.05
October	TM 132	36	32.62	32.4	32.7	0.03
November	TM 133	37	14.20	14.3	12.7	−1.06

6.2 QUALITY CONTROL (QC)

Again, regulatory guidelines mandate that QC samples be analyzed to verify sample analysis performance. For TDM assays, we use three levels of QC near medical decision points or at least in the subtherapeutic, therapeutic, and toxic range to span the AMR. QC multirules are then established in which the values must fall within plus/minus two SD. While one QC level may fall outside 2 SD, it must be within 3 SD. If the QC continues to fall outside 2 SD on consecutive runs/days, the run would fail and results would have to be repeated.

6.3 SYSTEM SUITABILITY SAMPLES

A nonextracted sample made up of standard/internal standard material in a suitable solvent at a known concentration is made up in a large batch and stored based on the stability study data. This sample is then commonly run before a run to determine if the MS is performing properly. These samples can also be run during a run, or after a run/batch. Our laboratory uses these samples after any maintenance to assess how things are operating. For this sample, the retention time, peak shape, height/width, and analyte/IS ratio is recorded and compared to previous day's runs.

For all TDM MS tests, the technologists also have to systematically review each run after analysis. The technologists first make sure that all peaks are properly integrated and the retention times are within ±0.1 min of the standards. The peak shape and integrity are reviewed to look for shouldering or tailing. If any interference is discovered, the sample must be repeated straight or at an appropriate dilution to try and resolve the interference. For the calibrators, the percentage accuracy for each standard must also fall within 10% and the $r^2 > 0.98$. The internal standard metric plot is also reviewed. The internal standard data is then exported into an Excel spreadsheet where the percentage mean of the internal standard is analyzed and flagged, if it falls outside the limits established during method validation. This allows the technologist to see if any ion enhancement or ion suppression may be affecting a result. Furthermore, the blank must not show any carryover and the dilution control (if diluted patient samples were analyzed) must also pass acceptance criteria. Together, these parameters verify the instrument and sample analysis performance and are an essential component of a post-implementation plan.

7 CONCLUSIONS

MS is powerful quantitative technique that can be used for TDM. MS offers clinical laboratories several advantages including its specificity, sensitivity, throughput, and cost-effective testing. However, MS is not without its challenges especially with the lack of standardization for TDM assays and the fear of additional regulations by the FDA on LDTs. However, many clinical laboratories are successfully using MS to perform TDM testing for the analytes discussed in this chapter. In the end, the MS applications are endless and offer clinical laboratories a way to meet the TDM needs of physicians and patients.

REFERENCES

[1] Oellerich M, Armstrong VW. The role of therapeutic drug monitoring in individualizing immunosuppressive drug therapy: recent developments. Ther Drug Monit 2006;28(6):720–5.

[2] Steimer W. Performance and specificity of monoclonal immunoassays for cyclosporine monitoring: how specific is specific? Clin Chem 1999;45(3):371–81.

[3] Taylor PJ, Tai CH, Franklin ME, Pillans PI. The current role of liquid chromatography-tandem mass spectrometry in therapeutic drug monitoring of immunosuppressant and antiretroviral drugs. Clin Biochem 2011;44(1):14–20.

[4] Russell JA, Kangarloo SB. Therapeutic drug monitoring of busulfan in transplantation. Curr Pharm Des 2008;14(20):1936–49.

[5] Tesfaye H, Branova R, Klapkova E, Prusa R, Janeckova D, Riha P, et al. The importance of therapeutic drug monitoring (TDM) for parenteral busulfan dosing in conditioning regimen for hematopoietic stem cell transplantation (HSCT) in children. Ann Transplant 2014;19:214–24.

[6] Napoli KL. Is microparticle enzyme-linked immunoassay (MEIA) reliable for use in tacrolimus TDM? Comparison of MEIA to liquid chromatography with mass spectrometric detection using longitudinal trough samples from transplant recipients. Ther Drug Monit 2006;28(4):491–504.

[7] Warnke C, Meyer zu Horste G, Hartung HP, Stuve O, Kieseier BC. Review of teriflunomide and its potential in the treatment of multiple sclerosis. Neuropsychiatr Dis Treat 2009;5:333–40.

[8] Fermiano M, Bergsbaken J, Kolesar JM. Glucarpidase for the management of elevated methotrexate levels in patients with impaired renal function. Am J Health Syst Pharm 2014;71(10):793–8.

[9] Voraxaze package insert. Available from: http://www.accessdata.fda.gov/drugsatfda_docs/label/2012/125327lbl.pdf; 2012

[10] Zhang Q, Simpson J, Aboleneen HI. A specific method for the measurement of tacrolimus in human whole blood by liquid chromatography/tandem mass spectrometry. Ther Drug Monit 1997;19(4):470–6.

[11] Karppi J, Akerman KK, Parviainen M. Suitability of collection tubes with separator gels for collecting and storing blood samples for therapeutic drug monitoring (TDM). Clin Chem Lab Med 2000;38(4):313–20.

[12] Koch TR, Platoff G. Suitability of collection tubes with separator gels for therapeutic drug monitoring. Ther Drug Monit 1990;12(3):277–80.

[13] Taylor PJ. Internal standard selection for immunosuppressant drugs measured by high-performance liquid chromatography tandem mass spectrometry. Ther Drug Monit 2007;29(1):131–2.

[14] Owen LJ, Keevil BG. Testosterone measurement by liquid chromatography tandem mass spectrometry: the importance of internal standard choice. Ann Clin Biochem 2012;49(Pt 6):600–2.

[15] Zheng N, Jiang H, Zeng J. Current advances and strategies towards fully automated sample preparation for regulated LC-MS/MS bioanalysis. Bioanalysis 2014;6(18):2441–59.

[16] Clarke WHA, Molinaro RJ, Iyer B, Bachmann LM, Kulasingam V, Bothelo JC, Mason DS, Cao Z, Rappold B, French D, Tacker DH, Garg S, Truscott DH, Gawoski JM, Yu CY, Grant RP, Zhu Y. C62-A liquid chromatography-mass spectrometry methods; approved guideline. Wayne, PA: Clinical and Laboratory Standards Institute; 2014.

[17] Levine DM, Maine GT, Armbruster DA, Mussell C, Buchholz C, O'Connor G, et al. The need for standardization of tacrolimus assays. Clin Chem 2011;57(12):1739–47.

[18] Armbruster DA, Pry T. Limit of blank, limit of detection and limit of quantitation. Clin Biochem Rev 2008;29(Suppl. 1):S49–52.

[19] Bartholomew M. Top 200 Drugs of 2012. Available from: http://www.pharmacytimes.com/publications/issue/2013/July2013/Top-200-Drugs-of-2012; 2013.

[20] Annesley TM. Ion suppression in mass spectrometry. Clin Chem 2003;49(7):1041–4.

VITAMIN D METABOLITE QUANTITATION BY LC-MS/MS

H. Ketha*, R.J. Singh**

*Department of Pathology, University of Michigan Hospital and Health Systems, Ann Arbor, MI, United States; **Department of Laboratory Medicine and Pathology, Mayo Clinic, Rochester, MN, United States

1 PHYSIOLOGICAL ROLE OF VITAMIN D

Rickets as a pathological entity has been known since the archaic times but the first scientific description of rickets dates back to the late 1600s [1]. It was not until the early 1920s that findings demonstrating that rickets could be cured with "accessory" food nutrients or by exposing affected children to ultraviolet (UV) light were published [2]. These groups of accessory food nutrients also shown at that time, to cure scurvy by feeding fresh vegetables to sea-faring sailors [3] were termed as vital amines or vitamins by Casimir Funk [4]. McCollum [5] and Edward Mellanby [6] classified vitamins into water or fat soluble compounds [7] and showed that cod liver oil contained the antirachitic compound that could cure experimental dogs made to acquire rickets by restricting sunlight exposure [5,8]. This fat-soluble antirachitic substance, shown to be different than vitamin A by McCollum [5], was named as vitamin D. Orr in 1923 demonstrated that vitamin D stimulates intestinal absorption of calcium [9], whereas Shipley and Holtrop's work in 1926 showed that bone mineralization is promoted by an increased plasma calcium and phosphate concentrations directly regulated by vitamin D [10,11]. The findings that vitamin D has to be metabolically activated to exert its calcemic effects and that it also plays a rather counter-intuitive role on calcium resorption from bone in presence of parathyroid hormone (PTH) were established during 1950–70 [1,12–14]. In the decades that followed the scientific literature witnessed an exponential growth in the understanding of vitamin D biochemistry along with the advent of advanced quantitative tools including radio immunoassays (RIAs), high pressure liquid chromatography (HPLC) and mass spectrometry (MS) that helped isolate and quantify vitamin D and its metabolites with high degree of accuracy.

Vitamin D has been classified as vitamin D_3 or D_2 based on the source of the vitamin. Vitamin D_3 is the mammalian form also called cholecalciferol whereas vitamin D_2, also called ergocalciferol, is the plant derived form of vitamin D. In this chapter, the subscripts will be used to designate each form when the distinction is necessary otherwise the metabolite will denote the total $(-D_3 + -D_2)$ metabolite concentration. Both vitamin D_3 and vitamin D_2 undergo similar metabolic processes in humans.

The vitamin D endocrine system and PTH are the principle regulators of calcium homeostasis in humans [15–17]. The biochemical pathway of vitamin D action has been shown in Fig. 9.1. Vitamin D_3 is synthesized from its precursor 7-dehydrocholesterol following a UV-B catalyzed photo isomerization. Vitamin D_3 formed in the skin or vitamin D_2 obtained from diet is converted to 25-hydroxyvitamin D_3 or

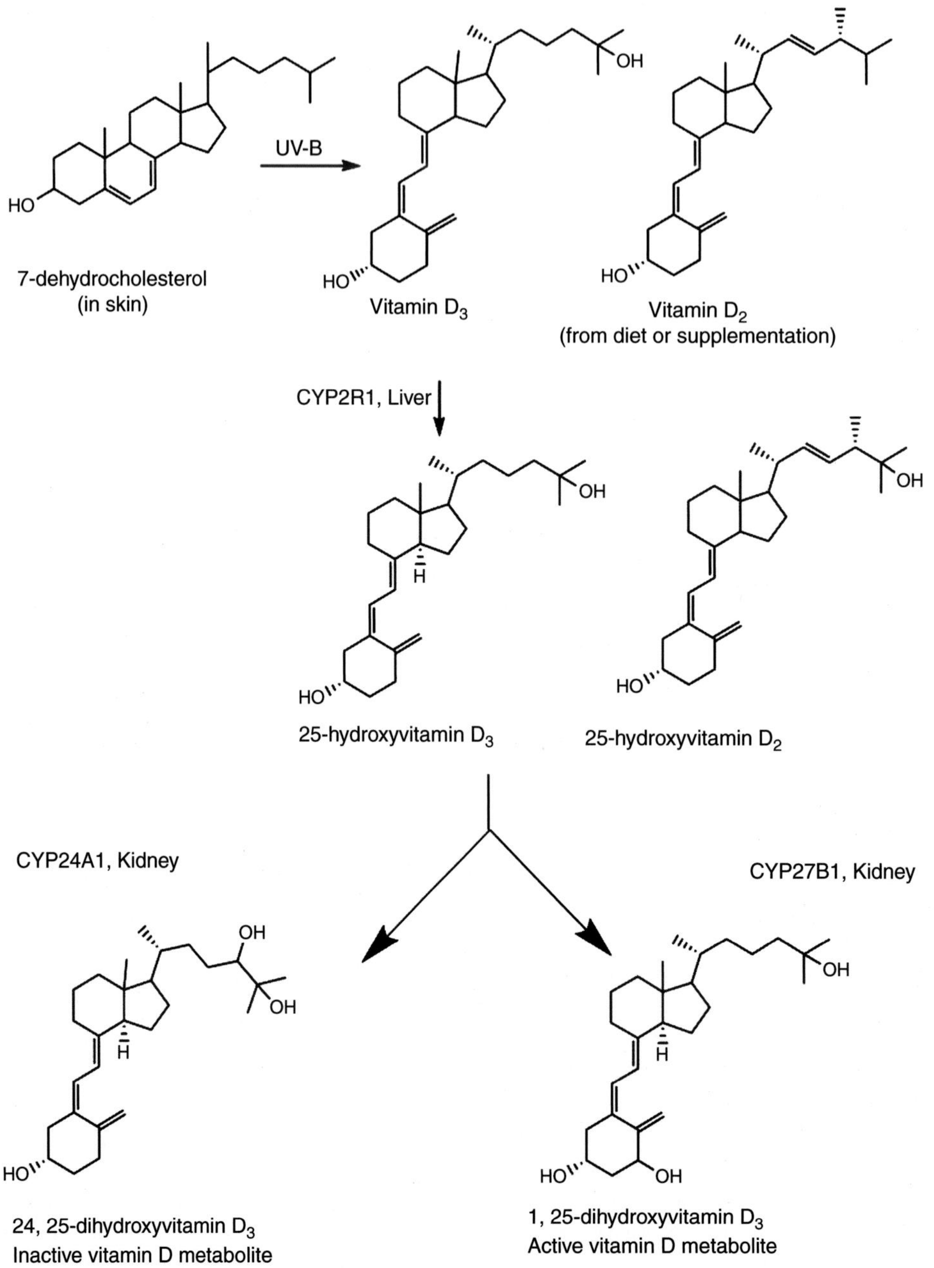

FIGURE 9.1 Vitamin D Metabolic Pathway

25-hydroxyvitamin D_2, the most abundant circulating form by vitamin D 25-hydroxylase also termed as cytochrome P450 (CYP) 2R1 in the liver. Vitamin D_3 and vitamin D_2 are transported to the liver by the vitamin D binding protein (VBP). VBP has a high affinity for vitamin D and its metabolites. Majority of the circulating 25(OH)D is bound to VBP in circulation [18]. PTH controls physiological calcium demand by regulating the conversion of 25(OH)D to the active hormone, 1,25-dihydroxyvitamin D (1,25(OH)$_2$D) by inducing the action of 25-hydroxyvitamin D_3-1-α-hydroxylase (CYP27B1). 24-hydroxylation of 1,25(OH)$_2$D by 25-hydroxyvitamin D_3-24-hydroxylase (CYP24A1) produces a trihydroxylated metabolite 1,24,25(OH)$_3$D which is finally excreted as calcitroic acid in urine [19]. Therefore, CYP24A1 deactivates the active metabolite of the vitamin D pathway. CYP24A1 also converts 25(OH)D to 24,25-dihydroxyvitamin D (24,25(OH)$_2$D), another inactive vitamin D metabolite [20]. The calcium regulatory effects of vitamin D metabolic pathway occur in conjunction with PTH. PTH functions in a seemingly paradoxical manner in the bone microenvironment where it stimulates bone resorption or bone formation, depending on its concentration and on the duration of exposure. Rapid calcium homeostasis is achieved by PTH by bone resorption whereas its long term effects function to achieve extreme systemic needs and maintain skeletal homeostasis [21]. The action of 1,25(OH)$_2$D is tightly regulated by several factors. This tight feedback regulates the amount of active calcemic hormone present in circulation and provides "protection" from excess 25(OH)D when present. For example, following prolonged sun exposure or high dose vitamin D supplementation circulating 25(OH)D levels increase significantly, but 1,25(OH)$_2$D levels do not change significantly and normal healthy adults do not experience hypercalcemia due to this tight feedback control on calcium homeostasis by PTH and vitamin D metabolism. The biochemistry and utility of quantifying clinically relevant vitamin D metabolites are discussed in later sections.

2 EVOLUTION OF ASSAYS FOR VITAMIN D METABOLITES

The vitamin D metabolites that are clinically relevant in the differential diagnosis of disorders of calcium homeostasis include 25(OH)D, C3-epi-25(OH)D, 1,25(OH)$_2$D, and 24,25(OH)$_2$D (Table 9.1). VBP is not commonly measured in clinical laboratories but its role in regulating the bioavailable 25(OH)D (fraction of 25(OH)D not bound to VBP) has caused an increased interest in measuring VBP [22,23]. Quantitative LC-MS/MS methods for VBP have been developed but will not be discussed here [22].

In the early 2000s, several clinical laboratories across United States, Europe, and Australia reported an enormous increase in testing related to vitamin D deficiency [27]. In Australia, an increase from 23,000 tests per year to about 2.5 million tests was reported from 2005 to 2010 [28]. An analysis of the trends in laboratory test volumes for Medicare Part B reimbursements in the United States from 2000 to 2010, showed that the number of vitamin D tests reimbursed per 10,000 tests had increased from less than a 100 in 2000 to a greater than 1500—an increase in reimbursement rate of 15-fold over 10 years [29]. The exponential increase in vitamin D testing is partly due to a revived public interest in vitamin D nutritional status following reports showing wide prevalence of vitamin D deficiency [30,31] along with the numerous studies exploring the physiological role of vitamin D and its metabolites in health and disease [32]. With the growing public demand for vitamin D testing and ensuing increasing testing volume, the interest in vitamin D assays grew and so did the number of assays available for routine use. Two methodologies that have received considerable attention for vitamin D metabolite quantitation are immunoassays and liquid chromatography mass spectrometry (LC-MS/MS).

Table 9.1 Clinical Utility of Vitamin D Metabolite Quantitation and Biochemical Effect on PTH and Serum Calcium in Various Pathologic States

Vitamin D Metabolite	Reference Interval [21]	Used in the Differential Diagnosis of	Effect on Vitamin D Metabolite[a]	PTH	Serum Calcium
25(OH)D	10–65 ng/mL [21] 25–162 nmol/L [21]	• Nutritional Rickets	↓	↑	↓
		• Vitamin D deficiency	↓	↑[a]	↓[a]
		• Vitamin D Toxicity	↑	↓	↑
		• Monitoring vitamin D supplementation	Variable	Normal	Normal
		• VDDR type I	↓	↑	↓
		• VDDR type II; (1,25(OH)2D is low	Normal	↑	↓
		• CYP24A1 mutations; has to be measured in conjunction with 24,25(OH)$_2$D.	Normal	↓	↑
C3-epi-25(OH)D	Variable percentage; dependent on 25(OH)D [24,25]	• To accurately determine the "native" 25(OH)D present since the two forms may have differential down-stream calcemic effects	Variable	Variable	Variable
1,25(OH)$_2$D	15–60 ng/mL [21] 36–144 nmol/L [21]	• Vitamin D deficiency	↑	↑	↓
		• Iatrogenic vitamin D toxicity	↑ or normal	↓	↑
		• Hypercalcemia due to malignancy, primary hyperparathyroidism, recurrant kidney stones, hypercalciuria	↑	↓	↑
		• Hypocalcemia in end stage renal disease	↓	↑	↓
24,25(OH)$_2$D and 25(OH) (25(OH) D/24,25(OH)$_2$D)	7–35 [26]	• CYP24A1 mutations; has to be measured in conjunction with 24,25(OH)$_2$D; elevated 1,25(OH)$_2$D, recurrant hypercalcemia and kidney stones are common	↑	↓	↑

[a]*Refers to the change in vitamin D metabolite.*

Majority of circulating (85%) of 25(OH)D is bound to VBP, a small portion (15%) bound to albumin, and only ~0.03% circulates as free form [18]. For the assay methodology to quantitate 25(OH)D to be successful, vitamin D and its metabolites need to be dissociated from VBP. The very first assays for 25(OH)D quantitation introduced in the 1970s were based on competitive binding with VBP. The sample preparation involved organic solvent extraction (allowing separation of VBP from vitamin D metabolites) followed by chromatographic separation of 25(OH)D from the dihydroxylated vitamin D metabolites and quantitation using a competitive binding assay with VBP [20,33]. With the development of an antibody against 25(OH)D in the 1980s a RIA was developed [34,35]. The early extraction assays were largely manual with tedious solvent extraction steps followed by labor intensive quantitation using a radiolabeled tracer. Turn-around time and throughput demands placed on the clinical laboratories led to immunoassays that were eventually fully automated to meet the clinical and operational needs of the institutions [36]. Automated immunoassays are a mainstay in clinical laboratories as these provide an optimum balance between specificity, sensitivity, and throughput for a variety of analytes. However, lack of specificity and wider intramethod variability especially for steroid quantitation using immunoassays is well known [36–38].

Over 40 metabolites of vitamin D have been reported to date [27,39]. Majority of these are not clinically useful as most have a very short half-life in circulation. However, from an analytical stand point, the presence of a large number of structurally related metabolites can be challenging due to potential interferences and cross reactivity. The nonextraction assays lack a "wash" or a separation step that removes structurally similar analogs or metabolites effectively before the quantitation step. However, nonextraction immunoassay methods may be susceptible to interferences and matrix effects especially due to the lipophilic nature of 25(OH)D and its high affinity to VBP [40]. Significant cross reactivity with $24,25\text{-}(OH)_2D_3$, $25,26\text{-}(OH)_2D_3$, and $25(OH)D_3\text{-}26,23$-lactone has been observed in several immunoassays [34]. While clinical significance of this interference has been debated, $24,25(OH)_2D$ circulates at about 7–35% of the 25OHD concentration [26] and its presence could potentially cause a misclassification of a patient as vitamin D sufficient. Additionally, commercially available immunoassays may differ greatly in their ability to measure $25(OH)D_2$ [41,42]. Today several manual and automated immunoassays for 25(OH) are available. Many are produced in a kit format that can be used in a manual or automated platforms [34].

Wide variability in 25(OH)D values of the same sample measured in different laboratories has been shown [41,43]. The mean concentration of the same sample in one study measured by different immunoassay methods ranged from 17 to 36 ng/mL [40]. In view of substantial between-method variability in 25(OH)D values, efforts to standardize 25(OH)D assays have been undertaken. National Institute of Standards and Technology (NIST) standard reference material for vitamin D metabolites is now available [44]. Vitamin D external quality assessment scheme (DEQAS) was formed in 1989 with a goal to achieve and maintain reliability in measurement of 25(OH)D and $1,25(OH)_2D$ by assays in use by clinical laboratories. Clinical laboratories accredited by College of American Pathologists can also use DEQAS as their primary proficiency testing for 25-OHD assays. The April, 2016 DEQAS reports for 25(OH)D shows 918 users performing 25(OH)D quantitation of quality-assurance samples by 28 different methods including LC-MS/MS (Table 9.2). Immunoassay methods are the predominant 81% whereas HPLC and LC-MS/MS combined constitute 19% of the users. Of note there has been an increase in LC-MS/MS users since 2012, when DEQAS reported 11% LC-MS/MS users [34].

Matrix effect refers to the impact of the sample components (e.g., proteins, salts, steroids, drugs, phospholipids, antibodies) other than the analyte of choice on the specificity and sensitivity of an analytical method. Matrix effects pose a significant challenge in immunoassays as they can lead to falsely elevated or low results [45]. Difference in the calibrator matrix and patient sample matrix, presence of high affinity binding proteins, coextraction of lipophilic components like phospholipids in the sample preparation step are all potential causes of matrix effects in immunoassays. In case of 25(OH)D assays, many immunoassays use a denaturing buffer to release 25(OH)D and other highly lipophilic vitamin D metabolites from VBP. The presence of other lipophilic components in the patient sample make the complete dissociation and capture of 25(OH)D difficult and nonspecific interaction of other vitamin D metabolites has proven to be problematic. Needless to say the effectiveness of the denaturing buffer will have a strong impact on assay specificity [46,47].

Table 9.2 DEQAS 2016 Report Showing 25(OH)D Method, the Number of Participating Laboratories, Mean of Reported Value by the Method Group, Standard Deviation (SD) in the Group and Percent Coefficient of Variation with a Methods Group (%CV)

Method	Number of Laboratories	Method Mean nmol/L	SD	%CV
All methods	918	33.6	5.1	15.1
Abbott Architect—New kit (5P02)	42	29.9	2.1	7.0
Abbott Architect—Old kit (3L52)	38	31.9	3.2	10.2
Automate IDS EIA	5	35.4	3.0	8.4
Beckman Access2 Total 25OHD	4	37.8	16.1	42.6
Beckman Unicel DXi Total 25OHD	31	33.7	4.6	13.7
bioMerieux 25OH Vitamin D Total	2	28.3	2.9	10.4
DiaSorin Liaison Total	237	30.4	3.0	9.9
DiaSorin RIA	2	33.8	4.8	14.3
DiaSource 25OH VitD Chem Analyzers	2	33.6	4.3	12.8
Diazyme 25OH VitD EIA	2	39.3	0.6	1.5
Euroimmun ELISA	11	32.2	12.4	38.6
Fujirebio Lumipulse D 25OH Vit D	3	35.1	3.5	10.0
HPLC	23	29.4	1.0	3.4
IDS EIA	17	36.8	6.4	17.9
IDS RIA	6	41.8	4.2	11.4
IDS-iSYS	88	35.7	5.1	12.2
LC-MS/MS	156	34.3	5.4	15.0
Ortho Total 25OHD	5	32.3	4.1	12.0
Roche Total 25OHD	160	34.9	5.4	16.7
Siemens Advia Centaur	65	40.5	4.4	12.5
Tosoh AIA	5	40.7	4.7	11.4
Others	9	32.0–54.5	0–12.4	0–5.3

All reported values represent total 25(OH)D = 25(OH)D$_2$ + 25(OH)D$_3$

Quantitation of 1,25(OH)$_2$D is particularly challenging mainly due to very low (1/1000th of 25(OH)D) circulating concentration. Similar to the earliest methods used, even with contemporary LC-MS/MS methods, appropriate sample preparation methods are crucial to quantitate 1,25(OH)$_2$D accurately. One of the first methods used 20 mL of plasma sample, which was extracted with methanol–chloroform followed by purification by liquid chromatography [48]. Quantitation was achieved by a protein binding assay using chick intestinal VDR. Then assays utilizing the calf thymus VDR, which was more stable compared to the chick intestinal VDR became available [49]. This assay was tedious to perform but was reliable and with a number of modifications made to the method over several years, was used widely [50]. The first RIA for 1,25(OH)$_2$D was introduced in 1978. The rabbit antibody used was nonspecific and showed cross reactivity to 25(OH). Due to poor specificity of the antibody, sample extraction and manipulation was very tedious and laborious. Then a RIA that used a simple acetonitrile extraction and a ^{125}I-labeled tracer that allowed gamma counting of the assay was introduced [51]. In one such method, use of sodium periodate treatment of the sample extract leading to the conversion of 24,25(OH)$_2$D$_3$ and 25,26(OH)$_2$D$_3$ to aldehyde and ketone forms greatly enhanced specificity. Immunodiagnostic Systems (Boldon, United Kingdom) developed a immunoenrichment based sample purification method that resulted in enhanced specificity towards 1,25(OH)$_2$D. The delipidated serum samples are treated with an antibody bound to a miniimmunocapsule. The antibody-bound 1,25(OH)$_2$D is eluted, eluant evaporated to dryness, then reconstituted prior to RIA quantitation [52]. Interferences in the Diasorin and IDS assays from 1,25(OH)$_2$D$_3$ 26,23-lactone, 1,24,25(OH)$_3$D$_3$, and 1,25,26(OH)$_3$D$_3$ has been demonstrated [35,51,53]. As obvious, 1,25(OH)$_2$D quantitation by immunoassay based methods have been problematic and laborious. The IDS's immunoenrichment based sample preparation approach has been coupled with LC-MS/MS to achieve highly sensitive and specific quantitation of 1,25(OH)$_2$D [54]. 1,25(OH)$_2$D measurement of quality-assurance samples has been shown in Table 9.3. The April, 2016 DEQAS reports for 1,25(OH)D shows 171 users performing quantitation of quality-assurance samples by nine different methods including LC-MS/MS (Table 9.3). Immunoassay methods are the predominant 92% whereas LC-MS/MS constitute 8% (compared to 3% in 2012) of the users.

Table 9.3 DEQAS 2016 Report Showing 1,25(OH)$_2$D Method, the Number of Participating Laboratories, Mean of Reported Value by the Method Group, Standard Deviation (SD) in the Group and Percent Coefficient of Variation with a Methods Group (%CV)

Method	Number of Laboratories	Method Mean pg/mL	SD	%CV
All methods	171	96.8	19.4	20.1
AMP RIA	2	71.1	27.6	38.9
Cusabio ELISA	1	50.4	0	0.0
DiaSorin Liaison XL	87	104.7	13.1	12.5
DiaSorin RIA	4	80.9	9.3	11.4
DIASource CT	3	84.9	3.8	4.5
IDS EIA	10	92.4	16.0	17.3
IDS RIA	18	102.7	15.0	14.6
IDS-iSYS	32	76.3	21.9	28.6
LC-MS/MS	14	99.8	24.5	24.5

All reported values represent total 1,25(OH)$_2$D = 1,25(OH)$_2$D$_2$ + 1,25(OH)$_2$D$_3$

3 CLINICAL UTILITY AND QUANTITATION OF VITAMIN D AND ITS METABOLITES BY LC-MS/MS

Accurate quantitation of 25(OH)D and 1,25(OH)$_2$D using nonextraction immunoassays has proven to be challenging as discussed in detail in the previous section. The limitations of immunoassays for accurate quantitation of vitamin D metabolites have been the motivators for clinical laboratories for adopting LC-MS/MS methods for vitamin D metabolite quantitation. In this section, we will discuss the key clinical, technical, and strategic considerations involved in the implementation of mass spectrometry for quantitation of vitamin D and its clinically relevant metabolites.

Performance characteristics of LC-MS/MS assays developed by various groups for vitamin D metabolites has been shown in Table 9.4.

3.1 VITAMIN D$_3$ AND VITAMIN D$_2$

3.1.1 Clinical Utility

Vitamin D$_3$ (or vitamin D$_2$), present in food, is absorbed via the intestinal lymphatics [1] where the vitamin exists in the chylomicron fraction. Approximately half of absorbed vitamin D in chylomicrons is transferred to VBP in blood. VBP is the principle chaperone protein that facilitates uptake of vitamin D by the liver while albumin also plays a minor role in transporting vitamin D to the liver for further metabolic conversion. Circulating serum concentration of vitamin D$_3$ or vitamin D$_2$ is not clinically relevant as they are converted rapidly to their 25-hydroxylated metabolite in the liver. The negative feedback of 25(OH)D on the activity of CYP2R1 and is not influenced by plasma calcium and phosphorus concentrations [16].

Table 9.4 Assay Performance Parameters for LC-MS/MS Assays for Clinically Useful Vitamin D Metabolites

Assay Parameter	25(OH)D [24,55–61]		1,25(OH)$_2$D [54,62]		24,25(OH)$_2$D [26,54,63,64]	
	25(OH)D$_2$	25(OH)D$_3$	1,25(OH)$_2$D$_2$	1,25(OH)$_2$D$_3$	24,25(OH)$_2$D$_2$	24,25(OH)$_2$D$_3$
Inter assay imprecision	5.0 16%	2.5–12%	11–13%	6-8%	3.1–10.1%	5.2–7.4%
Intra assay imprecision	4.5–11%	5.7–8.0%	9–11%	5.6–11	3.1–6.2%	7–9%
AMR	2.4–123 ng/mL	0.4–120 ng/mL	5.15–206 pg/mL	4.6–185 pg/mL	0.5–25 ng/mL	0.1–25 ng/mL
Mean recovery	89–110%	86–108%	112–110%	90–120%	94–100%	90–94%
LOD	0.3–2 ng/mL	0.3–2 ng/mL	2.0 pg/mL	2.7 pg/mL	0.5 ng/mL	0.03 ng/mL

The values are a compilation from several different laboratories.
Multiply by 2.5 to convert 25(OH)D$_3$ from ng/mL to nmol/L, by 2.41 to convert 25(OH)D$_2$ from ng/mL to nmol/L, by 2.40 to convert 1,25(OH)$_2$D$_3$ from pg/mL to pmol/L, by 2.33 to convert 1,25(OH)$_2$D$_2$ from pg/mL to pmol/L, by 2.40 to convert 24,25(OH)$_2$D$_3$ from ng/mL to nmol/L, by 2.33 to convert 24,25(OH)$_2$D$_2$ from ng/mL nmol/L.

3.1.2 Assay Procedure

LC-MS/MS methods have been focused at determining concentration of vitamin D_3 and/or vitamin D_2 in food matrices [65]. Quantitative LC-MS/MS methods used for vitamin D_3 and vitamin D_2 and their metabolites have many common steps particularly those involved in sample preparation and chromatographic separation. Most sample preparation methods described for 25(OH)D in this chapter will be amenable to vitamin D_3 and vitamin D_2 quantitation.

3.2 25-HYDROXYVITAMIN D

3.2.1 Biochemistry and Clinical Utility

Vitamin D_3, is converted to 25(OH)D in the liver [66]. The enzyme, CYP2R1 that catalyzes this reaction has been detected in liver microsomes and liver mitochondria.

Supplementation with large doses of vitamin D_3 or vitamin D_2 results in the parallel increase in circulating levels of total 25(OH)D (25(OH)D_3 + 25(OH)D_2). Since 25-hydroxylation is a poorly regulated process, the circulating level of 25(OH)D is a very good index of vitamin D reserve. On the contrary 1,25(OH)$_2$D production is highly regulated via CYP27B1 and does not provide information on the vitamin D nutritional status. 25(OH)D concentrations are not affected by phosphate but are affected by induction of CYP24A1, the enzyme responsible for degradation of 25(OH) and 1,25(OH)$_2$D. 25(OH)D quantitation alone is sufficient to identify underlying vitamin D deficiency in a patient. The half-life of 25(OH)D is approximately 1 month in humans [67]. 25(OH)D circulates bound to vitamin D binding protein (VBP) with total concentration in a healthy adult in the 5–80 ng/mL range [21]. At 25(OH)D concentrations above 200 ng/mL cause severe hypercalcemia and comorbidities associated with hypercalcemia including irritability, vomiting, and nephrocalcenosis can develop [68–70].

3.2.2 Assay Procedure

Several methods for 25(OH)D quantitation have been reported [55–60,71–73]. The sample preparation steps commonly used in clinical laboratories for LC-MS/MS quantitation of vitamin D and its metabolites are shown in Fig. 9.2. The key factors for achieving optimal sample clean-up for accurately quantitating total 25(OH)D are: (1) 25(OH)D has to be dissociated from VBP, (2) and other serum proteins, (3) interfering vitamin D metabolites and other phospholipids have to be chromatographically separated. Since 25(OH)D_3 differs from 25(OH)D_2 by 12 Da, the selectivity offered by the quadrupole for specific ion-transitions is sufficient to adequately separate them during MS analysis. The C3-epi-25(OH)D quantitation has been described in a later section. Dihydroxylated metabolites (1,25(OH)$_2$D and 24,25(OH)$_2$D) can be easily separated from 25(OH)D chromatographically whose retention times on reverse phase HPLC columns are considerably shorter than 25(OH)D.

Sample volume of 100–200 μL has been used to achieve optimal sensitivity. Deproteinization can be achieved by acetonitrile, acetonitrile/sodium hydroxide mixture, methanol or methanol:propanol mixture followed by solid phase extraction (SPE) with C8 or C18 solid phases [57,59–61]. A combination of deproteinization and extraction can be achieved by liquid–liquid extraction (LLE) with hexane or n-heptane [55,56]. An online SPE method (TurboFlow HPLC, formerly Cohesive Technologies, Thermo Fisher Scientific Group, Waltham, Massacheusetts, United States of America) has been used by several clinical laboratories including ours, to achieve an automated sample preparation that can combine sample clean-up and analytical separation [57,60]. TurboFlow technology uses a combination of diffusion, column chemistry, and size exclusion to accomplish online sample clean-up of complex

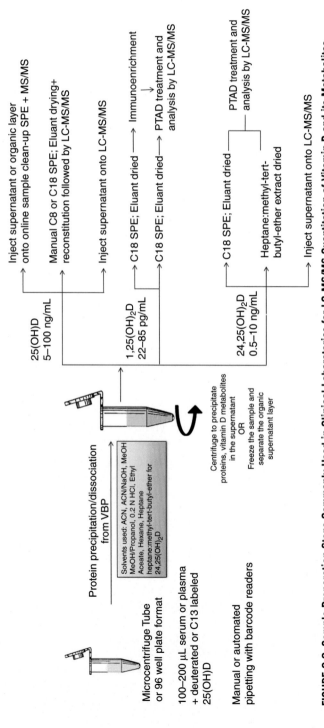

FIGURE 9.2 Sample Preparation Steps Commonly Used in Clinical Laboratories for LC-MS/MS Quantitation of Vitamin D and its Metabolites

The values shown with the analytes represent normal physiological concentrations. The circulating concentration in normal adults has been shown above the arrows. A PTAD based approach using differential mass tagging approach for 25(OH)D has been described in Fig. 9.5.

matrixes prior to HPLC or UHPLC separation, while improving selectivity for the analyte of choice. Online SPE involves injecting/loading the sample extract either directly from a LLE or SPE extraction or after drying/reconstitution from SPE or LLE step onto a C8, C18 or an equivalent cartridge or a TurboFlow column wherein higher flow rates can be facilitated allowing the matrix components to be effectively washed in a weakly organic mobile (loading) phase while retaining the analyte of choice. The sample is then transferred onto an analytical column followed by elution achieved by an increasing organic phase gradient (elution phase). Similar systems like 2-dimensional HPLC, capable of achieving comparable functionalities are offered by other vendors. Increased throughput can be achieved by a multiplexed HPLC or UPLC systems. One example of a multiplex HPLC system is the TurboFlow TLX4 HPLC system (now Transcend II System with Multi-Channel and TurboFlow Technology from Thermo Fisher Scientific Group, Waltham, Massacheusetts, USA). In this system, two HPLC columns are operated simultaneously with staggered sample elution and column regeneration steps to maximize mass spectrometer usage and reduce lag time between elution from columns (Fig. 9.3). Other innovative approaches to improve throughput have been undertaken [74]. Hoofnagel and coworkers developed a flexible rubber gasket capable of sealing two 96-well plates together and quantitatively transfer the organic layer contents of one plate to another. A workflow involving a 96-well plate format was developed by a combination of the transfer gasket and a dry-ice acetone bath to freeze the aqueous infranatant followed by LC-MS/MS analysis of the transferred organic phase [74].

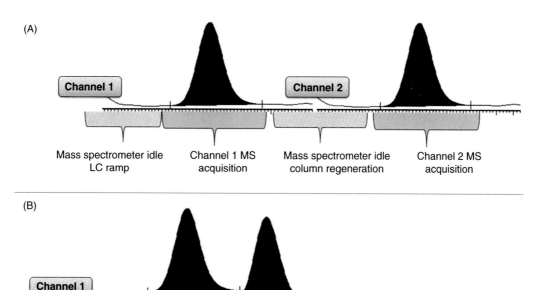

FIGURE 9.3 Multiplex Liquid Chromatography System

(A) Two LC systems operating in a sequential injection sequence with no overlap between the lag times (no multiplex) and (B) two parallel LC systems staggered to achieve minimal time lag between eluting analyte of interest coupled to a single mass spectrometer. Note that by multiplexing the LC channels, mass spectrometer idle-time is reduced and throughput is improved.

Mass detection can be achieved by atmospheric pressure chemical ionization (APCI) or electro-spray ionization (ESI) methods. Derivatization by Diels-Alder reaction with substituted phenyl-1,2,4-triazoline-3,5-diones (PTAD) improves sensitivity of 25(OH)D LC-MS/MS detection by a 100 fold [71,75]. PTAD addition to the 1,3-diene system common to vitamin D and its metabolites can lead to two isomeric products. Depending on the choice of the chromatographic column the different isomer may or may not separate. If no interferences are observed, choosing column that does not result in vita-min D-metabolite-PTAD isomer separation may help gain sensitivity in the method (Fig. 9.4).

A differential mass-tagging approach along with multiplex HPLC was used to improve throughput by fivefold by the Mayo Clinic reference laboratory [72]. In this method (Fig. 9.5), five different bar-coded patient samples are extracted and derivatized, each with a unique PTAD derivative. Then the five uniquely tagged patient samples are mixed and analyzed in a single analytical injection. Data output

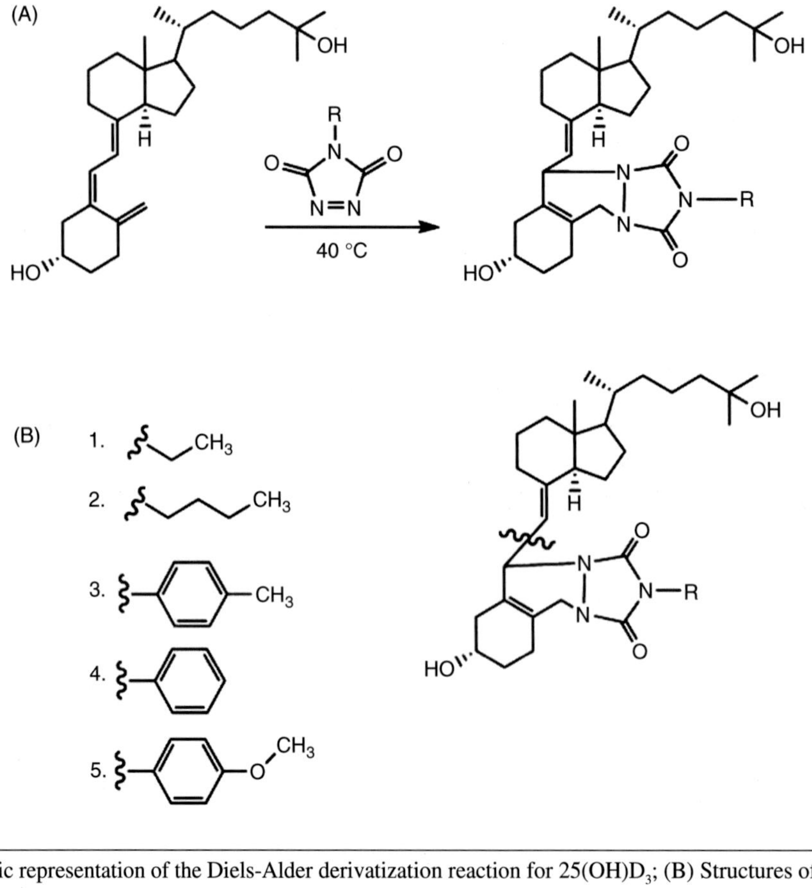

FIGURE 9.4

(A) Schematic representation of the Diels-Alder derivatization reaction for 25(OH)D$_3$; (B) Structures of PTAD derivatives used in the differential mass tagging approach.

Adapted from Netzel BC, Cradic KW, Bro ET, Girtman AB, Cyr RC, Singh RJ, Grebe SK. Increasing liquid chromatography-tandem mass spectrometry throughput by mass tagging: a sample-multiplexed high-throughput assay for 25-hydroxyvitamin D2 and D3. Clin Chem 2011;57:431–440. [72]

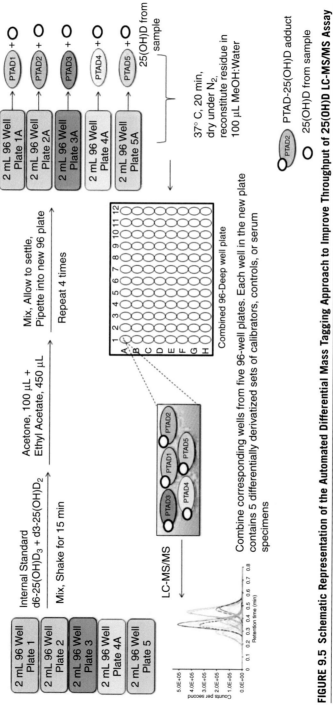

FIGURE 9.5 Schematic Representation of the Automated Differential Mass Tagging Approach to Improve Throughput of 25(OH)D LC-MS/MS Assay

Adapted from Netzel BC, Cradic KW, Bro ET, Girtman AB, Cyr RC, Bro ET, Singh RJ, Grebe SK. Increasing liquid chromatography-tandem mass spectrometry throughput by mass tagging: a sample-multiplexed high-throughput assay for 25-hydroxyvitamin D_2 and D_3. Clin Chem 2011;57:431–440. [72]

from five PTAD-25(OH)D adducts are assigned to the respective patient samples based on original bar-codes. In a high-volume clinical laboratory, differential mass-tagging combined with multiplex HPLC can provide significant gain in throughput.

NIST has developed a standard reference material (SRM972) for assay standardization purposes. Additionally, deuterated and C-13 labeled internal standards and certified reference materials are commercially available (Cerrilliant, Sigma-Aldrich, St. Louis, MO, USA) and have been used in several methods [61]. However other lipophilic compounds like tetrahydrocannabinol can be successfully used as an internal standard [56]. Calibrators and QC materials can be prepared from reference materials in drug free serum or in phosphate buffered saline with or without albumin [61]. Commercial ready-to-use calibrators are also available (Chromsystems Instruments and Chemicals, Gräfelfing, Germany).

3.3 C3-EPI-25(OH)D

3.3.1 Biochemistry and Clinical Utility

The C3-epi-25(OH)D$_3$ and C3-epi-25(OH)D$_2$ in infants was first reported by Singh and coworkers in 2006. Epimerization at the C3 position of 25(OH)D gives rise to a structurally identical molecule except for differential asymmetry on carbon 3. It has been hypothesized that epimerization to C3-epi-25(OH)D is a result of an immature hepatic vitamin D metabolism common in infants. Studies evaluating calcemic effects of C3-epi-1,25(OH)$_2$D, limited to invitro models, show that it can bind to VDR with a lower affinity compared to 1,25(OH)$_2$D and can stimulate downstream gene transcription. Wide variation in C3-epi-25(OH)D concentration of up to 9–60% of total 25(OH)D has been observed in infants [24] and a concentration of ~3.3 ng/mL has been reported in adolescents and adults aged 1–94 years. Failure to separate C3-epi-25(OH)D from 25(OH)D will result in an overestimation of "true" 25(OH)D concentration. An overestimation of 9% infants (<1 year) and it 3% adults (1–94 years) as vitamin D sufficient has been observed by methods that cannot separate 3-epi-25(OH)D$_3$ from 25(OH)D$_3$.

While the present clinical utility of measuring 3-epi-25(OH)D$_3$ has been debated, that the ongoing NHANES study group's decision to use methods that can separate the C3-epi-25(OH)D$_3$ from 25(OH)D underscores its potential clinical utility [25]. Since the LC-MS/MS 25(OH)D assays across many clinical laboratories share several analytical aspects including sample preparation methods and chromatographic parameters, it is important for clinical laboratorians to develop methods that can separate C3-epi-25(OH)D$_3$ from 25(OH)D especially in infants.

3.3.2 Assay Procedure

NIST has developed a traceable reference material containing 25(OH)D$_2$, 25(OH)D$_3$, and 3-epi-25(OH)D$_3$ [44]. The accurate quantitation of C3-epi-25(OH)D by LC-MS/MS can be achieved by longer chromatographic run time enabling separation of epimers [24,25,73]. The sample volume used in most methods range between 100 and 200 μL. The sample preparation steps are similar to the ones used for 25(OH)D (Fig. 9.2). A 50 mm column used for 25(OH)D can be replaced with a longer column (100 mm) to achieve optimal chromatographic separation. For example, 5-dinitrobenzoyl-(R)-phenylglycine column (ChirexPGLY and DNB 100 × 4.6 mm, 250 μm, Phenomenex, Torrance, CA), Restek PFP Propyl analytical column (100 × 3.2 mm, 5 μm) [25], or UPLC BEH Phenyl column (2.1 mm × 100 mm, 1.7 μm, Waters Corporation, Milford MA) [73] have been utilized successfully. Additionally, a significantly higher injection volume compared to 25(OH)D may be needed to achieve optimal sensitivity [24]. The calibration curve obtained for 25(OH)D$_2$ or 25(OHD)$_3$ were also calculated using the normalized 25(OH)D$_2$ and 25(OH)D$_3$ calibration curves, and the total C3-epi-25-(OH)D concentration is reported as the sum of C3-epi-25(OH)D$_2$ and C3-epi-25(OH)D$_3$ concentrations. The mass spectrometer can be operated in the positive ion, ESI or APCI mode.

3.4 1,25(OH)$_2$D

3.4.1 Biochemistry and Clinical Utility

The major physiological effects of vitamin D including increased active absorption of calcium from the proximal intestine and to its effect on bone mineralization are accomplished via 1,25(OH)$_2$D, the active form of vitamin D. Vitamin D has its most prominent effects in the intestinal tissue, its biochemical and cellular actions in the enterocyte are directly associated with the transport of calcium across the mucosa. Hydroxylation at the 1α-position of 25(OH)D by CYP27B1, results in the formation of 1,25(OH)$_2$D in the mitochondria of the proximal tubules of the nephron. CYP27B1 is a cytochrome P-450-like, mixed function oxidase that uses molecular oxygen as its source of oxygen. 1α-hydroxylation by CYP27B1 is subject to regulation by several factors including PTH, serum phosphate, serum calcium, and 1,25(OH)$_2$D itself [15]. 1,25(OH)$_2$D binds to the VDR resulting in the induction of calcium binding protein and several other proteins in many animal species. Structure activity relationship studies show that C-l and C-25 hydroxyl groups are critical for receptor binding followed by the C-3 hydroxyl. Binding of 1,25(OH)$_2$D to VDR results in its nuclear translocation from cytoplasm into the nucleus. "Free" receptors have been shown to be present in the cytoplasm whereas the bound receptors are localized on the nucleus [15]. This nuclear translocation of VDR results in increased calcium binding protein messenger RNA levels. The transport of calcium in the intestine mediated by calcium binding protein has been shown to be biphasic in nature involving cells located at the tip of villus and those located at the crypt. Additionally, significant effects of 1,25(OH)$_2$D on the calcium uptake by endoplasmic reticulum and Golgi apparatus and on the induction of cyclic AMP and cyclic GMP production in embryonal duodena have been described [76].

The biliary and fecal route of excretion play a major role in the elimination of 1,25(OH)$_2$D in humans. About 60–70% of 1,25(OH)$_2$D administered as a radioactively labeled compound is excreted in stool in the form of more polar metabolites. The elimination half-life of 1,25(OH)$_2$D is approximately 10 h. The production and degradation rates of dihydoxylated vitamin D metabolites are shown in Table 9.5 [77]. 1,25(OH)$_2$D$_3$ (and 1,25(OH)$_2$D$_2$) undergoes several metabolic changes as a part of it degradation process: (1) side chain oxidation to form an inactive metabolite, calcitroic acid [78]; (2) 24-hydroxylation to 1,24,25(OH)$_3$D$_3$ the kidney, the intestine, cartilage, and others [79]; (3) formation of 24-oxo-1,25(OH)$_2$D$_3$ [80]; (4) formation of 1,25(OH)$_2$D$_3$ 23,26-lactone [81]; (5) excretion as polar metabolites such as 1,25(OH)$_2$D$_3$ monoglucuronides in bile [82,83]; and (6) formation of 1,25,26(OH)$_3$D$_3$ [84,85].

Table 9.5 Metabolic Clearance Rate (MCR), Production Rate (PR), and Degradation Rates of 1,25(OH)$_2$D$_3$ and 24,25(OH)$_2$D$_3$ [15]

	MCR (Liter/day)	PR (µg/day)	Biliary Excretion (6–8 h)	Fecal Excretion at 6–7 day	Urinary Excretion
1,25(OH)$_2$D$_3$	44.6 ± 5.7	~1.5	15.6 ± 1%	54 ± 6%	14 ± 2% (6 day)
24,25(OH)$_2$D$_3$	9.2 ± 1.5	26.4 ± 7.2	15.3 ± 1.3%	49 ± 2.7%	7.4 ± 1.8% (2 day)

Quantitation of circulating concentrations of biologically active hormone, $1,25(OH)_2D$, is useful for the diagnosis and/or management of patients with chronic kidney disease (CKD) [16,86], oncogenic osteomalacia, or oncogenic hypophosphatemic osteomalacia syndrome [87], aberrations in phosphate homeostasis [87,88], and CYP24A1 mutations [26]. Oncogenic osteomalacia is associated with hypophosphatemia (low serum phosphate), elevated alkaline phosphatase, and low serum $1,25(OH)_2D$. Progressive loss of kidney function in advanced CKD is the most common cause of secondary hyperparathyroidism due to inadequate phosphate excretion and ineffective production of $1,25(OH)_2D$. This ultimately leads to hypocalcemia, increased PTH production and low serum $1,25(OH)_2D$. The National Kidney Foundation Kidney Disease Outcomes Quality Initiative (KDOQI) guidelines recommend that CKD patients with kidney failure (stage 5) receive an active vitamin D sterol (calcitriol, alfacalcidol, paricalcitol, or doxercalciferol), if the plasma levels of intact PTH are > 300 g/mL. Two well documented hereditary defects in vitamin D metabolic pathway are, vitamin D-dependent rickets type I (VDDR I) and type II (VDDR II). VDDR I is caused by a reduced function of the CYP27B1 leading to muscle weakness and rickets. Rickets in VDDR I can be treated by a normal physiologic dose of $1,25(OH)_2D_3$. VDDR II comprises a range of VDR gene mutations associated with an early onset of severe rickets and characteristic alopecia (spot baldness) [89]. Substantial doses of vitamin D analogs and calcium supplementation is usually required for the treatment; however, the response to therapy is sometimes variable. The response to $1,25(OH)_2D$ therapy can be monitored by measuring serum and urinary minerals and PTH. Loss of CYP24A1 function results in a build-up of the active hormone, $1,25(OH)_2D$. Persistent hypercalcemia with elevated serum $1,25(OH)_2D$ and suppressed or low serum PTH should prompt $25(OH)D/24,25(OH)_2D$ ratio analysis to rule out a CYP24A1 mutation [26]. Clinical utility of quantitation of vitamin D and its metabolites are summarized in Table 9.1. $1,25(OH)_2D$ circulates in very low concentrations (15–65 pg/mL) in normal subjects and can be challenging to quantify due to the high demand placed on sensitivity and specificity of the assay technique.

3.4.2 Assay Procedure

The sample preparation procedure for a commonly used method in our and other clinical laboratories [54] is very similar to the IDS immunoassay discussed earlier. Due to the relatively low circulating concentration of $1,25(OH)_2D$ immunoenrichment with a solid phase bound antibody greatly enhances the sensitivity and specificity of the method. Compared to $25(OH)D$ fourfold to fivefold larger sample volume (400–500 µL) of serum is usually used. The first step is the separation of the vitamin D metabolites from VBP. Solvents/solvent mixtures similar to that used in $25(OH)D$ quantitation are used (Fig. 9.2). SPE can also be used to obtain a delipidized, deproteinized sample extract. Methods may or may not use an immunoenrichment step for $1,25(OH)_2D$. PTAD derivatization has been used by almost all LC-MS/MS methods to enhance sensitivity. Improved sensitivity of contemporary mass spectrometers have enable successful quantitation of $1,25(OH)_2D$ without immunopurification [62]. In one method, a combination of SPE and fixed-charge derivitization using PTAD chemistry was found comparable to immunoextraction technique. C18 or equivalent chromatographic columns (50 mm) are used and the mass spectrometer is commonly operated in the ESI mode.

3.5 $24,25(OH)_2D$

3.5.1 Biochemistry and Clinical Utility

Circulating $1,25(OH)_2D$ constitutes only a minor fraction in picomolar concentrations of vitamin D related metabolites compared to $25(OH)D$, which circulates in nanomolar concentration in nonvitamin

D-deficient normal adults. Another dihydroxylated metabolite that is present approximately 7–35% of 25(OH)D concentration is 24,25(OH)$_2$D. 24,25(OH)$_2$D does not have a physiologically relevant role but is a surrogate marker for measuring CYP24A1 activity. The role of CYP24A1 enzyme in calcium homeostasis was demonstrated in CYP24A1 gene knock out mouse models. These animals developed severe hypercalcemia, hypercalciuria, and biochemical evidence of impaired vitamin D metabolism [90]. CYP24A1 biallelic mutations in humans leading to a nonfunctional CYP24A1 enzyme were first identified as the cause of idiopathic infantile hypercalcemia where an inability to metabolize the active hormone, 1,25(OH)$_2$D results in vitamin D supplementation associated toxicity [91]. Since then several more families with mutations in the CYP24A1 gene have now been described as the cause of hypercalcemia, elevated 1,25(OH)$_2$D, hypercalciuria, and nephrolithiasis commonly seen in these patients [26,63,92–98]. The prevalence of this recessive genetic defect is still unknown but appears to be higher than originally thought. Genotype-phenotype correlation studies suggest a variable penetrance and that haploinsufficiency is not associated with CYP24A1 deficiency [99].

A 25(OH)D/24,25(OH)$_2$D ratio of 99 or greater identifies patients with CYP24A1 mutations [26,63,99]. In patients presenting with hypercalcemia and vitamin D toxicity, 24,25(OH)$_2$D measurement can aide in differentiating between an iatrogenic versus genetic cause of hypercalcemia [26,69]. It is important to note that 24,25(OH)$_2$D measurement alone provides no useful information about CYP24A1 activity because of a strong linear correlation between 25(OH)D and 24,25(OH)$_2$D. Therefore 24,25(OH)$_2$D must be interpreted with a concomitant 25(OH)D measurement performed on LC-MS/MS platforms since current immunoassays cannot differentiate between 25(OH)D and 24,25(OH)$_2$D. Even a very low 24,25(OH)$_2$D value alone is not clinically sufficient to identify a patient with CYP24A1 mutation, since it could be merely indicative of low circulating 25(OH)D levels. As a hypothetical example in a patient with 25(OH)D = 5 ng/mL and 24,25(OH)$_2$D = 0.5 ng/mL shows normal CYP24A1 function, whereas in a patient with 25(OH)D = 80 ng/mL and 24,25(OH)2D = 0.5 ng/mL CYP24A1 mutation is highly likely. Of note, a 25(OH)D concentration < 20 ng/mL, an increase in 25(OH)D/24,25(OH)$_2$D is observed (Fig. 9.6) [63,99]. A role of 25(OH)D/24,25(OH)$_2$D as a biomarker of optimal response to vitamin D supplementation has been proposed [64]. However, based on currently available scientific information 25(OH)D/24,25(OH)$_2$D measurement is indicated only in patients presenting with persistently elevated 1,25(OH)$_2$D, hypercalcemia along with suppressed PTH with other causes of hypercalcemia having been ruled out [26,54,63,64,99,100].

3.5.2 Assay Procedure

Since 24,25(OH)$_2$D circulates at a much higher concentration compared to 1,25(OH)$_2$D, immunoenrichment is not required to accurately quantitate 24,25(OH)$_2$D. Sample preparation steps are similar to 25(OH)D (Fig. 9.2). Methods with and without PTAD derivatization have been described [64,100]. Clinical laboratories measuring 25(OH)D by LC-MS/MS currently should be able to simultaneously quantitate 25(OH)D and 24,25(OH)$_2$D by adjusting chromatographic times, optimizing ion-pairs and adding a set of calibrators and the corresponding deuterated analogs of 24,25(OH)$_2$D$_2$ and 24,25(OH)$_2$D$_3$. The calibrators can be prepared in drug free and charcoal tripped human serum or in phosphate buffered saline supplemented with 1–25 (w/vol) of albumin. It may be necessary to chromatographically separate the two isomers of 24,25(OH)$_2$D-PTAD adduct as an interference from a closely related, 25,26(OH)$_2$D that has the same fragmentation pattern as 24,25(OH)$_2$D, with major isomer has been observed in our laboratory. Interference from 1,25(OH)$_2$D-PTAD is unlikely to be a problem due to low circulating concentrations of 1,25(OH)$_2$D and fragmentation pattern different from 24,25(OH)$_2$D-PTAD. Chromatography can be performed on 50 mm high pressure liquid chromatography (HPLC) or

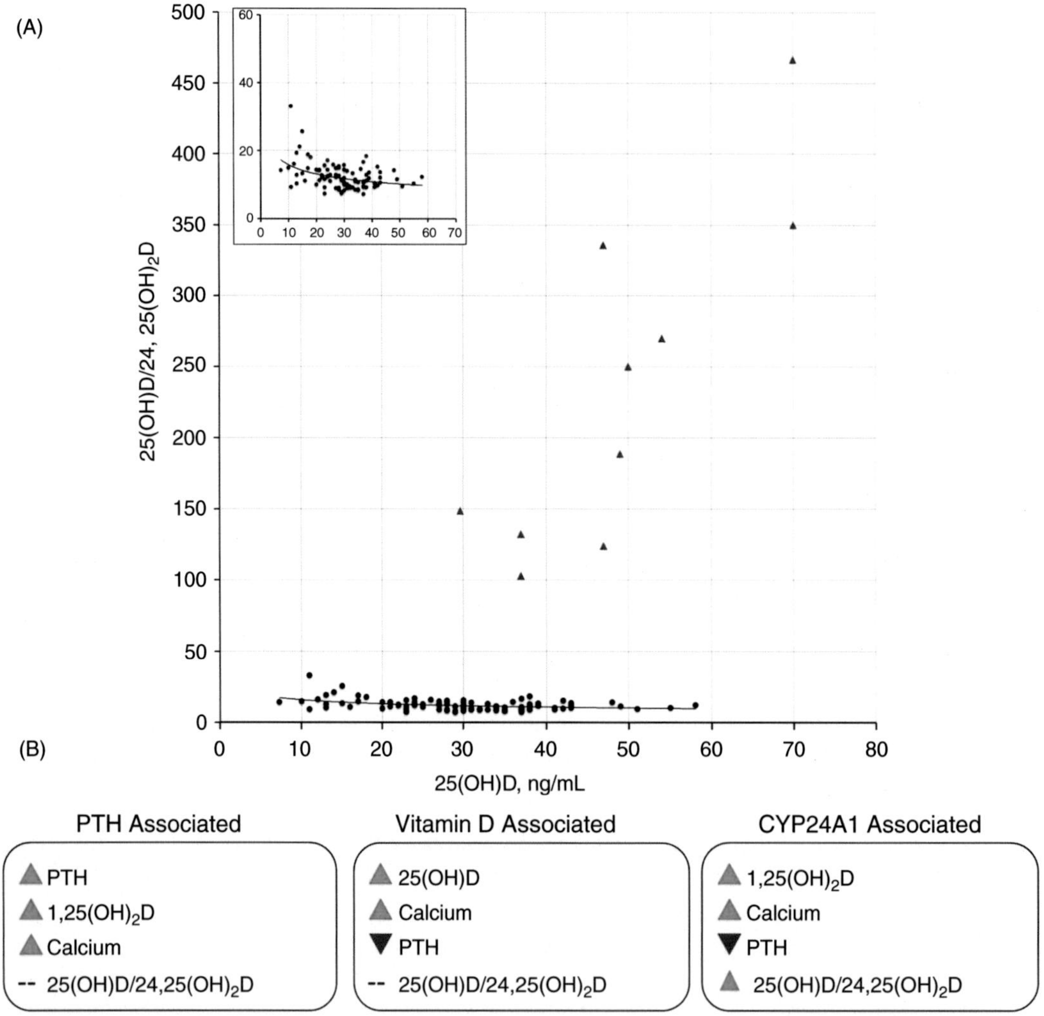

FIGURE 9.6

(A) Association between 25(OH)D/24,25(OH)$_2$D in normal subjects (*black circles*) and in patients with CYP24A1 mutations (*Red triangles*). A slight increase in the 25(OH)D/24,25(OH)$_2$D is observed at 25(OH)D <20 ng/mL. Even though 25(OH)D/24,25(OH)$_2$D in vitamin D deficient patients is higher, it distinguishes between unaffected and affected subjects. (B) Metabolic profiles observed in different forms of hypercalcemia.

Adapted from Ketha H, Kumar R, Singh RJ: LC-MS/MS for Identifying Patients with CYP24A1 Mutations. Clin Chem 2016; 62:236–242. [26]

ultra-pressure liquid chromatography (UPLC) reverse phase columns using water: methanol gradient methods. Examples of chromatographic columns that have been successfully used include BEH-Phenyl 1.7 μm, 2.1 × 50 mm UPLC column (Waters Corporation, Milford MA), XDB-C8, 2.1 × 50 mm column (Agilent Technologies, Santa Clara, CA), Eclipse C8 3.0 mm × 50 mm, 1.8 μm column. Other equivalent column chemistries can be easily utilized.

4 CONCLUSIONS

Vitamin D and PTH are the principle regulators of calcium homeostasis. Measurement of vitamin D metabolites is clinically valuable to investigate: (1) vitamin D nutritional deficiency; (2) hypocalcemia; (3) hypercalcemia; (3) vitamin D toxicity; (4) response to vitamin D supplementation; and (5) genetic causes of aberrant vitamin D metabolism (Table 9.1). 25(OH)D, 1,25(OH)$_2$D, C3-epi-25(OH)D, and more recently 24,25(OH)$_2$D are the clinically relevant vitamin D metabolites measured in clinical laboratories by LC-MS/MS—among which 25(OH)D is by far the most commonly utilized. With a growing interest in the role of vitamin D metabolites in health and disease and due to the limitations of immunoassay platforms to accurately quantitate vitamin D metabolites, the number of laboratories offering LC-MS/MS vitamin D assays has grown exponentially in the last decade. To say that several institutions in the past decade, commercial and academic alike, started implementing LC-MS/MS as a long-term and a financially viable option in their clinical laboratories primarily due to a steep increase in requests for vitamin D testing—certainly won't be an overstatement. LC-MS/MS is a useful and a financially viable option that has been tested widely and proven to provide an accurate quantitation of vitamin D metabolites in clinical laboratories. The decision to implement a LC-MS/MS platform for vitamin D testing by a clinical laboratory should be guided by an institution-specific balance of clinical, technical, financial, and strategic considerations.

It is evident from the era of the first vitamin D assays that used organic solvent extraction and competitive binding assay with VBP followed by the advent of manual RIAs and then finally automated RIAs and chemiluminescent immunoassays – that assay methodologies move through cyclical process of innovation, wide scale examination and implementation, and finally to complete automation. A key point to understand the evolution and current state of vitamin D metabolite assays is that the gains in throughput and turn-around time achieved by automated immunoassays was, in part, due to reduction in manual sample preparation steps (like antibody antigen complex separation from the matrix components prior to quantitation in an extraction RIA) that helped remove majority of the extraneous matrix components allowing an accurate final quantitation. While an automated approach resulted in platforms and methods that are amenable to clinical laboratories processing routine analytes in high volumes, the lack of accuracy and specificity especially for quantitation of low abundance steroid hormones have become apparent. It is interesting to note that while 25(OH)D extraction RIA shows very good agreement with LC-MS/MS methods, the nonextraction methods show poor correlation and wide intermethod variability [24,27]. The sample extraction process for LC-MS/MS assays is still essential to improve the specificity of vitamin D metabolite quantitation. It is not difficult to envision that in the coming decades academic and industry experts from a variety of fields including microfluidics, automated sample preparation platforms, and mass spectrometer manufacturing will collaborate and completely automated mass spectrometry platforms will become an integral part of clinical laboratories.

REFERENCES

[1] deLuca HF. The metabolism, physiology and funciton of vitamin D. Massachusetts, USA: Martinus Nijhoff Publishing; 1984.

[2] Hess A. The history of rickets. Philadelphia: Lee & Febiger; 1929.

[3] Holst A. Experimental studies relating to "ship-beri-beri" and scurvy. J Hyg (Lond) 1907;7:619–33.

[4] Funk C. On the chemical nature of the substance which cures polyneuritis in birds induced by a diet of polished rice. J Physiol 1911;43:395–400.

[5] McCollum EV, Simmonds N, Becker JE, Shipley PG. Studies on experimental rickets: xxi. An experimental demonstration of the existence of a vitamin which promotes calcium deposition. J Biol Chem 1922;53: 293–312.

[6] Mellanby E. An experimental investigation on rickets. Lancet 1919;193:407–12.

[7] McCollum EV, Davis M. The necessity of certain lipins in the diet during growth. J Biol Chem 1913;15: 167–75.

[8] Sallis JD, Deluca HF, Rasmussen H. Parathyroid hormone-dependent uptake of inorganic phosphate by mitochondria. J Biol Chem 1963;238:4098–102.

[9] Orr WJ, Holt LE Jr, Wilkins L, Boone FH. The calcium and phosphorus metabolism in rickets, with special reference to ultraviolet ray therapy. Am J Dis Child 1923;26:362–72.

[10] Shipley PG, Kramer B, Howland J. Studies upon calcification in vitro. Biochem J 1926;20:379–87.

[11] Holtrop ME, Cox KA, Clark MB, Holick MF, Anast CS. 1,25-dihydroxycholecalciferol stimulates osteoclasts in rat bones in the absence of parathyroid hormone. Endocrinology 1981;108:2293–301.

[12] Lund J, DeLuca HF. Biologically active metabolite of vitamin D3 from bone, liver, and blood serum. J Lipid Res 1966;7:739–44.

[13] Rasmussen H, Deluca H, Arnaud C, Hawker C, Vonstedingk M. The relationship between vitamin D and parathyroid hormone. J Clin Invest 1963;42:1940–6.

[14] Sallis JD, Deluca HF, Rasmussen H. Parathyroid hormone stimulation of phosphate uptake by rat liver mitochondria. Biochem Biophys Res Commun 1963;10:266–70.

[15] Kumar R. The metabolism and mechanism of action of 1,25-dihydroxyvitamin D3. Kidney Int 1986;30:793–803.

[16] Audran M, Kumar R. The physiology and pathophysiology of vitamin D. Mayo Clin Proc 1985;60:851–66.

[17] Kumar R, Tebben PJ, Thompson JR. Vitamin D and the kidney. Arch Biochem Biophys 2012;523:77–86.

[18] Bikle DD, Gee E, Halloran B, Kowalski MA, Ryzen E, Haddad JG. Assessment of the free fraction of 25-hydroxyvitamin D in serum and its regulation by albumin and the vitamin D-binding protein. J Clin Endocrinol Metab 1986;63:954–9.

[19] Esvelt RP, De Luca HF. Calcitroic acid: biological activity and tissue distribution studies. Arch Biochem Biophys 1981;206:403–13.

[20] Holick MF, Schnoes HK, DeLuca HF, Gray RW, Boyle IT, Suda T. Isolation and identification of 24,25-dihydroxycholecalciferol, a metabolite of vitamin D made in the kidney. Biochemistry 1972;11:4251–5.

[21] Burtis CA, Ashwood ER, Bruns DE. Tietz textbook of clinical chemistry and molecular diagnostics. UK: Elsevier Health Sciences; 2012.

[22] Hoofnagle AN, Eckfeldt JH, Lutsey PL. Vitamin D–binding protein concentrations quantified by mass spectrometry. New Engl J Med 2015;373:1480–2.

[23] Powe CE, Evans MK, Wenger J, Zonderman AB, Berg AH, Nalls M, Tamez H, Zhang D, Bhan I, Karumanchi SA, Powe NR, Thadhani R. Vitamin D–binding protein and vitamin D status of black Americans and white Americans. New Engl J Med 2013;369:1991–2000.

[24] Singh RJ, Taylor RL, Reddy GS, Grebe SK. C-3 epimers can account for a significant proportion of total circulating 25-hydroxyvitamin D in infants, complicating accurate measurement and interpretation of vitamin D status. J Clin Endocrinol Metab 2006;91:3055–61.

[25] Strathmann FG, Sadilkova K, Laha TJ, LeSourd SE, Bornhorst JA, Hoofnagle AN, Jack R. 3-epi-25 hydroxyvitamin D concentrations are not correlated with age in a cohort of infants and adults. Clin Chim Acta 2012;413:203–6.

[26] Ketha H, Kumar R, Singh RJ. LC-MS/MS for identifying patients with CYP24A1 mutations. Clin Chem 2016;62:236–42.

[27] Farrell CJ, Martin S, McWhinney B, Straub I, Williams P, Herrmann M. State-of-the-art vitamin D assays: a comparison of automated immunoassays with liquid chromatography-tandem mass spectrometry methods. Clin Chem 2012;58:531–42.

[28] Review of the funding arrangements for pathology services. Australian Department of health and Ageing. Australia, 2011.

[29] Shahangian S, Alspach TD, Astles JR, Yesupriya A, Dettwyler WK. Trends in laboratory test volumes for medicare part B reimbursements, 2000-2010. Arch Pathol Lab Med 2014;138:189–203.

[30] Rosen CJ. Clinical practice. Vitamin D insufficiency. N Engl J Med 2011;364:248–54.

[31] Holick MF. Vitamin D deficiency. New Engl J Med 2007;357:266–81.

[32] Zhang R, Naughton DP. Vitamin D in health and disease: current perspectives. Nutr J 2010;9:65.

[33] Haddad JG, Chyu KJ. Competitive protein-binding radioassay for 25-hydroxycholecalciferol. J Clin Endocrinol Metab 1971;33:992–5.

[34] Wallace AM, Gibson S, de la Hunty A, Lamberg-Allardt C, Ashwell M. Measurement of 25-hydroxyvitamin D in the clinical laboratory: current procedures, performance characteristics and limitations. Steroids 2010;75:477–88.

[35] Hollis BW, Napoli JL. Improved radioimmunoassay for vitamin D and its use in assessing vitamin D status. Clin Chem 1985;31:1815–9.

[36] Ketha H, Kaur S, Grebe SK, Singh RJ. Clinical applications of LC-MS sex steroid assays: evolution of methodologies in the 21st century. Curr Opin Endocrinol Diabetes Obes 2014;21:217–26.

[37] Grebe SK, Singh RJ. LC-MS/MS in the clinical laboratory—where to from here? Clin Biochem Rev 2011;32:5–31.

[38] Ketha H, Girtman A, Singh RJ. Estradiol assays—the path ahead. Steroids 2014;99:39–44.

[39] DeLuca HF, Schnoes HK. Vitamin D: recent advances. Annu Rev Biochem 1983;52:411–39.

[40] Binkley N, Krueger D, Cowgill CS, Plum L, Lake E, Hansen KE, DeLuca HF, Drezner MK. Assay variation confounds the diagnosis of hypovitaminosis D: a call for standardization. J Clin Endocrinol Metab 2004;89:3152–7.

[41] Carter GD, Carter R, Jones J, Berry J. How accurate are assays for 25-hydroxyvitamin D? Data from the international vitamin D external quality assessment scheme. Clin Chem 2004;50:2195–7.

[42] Zerwekh JE. Blood biomarkers of vitamin D status. Am J Clin Nutr 2008;87:1087s–91s.

[43] Glendenning P, Noble JM, Taranto M, Musk AA, McGuiness M, Goldswain PR, Fraser WD, Vasikaran SD. Issues of methodology, standardization and metabolite recognition for 25-hydroxyvitamin D when comparing the DiaSorin radioimmunoassay and the Nichols Advantage automated chemiluminescence protein-binding assay in hip fracture cases. Ann Clin Biochem 2003;40:546–51.

[44] Phinney KW, Bedner M, Tai SS, Vamathevan VV, Sander LC, Sharpless KE, Wise SA, Yen JH, Schleicher RL, Chaudhary-Webb M, Pfeiffer CM, Betz JM, Coates PM, Picciano MF. Development and certification of a standard reference material for vitamin D metabolites in human serum. Anal Chem 2012;84:956–62.

[45] Costelloe SJ, Woolman E, Rainbow S, Stratiotis L, O'Garro G, Whiting S, Thomas M. Is high-throughput measurement of 25-hydroxyvitamin D3 without 25-hydroxyvitamin D2 appropriate for routine clinical use? Ann Clin Biochem 2009;46:86–7.

[46] Beastall G, Rainbow S. Vitamin D reinvented: implications for clinical chemistry. Clin Chem 2008;54: 630–2.

[47] Hollis BW. Editorial: the determination of circulating 25-hydroxyvitamin D: no easy task. J Clin Endocrinol Metab 2004;89:3149–51.

[48] Brumbaugh PF, Haussler DH, Bressler R, Haussler MR. Radioreceptor assay for 1 alpha,25-dihydroxyvitamin D3. Science 1974;183:1089–91.

[49] Clemens TL, Hendy GN, Graham RF, Baggiolini EG, Uskokovic MR, O'Riordan JL. A radioimmunoassay for 1,25-dihydroxycholecalciferol. Clin Sci Mol Med 1978;54:329–32.

[50] Fraser WD, Milan AM. Vitamin D assays: past and present debates, difficulties, and developments. Calcif Tissue Int 2013;92:118–27.

[51] Hollis BW, Kamerud JQ, Kurkowski A, Beaulieu J, Napoli JL. Quantification of circulating 1,25-dihydroxyvitamin D by radioimmunoassay with 125I-labeled tracer. Clin Chem 1996;42:586–92.

[52] Fraser WD, Durham BH, Berry JL, Mawer EB. Measurement of plasma 1,25 dihydroxyvitamin D using a novel immunoextraction technique and immunoassay with iodine labelled vitamin D tracer. Ann Clin Biochem 1997;34(Pt 6):632–7.

[53] Hollis BW. 1,25-Dihydroxyvitamin D3-26,23-lactone interferes in determination of 1,25-dihydroxyvitamin D by RIA after immunoextraction. Clin Chem 1995;41:1313–4.

[54] Strathmann FG, Laha TJ, Hoofnagle AN. Quantification of 1alpha,25-dihydroxy vitamin D by immunoextraction and liquid chromatography-tandem mass spectrometry. Clin Chem 2011;57:1279–85.

[55] Maunsell Z, Wright DJ, Rainbow SJ. Routine isotope-dilution liquid chromatography-tandem mass spectrometry assay for simultaneous measurement of the 25-hydroxy metabolites of vitamins D2 and D3. Clin Chem 2005;51:1683–90.

[56] Saenger AK, Laha TJ, Bremner DE, Sadrzadeh SM. Quantification of serum 25-hydroxyvitamin D(2) and D(3) using HPLC-tandem mass spectrometry and examination of reference intervals for diagnosis of vitamin D deficiency. Am J Clin Pathol 2006;125:914–20.

[57] Chen H, McCoy LF, Schleicher RL, Pfeiffer CM. Measurement of 25-hydroxyvitamin D3 (25OHD3) and 25-hydroxyvitamin D2 (25OHD2) in human serum using liquid chromatography-tandem mass spectrometry and its comparison to a radioimmunoassay method. Clin Chim Acta 2008;391:6–12.

[58] Knox S, Harris J, Calton L, Wallace AM. A simple automated solid-phase extraction procedure for measurement of 25-hydroxyvitamin D3 and D2 by liquid chromatography-tandem mass spectrometry. Ann Clin Biochem 2009;46:226–30.

[59] Tsugawa N, Suhara Y, Kamao M, Okano T. Determination of 25-hydroxyvitamin D in human plasma using high-performance liquid chromatography—tandem mass spectrometry. Anal Chem 2005;77:3001–7.

[60] Vogeser M, Kyriatsoulis A, Huber E, Kobold U. Candidate reference method for the quantification of circulating 25-hydroxyvitamin D3 by liquid chromatography-tandem mass spectrometry. Clin Chem 2004;50: 1415–7.

[61] Vogeser M. Quantification of circulating 25-hydroxyvitamin D by liquid chromatography–tandem mass spectrometry. J Steroid Biochem Mol Biol 2010;121:565–73.

[62] Chan N, Kaleta EJ. Quantitation of 1alpha,25-dihydroxyvitamin D by LC-MS/MS using solid-phase extraction and fixed-charge derivitization in comparison to immunoextraction. Clin Chem Lab Med 2015;53: 1399–407.

[63] Kaufmann M, Gallagher JC, Peacock M, Schlingmann K-P, Konrad M, DeLuca HF, Sigueiro R, Lopez B, Mourino A, Maestro M, St-Arnaud R, Finkelstein JS, Cooper DP, Jones G. Clinical utility of simultaneous quantitation of 25-hydroxyvitamin D and 24,25-dihydroxyvitamin D by LC-MS/MS involving derivatization with DMEQ-TAD. J Clin Endocrinol Metab 2014;99:2567–74.

[64] Wagner D, Hanwell HE, Schnabl K, Yazdanpanah M, Kimball S, Fu L, Sidhom G, Rousseau D, Cole DE, Vieth R. The ratio of serum 24,25-dihydroxyvitamin D(3) to 25-hydroxyvitamin D(3) is predictive of 25-hydroxyvitamin D(3) response to vitamin D(3) supplementation. J Steroid Biochem Mol Biol 2011;126:72–7.

[65] Adamec J, Jannasch A, Huang J, Hohman E, Fleet JC, Peacock M, Ferruzzi MG, Martin B, Weaver CM. Development and optimization of an LC-MS/MS-based method for simultaneous quantification of vitamin D2, vitamin D3, 25-hydroxyvitamin D2 and 25-hydroxyvitamin D3. J Sep Sci 2011;34:11–20.

[66] Ponchon G, Deluca HF. The role of the liver in the metabolism of vitamin D. J Clin Invest 1969;48:1273–9.

[67] Vieth R. Vitamin D supplementation, 25-hydroxyvitamin D concentrations, and safety. Am J Clin Nutr 1999;69:842–56.

[68] Jacobus CH, Holick MF, Shao Q, Chen TC, Holm IA, Kolodny JM, Fuleihan GE, Seely EW. Hypervitaminosis-D associated with drinking milk. New Engl J Med 1992;326:1173–7.

[69] Ketha H, Wadams H, Lteif A, Singh RJ. Iatrogenic vitamin D toxicity in an infant—a case report and review of literature. J Steroid Biochem Mol Biol 2015;148:14–8.

[70] Jones G. Pharmacokinetics of vitamin D toxicity. Am J Clin Nutr 2008;88:582S–6S.

[71] Aronov PA, Hall LM, Dettmer K, Stephensen CB, Hammock BD. Metabolic profiling of major vitamin D metabolites using Diels-Alder derivatization and ultra-performance liquid chromatography-tandem mass spectrometry. Anal Bioanal Chem 2008;391:1917–30.

[72] Netzel BC, Cradic KW, Bro ET, Girtman AB, Cyr RC, Singh RJ, Grebe SK. Increasing liquid chromatography-tandem mass spectrometry throughput by mass tagging: a sample-multiplexed high-throughput assay for 25-hydroxyvitamin D2 and D3. Clin Chem 2011;57:431–40.

[73] Yu S, Zhou W, Zhang R, Cheng X, Fang H, Han J, Cheng Q, Su W, Qiu L. Validation and comparison of a rapid liquid chromatography tandem mass spectrometry method for serum 25OHD with the efficiency of separating 3-epi 25OHD. Clin Biochem 2016;49(13–14):1004–8.

[74] Hoofnagle AN, Laha TJ, Donaldson TF. A rubber transfer gasket to improve the throughput of liquid-liquid extraction in 96-well plates: application to vitamin D testing. J Chromatogr B Analyt Technol Biomed Life Sci 2010;878:1639–42.

[75] Higashi T, Awada D, Shimada K. Simultaneous determination of 25-hydroxyvitamin D2 and 25-hydroxyvitamin D3 in human plasma by liquid chromatography-tandem mass spectrometry employing derivatization with a Cookson-type reagent. Biol Pharm Bull 2001;24:738–43.

[76] MacLaughlin JA, Weiser MM, Freedman RA. Biphasic recovery of vitamin D-dependent Ca2+ uptake by rat intestinal Golgi membranes. Gastroenterology 1980;78:325–32.

[77] Gray RW, Caldas AE, Wilz DR, Jacob Lemann J, Smith GA, Deluca HF. Metabolism and excretion of 3H-1,2 5-(OH)2-vitamin D3 in healthy adults. J Clin Endocrinol Metab 1978;46:756–65.

[78] Harnden D, Kumar R, Holick MF, Deluca HF. Side chain metabolism of 25-hydroxy-[26,27-14C] vitamin D3 and 1,25-dihydroxy-[26,27-14C] vitamin D3 in vivo. Science 1976;193:493–4.

[79] Holick MF, Kleiner-Bossaller A, Schnoes HK, Kasten PM, Boyle IT, DeLuca HF. 1,24,25-trihydroxyvitamin D3. A metabolite of vitamin D3 effective on intestine. J Biol Chem 1973;248:6691–6.

[80] Mayer E, Bishop JE, Chandraratna RA, Okamura WH, Kruse JR, Popjak G, Ohnuma N, Norman AW. Isolation and identification of 1,25-dihydroxy-24-oxo-vitamin D3 and 1,23,25-trihydroxy-24-oxo-vitamin D3. New metabolites of vitamin D3 produced by a C-24 oxidation pathway of metabolism for 1,25-dihydroxyvitamin D3 present in intestine and kidney. J Biol Chem 1983;258:13458–65.

[81] Kumar R, Harnden D, DeLuca HF. Metabolism of 1,25-dihydroxyvitamin D3: evidence for side-chain oxidation. Biochemistry 1976;15:2420–3.

[82] Litwiller RD, Mattox VR, Jardine I, Kumar R. Evidence for a monoglucuronide of 1,25-dihydroxyvitamin D3 in rat bile. J Biol Chem 1982;257:7491–4.

[83] Wiesner RH, Kumar R, Seeman E, Go VL. Enterohepatic physiology of 1,25-dihydroxyvitamin D3 metabolites in normal man. J Lab Clin Med 1980;96:1094–100.

[84] Tanaka Y, Schnoes HK, Smith CM, DeLuca HF. 1,25,26-trihydroxyvitamin D3: isolation, identification, and biological activity. Arch Biochem Biophys 1981;210:104–9.

[85] Reinhardt TA, Napoli JL, Beitz DC, Littledike ET, Horst RL. 1,24,25-trihydroxyvitamin D3: a circulating metabolite in vitamin D2-treated bovine. Arch Biochem Biophys 1982;213:163–8.

[86] Kumar R, Abboud CF, Riggs BL. The effect of elevated prolactin levels on plasma 1,25-dihydroxyvitamin D and intestinal absorption of calcium. Mayo Clin Proc 1980;55:51–3.

[87] Berndt T, Kumar R. Phosphatonins and the regulation of phosphate homeostasis. Ann Rev Physiol 2007;69:341–59.

[88] Audran M, Gross M, Kumar R. The physiology of the vitamin D endocrine system. Semin Nephrol 1986;6: 4–20.

[89] Kato S, Takeyama K, Kitanaka S, Murayama A, Sekine K, Yoshizawa T. In vivo function of VDR in gene expression-VDR knock-out mice. J Steroid Biochem Mol Biol 1999;69:247–51.

[90] St-Arnaud R, Arabian A, Travers R, Barletta F, Raval-Pandya M, Chapin K, Depovere J, Mathieu C, Christakos S, Demay MB, Glorieux FH. Deficient mineralization of intramembranous bone in vitamin D-24-hydroxylase-ablated mice is due to elevated 1,25-dihydroxyvitamin D and not to the absence of 24,25-dihydroxyvitamin D. Endocrinology 2000;141:2658–66.

[91] Schlingmann KP, Kaufmann M, Weber S, Irwin A, Goos C, John U, Misselwitz J, Klaus G, Kuwertz-Broking E, Fehrenbach H, Wingen AM, Guran T, Hoenderop JG, Bindels RJ, Prosser DE, Jones G, Konrad M. Mutations in CYP24A1 and idiopathic infantile hypercalcemia. N Engl J Med 2011;365:410–21.

[92] Jacobs TP, Kaufman M, Jones G, Kumar R, Schlingmann KP, Shapses S, Bilezikian JP. A lifetime of hypercalcemia and hypercalciuria, finally explained. J Clin Endocrinol Metab 2014;99:708–12.

[93] Dauber A, Nguyen TT, Sochett E, Cole DE, Horst R, Abrams SA, Carpenter TO, Hirschhorn JN. Genetic defect in CYP24A1, the vitamin D 24-hydroxylase gene, in a patient with severe infantile hypercalcemia. J Clin Endocrinol Metab 2012;97:E268–74.

[94] Dinour D, Beckerman P, Ganon L, Tordjman K, Eisenstein Z, Holtzman EJ. Loss-of-function mutations of CYP24A1, the vitamin D 24-hydroxylase gene, cause long-standing hypercalciuric nephrolithiasis and nephrocalcinosis. J Urol 2013;190:552–7.

[95] Dowen FE, Sayers JA, Hynes AM, Sayer JA. CYP24A1 mutation leading to nephrocalcinosis. Kidney Int 2014;85:1475.

[96] Sayers J, Haynes AM, Rice SJ, Hogg P, Sayer JA. Searching for CYP24A1 mutations in cohorts of patients with calcium nephrolithiasis. OA Nephrol 2013;1:6.

[97] Nesterova G, Malicdan MC, Yasuda K, Sakaki T, Vilboux T, Ciccone C, Horst R, Huang Y, Golas G, Introne W, Huizing M, Adams D, Boerkoel CF, Collins MT, Gahl WA. 1,25-(OH)2D-24 Hydroxylase (CYP24A1) deficiency as a cause of nephrolithiasis. Clin J Am Soc Nephrol 2013;8:649–57.

[98] O'Keeffe DT, Tebben PJ, Kumar R, Singh RJ, Wu Y, Wermers RA. Clinical and biochemical phenotypes of adults with monoallelic and biallelic CYP24A1 mutations: evidence of gene dose effect. Osteoporos Int, 2016.

[99] Molin A, Baudoin R, Kaufmann M, Souberbielle JC, Ryckewaert A, Vantyghem MC, Eckart P, Bacchetta J, Deschenes G, Kesler-Roussey G, Coudray N, Richard N, Wraich M, Bonafiglia Q, Tiulpakov A, Jones G, Kottler ML. CYP24A1 mutations in a cohort of hypercalcemic patients: evidence for a recessive trait. J Clin Endocrinol Metab 2015;100:E1343–52.

[100] Cashman KD, Hayes A, Galvin K, Merkel J, Jones G, Kaufmann M, Hoofnagle AN, Carter GD, Durazo-Arvizu RA, Sempos CT. Significance of serum 24,25-dihydroxyvitamin D in the assessment of vitamin D status: a double-edged sword? Clin Chem 2015;61:636–45.

STEROID HORMONES

10

J.C. Cook-Botelho*, L.M. Bachmann**, D. French†

*Clinical Chemistry Branch, Division of Laboratory Sciences, Centers for Disease Control and Prevention, Atlanta, GA, United States; **Department of Pathology, Virginia Commonwealth University, Richmond, VA, United States; †Department of Laboratory Medicine, University of California San Francisco, San Francisco, CA, United States

1 STEROID HORMONE APPLICATIONS FOR CLINICAL USE

Steroid hormone measurements are widely used in clinical research, public health assessments, and patient care. The major classes of steroid hormones considered in this chapter include androgens, estrogens, progestogens, mineralocorticoids, and glucocorticoids.

Steroid hormones have three six-member rings (A, B, C) and one five-member ring (D) called a cyclopentanophenophenanthrene ring system (Fig. 10.1). All hormones have an oxygen at C3 on the A ring either as a double bonded oxygen ($=O$) or a hydroxyl group ($-OH$). Methyl groups ($-CH_3$) are typically found at C10 and C13. The functional groups on C17 upon the D ring vary significantly. Many steroid hormones have the same molecular weight but different functions due to the six centers of asymmetry resulting in 64 possible stereoisomers (Fig. 10.2). The ability to distinguish between these differences is critical for reporting the correct results. The functionality of both the mass spectrometry and chromatography is critical.

Measurement of steroid hormones by mass spectrometry does provide distinct advantages. First due to the flexibility of the instrument multiple analytes can be measured in a single run which allows for reporting of several analytes in one sample [1]. Additionally, the use of mass spectrometry can support the large dynamic range typically seen in steroid hormones, where concentration ranges within analytes can vary significantly (e.g., testosterone concentrations < 10 ng/dL can be observed in women and children and concentrations over 1000 ng/dL can be seen in men) [2]. Additionally, it can support differences in the concentration range between analytes; for example, testosterone in men as high as 1000 ng/dL can be measured in the same sample with an estradiol concentration as low as 10 pg/mL [1].

The goal of the chapter is to provide some basic information on the development, validation, and routine use of mass spectrometry methods for steroid hormones.

2 CALIBRATION AND QUALITY CONTROL REQUIREMENTS

2.1 STANDARD MATERIALS

Materials used to prepare calibrators should be carefully considered to ensure that accurate results are obtained [3,4]. Primary reference materials or materials traceable to primary reference materials should be considered [5,6]. These materials are available through international metrological institutes, such

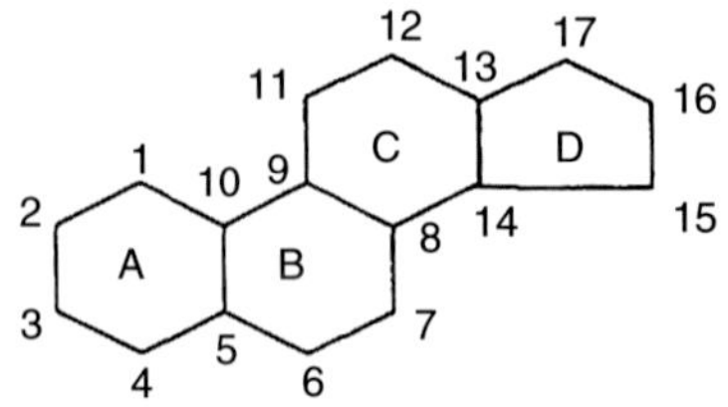

FIGURE 10.1 Illustration of the Cyclopentanophenophenanthrene Ring System Common to the Steroid Hormones

as the National Institute of Standards and Technology (NIST) in the United States and the Institute for Reference Materials and Measurements (IRMM) in Europe. A database of acknowledged reference materials can be found through the International Bureau of Weights and Measures (BIPM) and Joint Committee Traceability in Laboratory Medicine (JCTLM) website, www.bipm.org/jctlm/. If primary reference materials cannot be obtained due to limited availability, cost, or legal restrictions (some steroid hormones require a DEA license to obtain), use of commercially available material traceable to primary reference material is recommended.

The purity of the material used to prepare calibrators is important. A purity of > 99% is preferred when available. Careful evaluation of materials, used both for the calibrator and internal standard should be assessed when methods of multiple steroid hormones are used thus ensuring that impurities from the standards do not include other steroid hormones evaluated in the method. For example, it has been anecdotally noted that in estrone (E1) standards can contain trace amounts of estriol (E3). In a method evaluating E1 and E3 together, a positive bias could be introduced into the E3 results due to the presence of E1 in the E3 standard.

Steroid hormones are nonpolar and hydrophobic; nevertheless, there can be an issue with crystalline powders taking on moisture during storage. The storage of crystalline materials should be carefully considered to avoid the addition of moisture. Materials should be stored in line with the certificate of analysis (COA) provided by the vendor. Consultation with vendor and evaluation of moisture content on a routine base is needed if long-term storage is necessary.

2.2 STANDARDS PREPARATION

Errors in the preparation of standard solutions can result in significant bias in final patient results, and as a consequence certain parameters should be careful controlled [3]. Many considerations during preparation of standard solutions are universal and considered good laboratory practice. This includes, but is not limited to: the use of gravimetric measurements, class A glassware, calibrated pipettes, and restricted use of serial dilutions. Concerns specific to steroid hormones, such as solubility of crystalline powder materials, should be carefully considered. As steroid hormones are mostly nonpolar molecules, most are hydrophobic and poorly soluble in water. As such, the solution used to reconstitute a powder should be considered to ensure that the compound is completely dissolved prior to use; typically, an organic solvent, such as methanol or ethanol is used. After reconstitution, ample time with gentle shaking or rocking should be allowed to ensure that the solid is completely dissolved prior to the preparation of additional solutions from this solution. Incomplete incorporation of the crystalline powder into solution prior to future transfers and dilutions will result in inconsistent and inaccurate concentrations of dilutions of the stock solution.

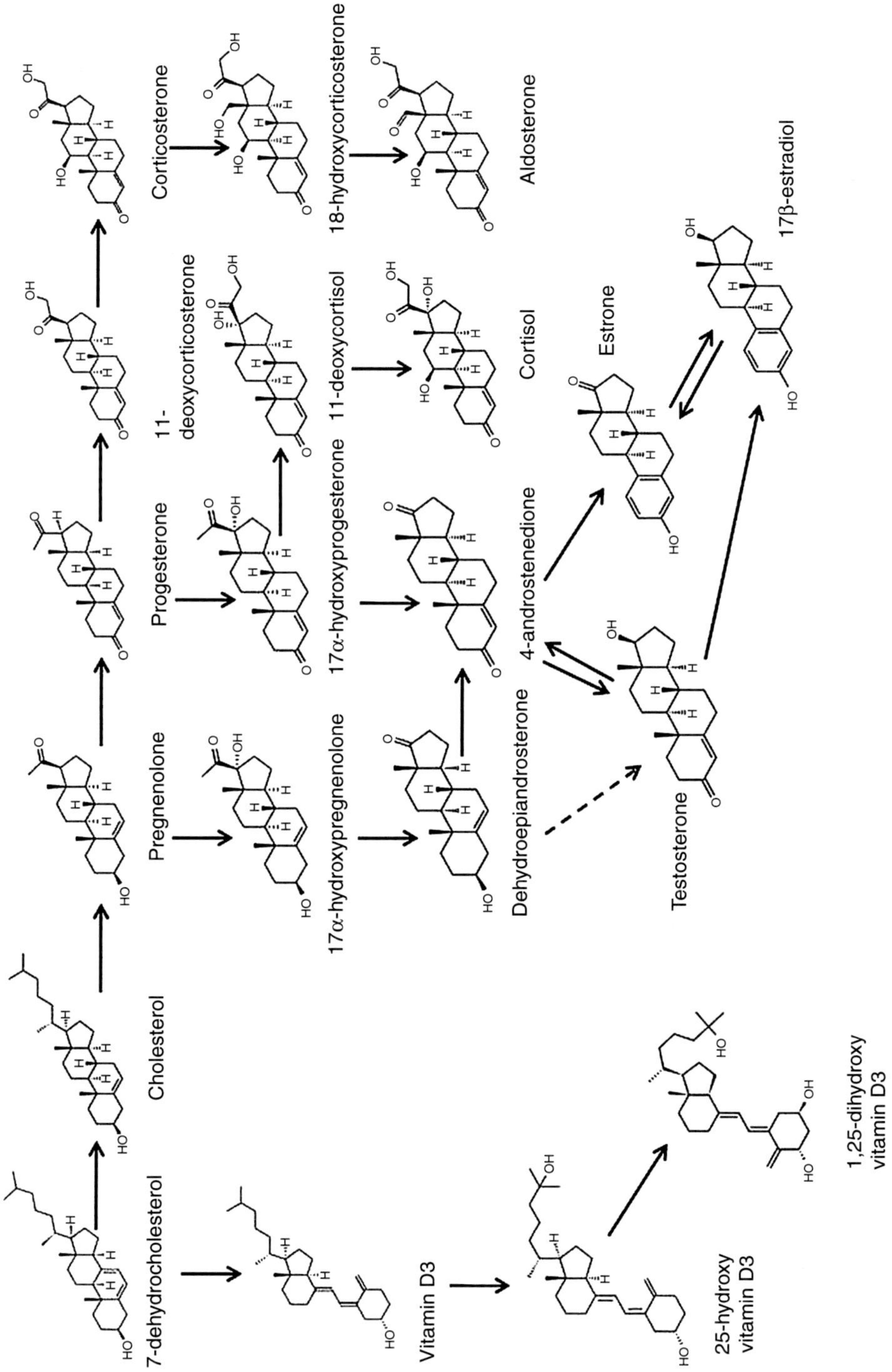

FIGURE 10.2 Steroid Hormone Pathway

Careful control of environmental conditions is needed, as temperature-dependent volume changes of solvents can occur especially with the use of volatile solvents that are used to prepare standard solutions. Controlling the temperatures of the standard and working solutions will help to ensure that proper volumes of solutions are transferred. A constant temperature, typically 20°C, consistent with the rating of the glassware and pipettes, should be maintained. This is executed through the use of a carefully monitored and maintained water bath. Additionally, due to the volatility of the solvents used in preparation of the standards, steps should be taken to minimize, and if possible avoid, evaporation of the solvents, as evaporation will alter the concentration of standard solutions and potentially introduce a measurement bias. Materials should be stored in containers with caps that create a sufficient seal to avoid evaporation and should be left uncapped for the minimal time needed to make a transfer. If materials are to be stored long-term in volatile solvents, gravimetric measurements of the containers with the solution should be considered both before and after storage as a way to monitor evaporation loss during storage and determine if concentration corrections are necessary. If standards are light-sensitive, (e.g., estradiol), amber glassware should be used and steps should be taken to avoid light exposure.

2.3 CALIBRATORS

The selection of the calibrator concentration range should be tailored to the clinical application of the selected method, and depending on the application, this could be significantly different [3,4]. Careful consideration should be taken to ensure that the concentration range applicable to the clinical application is adequate and should extend beyond the lower and upper physiological range. Suggestions for the number of calibration points and spacing are the same with any other clinical method as previously discussed in earlier chapters [4]. However, with steroid hormones the concentration range typically covers several orders of magnitude; for example, estradiol can vary from < 1 pg/mL in postmenopausal women on aromatase inhibitor treatment for breast cancer versus pregnant women and women on IVF with values greater than 4000 pg/mL. Additional calibration points may be advantageous beyond the minimum recommendation to adequately cover the large dynamic range required due to the varied clinical applications of steroid hormone measurement. Additionally, a validated dilution procedure for samples beyond the established calibration curve should be considered for the occasional sample that will fall outside the established concentration range. Due to the large dynamic range, weighted regression may be needed to avoid the highest calibrator points dominating the slope of the curve and should be evaluated during method development [4]. If panels of steroids are measured in one method, the dynamic range not only within an analyte of interest but also between analytes has to be considered. For example, testosterone is measured in ng/dL and estradiol in pg/mL, and even with these significant differences in concentration, a single calibration curve with all analytes of interest in the same mixture is encouraged to minimize sampling time spent on individual calibrators.

It has been recommended that preparation of calibrators should be done in matrix-matched materials to avoid introduction of a bias resulting from matrix differences among patient samples and calibrators [3,4]. Typical surrogate matrices used in steroid hormone analysis in serum include, but are not limited to, bovine serum albumin (BSA) and charcoal-stripped serum. Commercially available synthetic urine is typically used in analysis of steroid hormones in urine. Careful evaluation is needed, as problems have been noted with some surrogate matrices. Charcoal-stripped serum is prepared with the addition of activated charcoal to a serum pool, and then the charcoal is removed through filtration.

Bias can be introduced if residuals of charcoal are present after filtration. Residual charcoal can absorb the spiked standard material, thereby resulting in an inaccurate concentration of the steroid hormones in the calibrator.

Additionally, if BSA is used as a surrogate matrix it should be evaluated for residual steroid hormones. Bioavailable steroid hormones also bind to albumin, and as a result, trace amounts of steroid may be present in BSA. Careful evaluation of a surrogate matrix should be made prior to use to ensure that it will not introduce unaccounted for the concentrations of the analyte of interest or deplete the added concentration. Additionally, it should not be assumed that the surrogate matrices will remain consistent between manufactured lots. Evaluation should be made each time a new lot is used to prepare calibrators. Calibrator preparations using matrices other than unadulterated native serum should be evaluated and compared to clinical samples for differences in recovery and ion suppression [7]. Depending on the extent of sample preparation and extraction of the analyte of interest from the matrix prior to analysis by mass spectrometry, a matrix match may not be necessary but rigorous recovery and ion suppression testing is needed to confirm the absence of a matrix bias of a set of neat calibrators, for example, prepared in solvent. In addition to recovery and ion suppression experiments, matrix bias for the calibrator matrix can be assessed using a commutability evaluation strategy as detailed in CLSI EP-14 [8].

2.4 INTERNAL STANDARD

Isotope-dilution should be considered for steroid hormone MS methods [4]. This approach, as previously described, can help provide the most accurate and precise results. Isotopically-labeled internal standards (IS) are commercially available for most steroid hormones (typically deuterium and/or C^{13} labeled forms). Standards with an incorporation of a minimum of three additional mass units are recommended to avoid overlap between the isotopic envelope of the analyte and the internal standard. A full scan evaluation of an internal standard is suggested to ensure that incomplete isotopic incorporation, or that isotopic exchange, does not overlap either directly with the analyte of interest or the isotopic envelope. Careful evaluation should be carried out in the selection of an isotopically labeled standard to ensure minimal isotopic exchange occurs. While deuterium standards may provide a more affordable approach, the standard can undergo hydrogen/deuterium exchange depending on the location of the deuterium. Isotopic exchange can also occur as a result of the environmental conditions. Evaluation of the sample preparation procedure (e.g., solvents used and pH), and mass spectrometry parameters (e.g., voltages, temperatures, and solvents) on the isotopic stability of the standard is recommended. Storage stability of internal standard materials should also be considered and evaluated, as hydrogen/deuterium exchange may occur during long-term storage of materials.

In the selection of the concentration of the internal standard, the approximate mid-point of the calibration curve or at close proximity to a medical decision point has been recommended [4]. This can be somewhat extreme in situations where the dynamic range is large. The internal standard should be added to all calibrators, quality control (QC) materials, and samples prior to further sample preparation. In addition, ample time for equilibration of the internal standard is important with steroid hormones that are associated with binding proteins, such as sex-hormone binding globulin (SHBG) and albumin. If added and equilibrated appropriately, the use of the internal standard will improve the accuracy and reproducibility of the measurements by accounting for any sample loss or incomplete recovery of the analyte.

2.5 QUALITY CONTROL MATERIALS

QC preparation and use should follow criteria discussed in previous chapters [4]. Again, with the large dynamic range measurements typically seen in steroid hormone analysis, additional QCs may be needed to ensure the appropriate concentrations for the clinical application are included. For steroid hormones, bulk materials can be obtained from commercial sources and pooled. Ideally, native materials without any enhancement or depletion are preferred. However, the use of only pooled, unaltered materials may be difficult especially for the analysis of several steroid hormones at one time that require specific concentrations. As a result, spiking or dilution of the matrix may be required. Pooling, spiking, and/or diluting will alter the materials and could affect the binding protein profiles in the serum. This alteration may result in the final concentrations being significantly different than expected (especially with the analysis of free steroid hormones) and should be evaluated during preparation.

3 SAMPLE PREPARATION

The first step, prior to any manipulation of the sample, is to add the IS to all samples, calibrators, and QC materials. Accuracy is critical with the addition of IS solution. Volumes greater than 50 µL, dispensed by properly calibrated pipettes are recommended. For sample preparation procedures that do not include an initial protein precipitation (PPT) step [e.g., liquid–liquid extraction (LLE), solid phase extraction (SPE), or supported liquid extraction(SLE)], care should be taken to minimize the amount of organic solvent contributed by the IS solution to avoid PPT. The precipitated protein could sequester protein-bound steroids away from subsequent extraction steps, resulting in falsely low results. Additionally, ample time should be allowed for the equilibration of the IS solution with any protein-bound steroids. Testosterone has been reported to reach equilibration with the internal standard in 30 min [9]. With proper addition and equilibration time, a stable IS signal should be observed from the MS. If large variations in signal response are observed, incomplete equilibration or inaccurate addition of the internal standard solution could be the cause.

Steroid hormones are bound to albumin with low affinity and to SHBG with high affinity in serum. Only a small percentage of steroids are not protein-bound, and these represent the free fraction. Due to the lower affinity of binding to albumin, these steroids are also available in vivo. Therefore, the free fraction and the albumin-bound fractions make up the bioavailable steroid hormones. Steroid hormones also form conjugates with sulfates and glucuronides, especially in urine, and are found at higher concentrations in some cases than the free steroid.

Free steroid hormones can be directly measured by liquid chromatography-tandem mass spectrometry (LC-MS/MS) but require additional procedures to isolate the free steroids while taking care not to effect the equilibrium of free and bound steroids [10]. Procedures, such as equilibrium dialysis and ultracentrifugation have been reported in literature [11–13]. Assays for free steroids have to be more sensitive due to the low concentrations that are observed in free steroid measurements. To measure total steroid hormone concentrations, the analyte of interest must be released from the binding protein or conjugate. Protein dissociation can be achieved with protein denaturing. The denaturation should break the bond between the steroid and the protein(s) without altering the analyte of interest. For testosterone it has been reported that this can be achieved at a pH of 5.5 for 30 min [9]. A hydrolysis step is needed to deconjugate the glucuronide group from the steroid hormones if total hormone concentration is required. In most cases, this is achieved with β-glucuronidase [14].

Isolation of the analyte of interest from the matrix is recommended to reduce the complexity of the sample that is to be analyzed. Proteins, peptides, small molecules, salts, phospholipids, and other compounds can suppress or interfere with the measurements of steroid hormones by LC/MS, resulting in insufficient resolution, loss of sensitivity, reproducibility, and robustness due to ion suppression [15]. Extraction can be performed offline or online to allow for higher sample throughput. Automation steps can be considered as previously discussed.

Common sample extraction procedures used in steroid hormone analysis include; PPT, LLE, SPE, and SLE. Examples of publications with different sample preparation techniques are referenced in Table 10.1. For steroid hormone analysis, PPT is most often performed by adding a high concentration of organic solvent or organic solvent plus a salt solution to the sample to separate the steroids from binding proteins and other matrix proteins. PPT is inexpensive and easily automated, however ion suppression and instrument contamination due to insufficient matrix removal may prevent the ability to achieve sufficient steroid assay sensitivity. SPE is a rapid preparation technique that employs partitioning between a mobile phase and a solid phase, similar to conventional liquid chromatography (LC). SPE cartridges containing reversed-phase sorbent are commonly used for steroid analysis. The relatively large particle size of the SPE sorbent allows for fast sample preparation. SPE is easily automated and can also be performed online, where samples interact with a SPE column immediately prior the analytical column. LLE utilizes partitioning between two immiscible liquids (e.g., aqueous and organic solvent) and is a common approach for steroid analysis, as LLE exhibits good extraction efficiency for nonpolar analytes and is more selective than PPT. Modification of the pH in LLE can be used to change the selectivity of the extraction and improve recovery. In SLE, a solid support material with a hydrophilic surface is used to retain droplets of the aqueous serum sample containing the steroids to be measured. The supported aqueous phase is then exposed to an organic solvent for extraction of the steroids, similar to standard LLE. SLE reduces the number of steps required for the preparation compared to LLE and SPE, and allows faster sample processing [16]. However, one drawback of SLE is reduced mixing and surface area for extraction as compared to LLE.

The choice of sample preparation technique depends on required assay sensitivity, selectivity, and labor restraints. Many sample preparation techniques include use of multiple combinations of sample preparation approaches (Table 10.1), and most are amenable to some level of automation, such as automated SPE/SLE or use of an automated liquid handler.

Many steroids are difficult to resolve chromatographically and they exhibit poor ionization in the mass spectrometer. Chemical derivatization during the sample preparation process may be used to improve chromatographic separation, increase robustness in MS fragmentation, and achieve greater sensitivity by improving ionization efficiency. Common derivatizing agents used in steroid hormone analysis include dansyl chloride, pentafluorobenzyl, picolinoyl, pyridyl, and piperazinyl. Through the addition of the derivative, ionization and chromatographic separation may improve and the molecular weight of the analyte is increased, which may result in more robust precursor ions and availability of additional mass transitions. Although derivatization can improve assay sensitivity, use of derivatization procedures can induce the formation of isomers and nonspecific product ions. Therefore, the specificity of the method will increase if the fragment ion selected for quantification is not the loss of the derivative (i.e., just the mass of the analyte you are measuring). The sensitivity and robustness of a method may be improved with the removal of derivatization reagents via an additional cleanup step. Routine clinical laboratories may choose to avoid methods that employ derivatization procedures because derivatization often increases assay labor requirements and reagent expense.

Table 10.1 Examples of Published Transitions for Selected Reaction Monitoring of Nonderivatized Steroid Hormones by Liquid Chromatography Tandem Mass Spectrometry

Steroid Hormone	Ion	Mass Transition	Ionization Source	Number of Steroids	LoQ (ng/L)	Analytical Column	Sample Preparation	References
Androstene-dione	[M + H]$^+$	287/97	ESI	10	404	C18	PP	[18]
		287/109; 97	APCI	3	100	C18	PP + SPEOL	[19]
		287/97	ESI	3	30	C18	SPE	[20]
		287/97	ESI	2	72	C18	LLE	[21]
		287/97; 109	APCI	9	20	C18	PP + SPE	[22]
DHEA[a]	[M−H$_2$O + H]$^+$	271/197; 213	APCI	9	20	C18	PP + SPE	[23]
DHEAS	[M − H]$^-$	367/97	ESI	1	3.6e^5	C18	PP	[23]
Testosterone	[M + H]$^+$	289/97; 109	ESI	2	20	C18	LLE	[24], [25]
		289/97	ESI	1	50	C18	LLE	[26]
		289/97; 109	APCI	1	100	C18	PP + SPEOL	[27]
		289/109	ESI	10	600	C18	PP	[18]
		287/109; 97	APCI	3	100	C18	PP + SPEOL	[19]
		287/109; 97	APCI	1	3	C12	SPEOL	[28]
		289/97	ESI	1	61	C18	PP + LLE	[29]
		289/97	ESI	1	87	C18	PP	[30]
		289/97	ESI	3	30	C18	SPE	[20]
		289/109	ESI	2	72	C18	LLE	[21]
		289/109; 97	APCI	9	20	C18	PP + SPE	[22]
		289/97; 109	ESI	1	10	C18	LLE	[31]
		289/97	APCI	8	20	C18	LLE	[32]
		289/97; 109	ESI	1	20	C18	LLE	[33]
DHT	[M + H]$^+$	291/255	ESI	2	20	C18	LLE	[24], [25]
		291/255	ESI	10	854	C18	PP	[18]
		291/255	ESI	3	30	C18	SPE	[20]
Estradiol	[M − H]$^-$	271/145; 183	ESI	4	2[b]	C8	PP	[34]
Estrone	[M − H]$^-$	269/145; 143	ESI	4	1[b]	C8	PP	[34]
Estrone-sulfate	[M-SO$_3^-$]$^-$	349/269	ESI	1	8	C18	PP	[35]
		349/269	ESI	1	80	C18	SPE	[36]

NOTE: The m/z of the tenth decimal position may vary slightly, based on instrument tuning. Therefore, values for m/z are given to the nearest whole number.

Abbreviations: APCI, atmospheric pressure chemical ionization; CV, coefficient of variation; DHEA, dehydroepiandrosterone; DHEAS, dehydroepiandrosterone sulfate; DHT, dihydrotestosterone; ESI, electrospray ionization; LLE, liquid–liquid extraction; LoD, limit of detection; LoQ, limit of quantitation; PP, protein precipitation; SPE, solid phase extraction; SPEOL, solid phase extraction online.

[a]Water-loss ion.

[b]Value stated as "LOD" with CV < 10%.

Adapted from Kushnir MM, Rockwood AL, Roberts WL, Yue B, Bergquist J, Meikle AW. Liquid chromatography tandem mass spectrometry for analysis of steroids in clinical laboratories. Clin Biochem 2011; 44(1):77–88, [17] with permission from Elsevier.

4 INSTRUMENTATION

4.1 CHROMATOGRAPHY

Chromatography is an analytical technique used to separate compounds in a complex mixture. In clinical laboratories, this complex mixture is often serum, plasma, or urine, among others. There are two commonly used types of chromatography that are used before mass spectrometry analysis; GC and LC.

4.1.1 Gas Chromatography

As the name suggests, with GC, volatile analytes with a boiling point of < 300°C are measured in the gaseous phase. Separation occurs when the analytes, carried in a gaseous mobile phase (usually helium or nitrogen), pass over a liquid stationary phase, which lines the inside of the column. Analytes are separated in time based on interaction with the column, due to differences in vapor pressure or boiling point. To remove matrix components, patient samples must undergo a sample preparation procedure prior to GC analysis which can involve LLE, SPE, and others, as mentioned previously. In order to increase volatility, analytes may also be derivatized which adds another layer of complexity to the sample preparation process. After extraction, the samples are dissolved in a volatile solvent (e.g., acetonitrile, ethyl acetate) and then injected onto the GC system.

4.1.2 Liquid Chromatography

GC is a technique amenable to the analysis of volatile analytes. However, the advent and simplification of LC has increased the popularity of this technique for clinical laboratory analysis, especially since it is applicable to a wider range of analytes, including polar and nonvolatile analytes. In LC, separation occurs when analytes are carried in a liquid mobile phase and pass over a solid stationary phase on the inside of the column. Analytes are separated in time based upon the interaction they have with the column and mobile phase due to differences in polarity. To remove matrix components, patient samples also have to undergo sample preparation prior to LC analysis, but it can be simpler than for GC analysis. Also, derivatization is required much less frequently as analytes do not need to be volatile and additionally, if an improvement in sensitivity is required, the injection volume can be increased.

The majority of LC systems in clinical laboratories used for steroid hormone analysis are high performance LC (HPLC) instruments [4]. In this technique, two pumps produce high and reproducible flow rates of mobile phase to move the analyte through the column and into the mass spectrometer. For steroid hormones, reversed-phase chromatography is typically utilized therefore one of the LC pumps is used to pump an aqueous mobile phase (e.g., water) and the other, an organic mobile phase (e.g., methanol or acetonitrile) up to flow rates of about 1 mL/min and a maximum allowable pressure of around 6000 psi (~400 bar). Ultra-high pressure LC (UHPLC) systems exist that have maximum allowable pressure of around 18,000 psi (~1200 bar). UHPLC allows the user to implement shorter columns with a smaller particle size and higher mobile phase flow rates for faster analysis while maintaining the required separation of analytes on the column [37].

The type of stationary phase inside the chromatography column is important in determining the separation of analytes in time on the column [4]. The most common stationary phase used in reversed-phase chromatography implemented for steroid hormone analysis is an octadecyl (C18) bonded phase (Table 10.1). Gradient elution on this column begins with a high percentage of aqueous mobile phase and a low percentage of organic mobile phase (e.g., 95% water; 5% methanol) and during the analysis, this incrementally changes to a high percentage organic mobile phase and a low percentage aqueous mobile phase (e.g., 5% water; 95% methanol), in order to elute the nonpolar steroid hormones from the chromatography column. Gradient elution is most commonly used for steroid hormone analysis as it allows the user to control the separation of analytes with the same molecular formula, and potential interferences, more easily by varying the percentage of organic mobile phase over the length of the chromatographic run time [38]. In addition to water or organic solvent, mobile phases for mass spectrometry analysis of steroid hormones often include additives, such as formic acid (e.g., 0.1%) and ammonium formate (0.01 M) in order to improve ionization [4].

4.2 MASS SPECTROMETRY

In the context of this book, mass spectrometers essentially act as detectors for either GC or LC analysis. Earlier chapters have described the technique of mass spectrometry and so here only steroid hormone specific considerations will be discussed.

4.2.1 Gas Chromatography-Mass Spectrometry

Gas chromatography-mass spectrometry (GC-MS) has historically been the method of choice for measuring steroid hormones in clinical samples. This technique is particularly useful for the detection of low concentration analytes, such as testosterone and estradiol in certain patient populations [39]. Due to the lipophilic or nonpolar nature of the steroid hormones, they are well suited to GC-MS analysis. By definition, a GC-MS has one quadrupole and is most commonly used for selected ion monitoring or SIM analysis where after mass fragmentation, three of the most prominent or reproducible ions are measured and the relative abundance must fall within predetermined ratios [38]. As many steroid hormones share common fragment ions, they must be separated chromatographically before they enter the mass spectrometer, and this can lead to long chromatography run times. Additionally, a number of steroid hormones are present in clinical samples as sulfate or glucuronide conjugates, which are too polar to be measured by GC-MS analysis. These conjugates have to be removed by either acidic or enzymatic hydrolysis before being analyzed as the parent compound.

4.2.2 Liquid Chromatography-Tandem Mass Spectrometry (LC-MS/MS)

As mentioned earlier, for GC-MS analysis, extensive sample preparation is required, as is derivatization in order to make the compounds more volatile. Therefore, most clinically laboratories have turned to LC-MS/MS for analysis of steroids, as it is typically less labor intensive. In LC-MS/MS, there are three quadrupoles and the most commonly used mode for quantitative analysis is selected reaction monitoring (SRM). In this mode, the precursor ion is selected in the first quadrupole, specific voltages are applied in the second quadrupole to fragment the compound of interest, and then the specific product ions are selected in the third quadrupole. The combination of a precursor and product ion is called a mass transition. It is recommended that common losses that occur in the mass spectrometer source should be avoided when selecting mass transitions (e.g., a loss of 18 Da which would be the loss of a water molecule). These losses are neither specific nor selective and should not be monitored. Commonly used mass transitions for steroid hormones can be found in Table 10.1. It is best practice to use two mass transitions per steroid hormone and per internal standard allowing calculation of the ratios of the peak areas between the two product ions [4]. This adds an additional layer of quality assurance to the method. If the ion ratios in a particular patient sample are not consistent with normal patient samples and calibrators, there could be an interference and the result should not be reported. The patient sample with aberrant ion ratios should be run on a different methodology for that analyte. Tandem mass spectrometry is very selective for the analyte of interest, but as mentioned earlier, many steroid hormones have common fragment ions and so compounds that have the same molecular formula must be separated in time on the chromatography column before they enter the mass spectrometer. However, as LC-MS/MS is capable of measuring compounds with a range of polarities, and the analytes do not have to be volatile, a derivatization step is not necessarily required, and the conjugated analytes can be measured by this technique without prior hydrolysis. The following sections will discuss some steroid hormone specific considerations for LC-MS/MS analysis.

4.2.2.1 Sensitivity

Due to the physiological relevance of low concentrations of steroid hormones, a pg/mL (~2 pg/mL for estradiol) or ng/mL limit of quantification is required for most steroid assays. The limit of quantitation (LOQ) for some published LC-MS/MS methods used to measure certain steroid hormones can be seen in Table 10.1. Sensitivity depends on the combination of numerous assay components, such as the quality of the sample preparation, the efficiency of the chromatographic separation, and the inherent sensitivity of the mass spectrometer. As described in Section 3, the sample preparation strategy should be selected to ensure adequate recovery and to minimize the effects of ion suppression. The separation step should be selected to optimize chromatographic resolution and to minimize band width. It is also necessary to have a mass spectrometer that is capable of the required sensitivity. Steroid hormone analysis will typically require a tandem mass spectrometer that is in mid-, to top-of-the quality range. If the LC-MS/MS method is developed solely to measure one analyte, it is far easier to optimize the measurement. However, if it is a steroid panel, that is being developed, the method should be first optimized for the lowest concentration steroid hormones and the higher concentration steroids can be added in subsequently.

4.2.2.2 Ionization

Atmospheric pressure ionization is the most common type of ionization in LC-MS/MS instruments. There are three different ionization types: electrospray ionization (ESI), atmospheric pressure chemical ionization (APCI), and atmospheric pressure photoionization (APPI) [4]. ESI and APCI are most commonly used in clinical laboratories for steroid hormone analysis. The advantage of ESI is that it is normally the default ionization source in LC-MS/MS systems and due to the formation of protonated or deprotonated ions with little fragmentation, it makes it ideal for selection of precursor and product ions for SRM transitions. The major disadvantage of ESI is that this technique is more prone to matrix effects or ion suppression than the other types of ionization. In published methods, estradiol, estriol, estrone, DHEAS, and dihydrotestosterone are usually measured by ESI [18,23,24,34–36]. The advantage of APCI is that matrix effects are less pronounced using this type of ionization and it is more suited to analysis of nonpolar analytes, such as the steroid hormones. The disadvantage is that it is a little harder to use than ESI due to issues with preferential ionization of solvents or additives that have relatively higher proton affinities making solvent selection a challenge. Additives at higher concentrations are also necessary in the mobile phase with APCI ionization in order to help control the ionization process. In publications to date, testosterone and androstenedione have been measured with ESI or APCI [18–22] and DHEA has been measured with APCI [22]. APPI is not used very often in clinical laboratories, even though it was developed to be more efficient at ionization of steroid hormones. However, there is a report of APPI being used to measure a panel of 12 steroid hormones and another measuring testosterone, dihydrotestosterone, estradiol, and estrone in one run [1,40].

These ionization techniques are "soft" ionization techniques with little fragmentation occurring and so the resulting mass spectra typically show ions that are protonated or deprotonated, for example, $[M + H]^+$ or $[M - H]^-$, where M is the mass of the analyte. The choice of whether to run the MS in positive versus negative mode is dependent upon the polarity of the analytes and the LC conditions (Table 10.1). The steroid hormones that are most commonly detected in positive mode are testosterone and dihydrotestosterone [18,20], whereas underivatized estradiol, estrone, and the sulfate conjugates are commonly detected in negative mode [34–36]. It should be noted that these steroid hormones can be detected in either positive or negative mode and the choice of ionization source is dependent upon each laboratory and whether the steroid hormones will be analyzed in a panel or individually. It is also possible with most LC-MS/MS systems to do positive/negative polarity switching during a method. It should be noted that positive/negative polarity switching can result in loss of sensitivity. In order to counteract the sensitivity loss, it is recommended to perform timed switching wherever possible, where the chromatogram is divided into sections based on the retention time of the analytes.

4.2.2.3 Other MS Parameters

4.2.2.3.1 Source temperature. During method development, source parameters are optimized and one of these is source temperature. The ion source in LC-MS/MS analysis is heated in order to help the ionization of the analyte as well as limiting the condensation of the aqueous component of the mobile phase from the LC system. In general, increasing the source temperature improves analyte ionization but with steroid hormones, if the temperature is too high, there can be water loss in the source of both the analyte and the internal standard, which will decrease the sensitivity of the method since the precursor ion mass has changed [4].

4.2.2.3.2 Cone or Curtain gas. The flow of cone or curtain gas is another source parameter that is optimized during method development. This is a flow of gas that is in the direction opposite to the spray from the ion source. The purpose of the curtain gas is to prevent water, solvent or nonionized matrix components from entering the mass spectrometer. Increasing the flow of curtain gas will limit contamination of the mass spectrometer with these entities, but it will also decrease the number of analyte ions that enter the mass spectrometer thereby reducing the sensitivity of the method [4].

4.2.2.3.3 Dwell time. The dwell time of a mass spectrometer describes the amount of time the instrument is collecting data for a specific SRM mass transition (in milliseconds). This too should be optimized during method development. Increasing the dwell time will increase the sensitivity and precision of the method, but it will reduce the number of SRM mass transitions that can be monitored in a given assay. It is best practice to optimize the dwell time so that 15–20 data points are collected across a chromatographic peak. The minimum number of points required to adequately define a peak is 10 data points. It is also possible to set time windows during which the mass spectrometer only detects a subset of the analytes in a given assay, for example, a steroid panel [4].

4.2.2.4 Other Types of Mass Spectrometry

4.2.2.4.1 Field asymmetric ion mobility spectrometry. Field asymmetric ion mobility spectrometry (FAIMS) is a technique based on gas phase separations on a millisecond timescale at atmospheric pressures and ambient temperature. It separates ions based on their differential mobility in high and low electric fields which is a function of mass, charge, size, and shape. It can be used after LC but before MS/MS in order to add another layer of selectivity to a LC-MS/MS method. This technology can separate isomeric analytes, such as testosterone and DHEA, or 11-deoxycortisol and corticosterone, and increase the signal to noise ratio, therefore increasing the sensitivity of a method for steroid hormones [41].

4.2.2.4.2 *High resolution mass spectrometry.* High resolution mass spectrometry (HR-MS) allows detection of analytes to the nearest 0.001 atomic mass units. Examples of HR-MS instruments are time-of-flight (TOF) and Fourier transform ion cyclotron resonance (FTICR), which forms the basis of the orbitrap technology. Generally, these instruments measure the exact mass of analytes without fragmentation, however, they can be combined with a quadrupole in which case fragmentation is also possible and can add more selectivity to the method [42]. The major advantage of this technique is that it is very selective since it measures the exact mass of a compound allowing even minor changes in structure to be distinguished. However, just as with LC-MS/MS, testosterone and DHEA would not be distinguished since they have the same molecular formula and therefore the same exact mass. They would require separation by LC before they enter the HR-MS.

5 VALIDATION

The requirements for validating a mass spectrometry assay have been discussed previously, and guidelines have been published [3,4]. This section will focus on some steroid hormone-specific nuances that are important to consider during validation. Biological variation, clinical guidelines, and regulatory requirements should guide performance requirements, and acceptable criteria should be established prior to validation to fit the intended use of the assay.

5.1 ACCURACY

In order to ensure that a mass spectrometry method is accurate, a laboratory can compare the method to a reference measurement procedure using patient samples following appropriate guidelines [43,44] or confirm calibration using certified reference materials [44]. The Joint Committee for Traceability in Laboratory Medicine maintains a database of acknowledged higher order reference measurement procedures and reference materials. A section for nonpeptide hormones includes reference measurement procedures or materials for aldosterone, estriol, estrone, estradiol, progesterone, 17-hydroxyprogesterone, and testosterone (www.bipm.org/jctlm/). In addition, standardization and certification programs are currently available for testosterone and estradiol from the Centers for Disease Control Hormone Standardization Program (CDC HoST) (http://www.cdc.gov/labstandards/hs.html). Certified reference materials are currently available for cortisol, progesterone, and testosterone from the National Institutes of Science and Technology (NIST) (http://www.nist.gov/srm/). Evaluations using acknowledged higher order reference measurement procedures and materials demonstrate method traceability; in other words, the method has been directly compared through an unbroken chain to a known method or material in terms of the accuracy, bias, and precision of the measurement. The maximum acceptable total bias from the true value should be $\leq 15\%$ [45], but accuracy criteria derived from biological variability [46] or clinical guideline requirements are preferred.

If access to the higher-order measurement procedures or reference materials is cost prohibitive, or if they are not available for the appropriate analyte, accuracy can be assessed by the recovery of a known spiked amount of each analyte into a blank matrix.

It should be noted that sample exchange between laboratories, while important and informative, does not provide information about accuracy but rather difference (bias) between two laboratories.

5.2 IMPRECISION

The imprecision of a mass spectrometry method should be evaluated within each run or batch, between runs, and on different days, using different technologists and, if relevant, using each instrument that will be used to run the method. The samples used to test imprecision should be representative of the samples that will be run routinely on the method in the laboratory (e.g., serum, urine, etc.) and include quality control samples. The entire testing method should be included during imprecision testing: from aliquoting the sample through the extraction method and derivatization (if applicable) to the chromatography and the mass spectrometry method. Low concentration measurements are needed for some steroid hormones, [e.g., testosterone and estradiol in certain patient populations (see Section 6)]. Therefore, the imprecision should be determined at the lower limit of quantitation (LLOQ) of these assays and at medical decision points where applicable. Guidance on imprecision testing has been previously published [45,47], but it is generally accepted that the coefficient of variation (%CV) at the LLOQ should be $\leq 20\%$ and throughout the rest of the linear range of the assay should be $\leq 15\%$. However, the total bias of the measurement compared to a true value (e.g., that of a reference method or certified reference material) should also be determined (see earlier section), and it is recommended to be $\leq 15\%$ [45]. Imprecision acceptability criteria can also be determined based upon known biological variation of each independent analyte (https://www.westgard.com/biodatabase1.htm) [48,49].

5.3 SENSITIVITY

The sensitivity required of a mass spectrometry method for measurement of steroid hormones depends on the clinical need of the patients that the laboratory serves. For the purposes of quantitative mass spectrometry methods, there are two important sensitivity parameters that should be determined: the limit of detection (LOD) and the LLOQ. The LOD is important in determining the LLOQ of the method, which is the direct measurement of assay sensitivity. It is recommended that for quantitative mass spectrometry methods, no value below the LLOQ of the assay is reported. As mentioned earlier, the LLOQ should meet acceptable precision and total bias limits.

5.3.1 Total Testosterone, Free Testosterone, Dihydrotestosterone, and Androstenedione

Measurement of the androgens in adult, nonhypogonadal males is easily achievable by mass spectrometry methods as the concentrations are reasonably high [19,24,50]. However, in female, pediatric, and hypogonadal male populations, the concentrations can be significantly lower, requiring highly sensitive mass spectrometry methods to detect them [51,52]. In order to achieve such high sensitivity, extensive sample clean-up and high-quality instrumentation are required (see earlier sections). For total testosterone, dihydrotestosterone, and androstenedione measurement in pediatric and female patients, a LLOQ of 2 ng/dL is recommended [2,22,24,49,52], whereas for free testosterone, a LLOQ of 0.1 pg/mL is recommended [12].

5.3.2 Dehydroepiandrosterone (DHEA) and Dehydroepiandrosterone Sulfate (DHEAS)

In order to measure DHEA and DHEAS accurately in pediatric populations with adrenal insufficiency and congenital adrenal hyperplasia, LLOQs of < 10–20 ng/dL and < 5 µg/dL respectively should be achieved by mass spectrometry methods [1,2].

5.3.3 Estrogens

In premenopausal adult females, highly sensitive methods to measure estradiol and estrone are not necessarily required due to reasonably high serum concentrations of these steroid hormones [53]. However, in males, pediatric patients, postmenopausal females, and postmenopausal females undergoing aromatase inhibitor treatment for breast cancer, highly sensitive methods are absolutely necessary [53]. For estradiol and estrone measurement in these patients, a LLOQ of < 2 pg/mL is required [34,53], and for free estradiol, a LLOQ of 0.5 pg/mL is required [11].

5.3.4 Aldosterone

Excluding neonates, the serum aldosterone concentration in the general population is generally low compared to that of other steroids. The reference ranges have a lower limit of 2 ng/dL, and therefore methods used to measure serum aldosterone have to be highly sensitive in order to measure this steroid hormone accurately at this level [54].

5.4 LINEARITY

Strategies for the evaluation of linearity have been previously published [55] and discussed in earlier chapters. The lower and upper limits of the analytical measurement range need to be determined. To demonstrate a linear response over the analytical measurement range, a minimum of six concentration points between these limits should be included for the steroid hormones, since measurement requires accurate quantification (calibrators; see earlier section). Linearity experiments should be carried out at a minimum in quadruplicate in order for the precision of each measurement to be calculated. Additionally, each sample matrix tested should have linearity evaluated [3,4].

During method validation, a procedure should be determined for patient samples with concentrations that fall outside the analytical measurement range of the assay. The result could be reported as "less than the lower limit of quantitation (< LLOQ)" or "greater than the upper limit of quantitation (> ULOQ)." Alternatively, patient samples may be concentrated or diluted in order to report an actual value if clinically applicable, and each part of these procedures requires verification during the method validation process (e.g., the diluent and the concentration device). Acceptable criteria must be set in advance; for example, the diluted or concentrated patient sample result should have a recovery of $100 \pm 15\%$ and experiments should be conducted in at least five separate replicates [4,55]. If the steroid concentration in the patient sample is very high, a procedure should be in place to evaluate the next sample(s) for potential carryover (see Section 6). Steroid hormones are often measured in neonates and to that end, dilutions may also be carried out on specimens that have low volume [4]. Mass spectrometry assays of steroid hormones may require as much as 500 µL of patient sample to obtain the required sensitivity, and in neonates that is a significant volume.

5.5 CARRY OVER

Since mass spectrometry systems are continuous flow, they are more susceptible to carryover than random access systems. Therefore, for each method, the potential carryover must be evaluated. For steroid hormones in particular, it must be determined that a high concentration sample does not affect the next sample that could have a very low concentration. Methods for evaluation of carryover have been published [45,56]. At a minimum, a very high concentration sample (far above the measurement range of the assay) should be injected followed by a number of low concentration or blank samples (extracted matrix) to determine if there is carryover. If there is carryover, the number of samples after the high sample that are affected has to be determined. A rule of thumb is that the carryover from a high sample into a blank or low concentration sample should be no more than 25% of the LLOQ concentration [4]. Some ways to reduce carryover in mass spectrometry systems are to optimize the needle wash solvent for the analytes that you are measuring, to have a wash method for the LC system (including the tubing and column), and to make sure the front end of the MS is regularly cleaned or maintained.

5.6 MATRIX EFFECTS/ION SUPPRESSION

Matrix effects or ion suppression can affect the sensitivity or the recovery of an assay [4,15]. It is critical to evaluate how matrix effects affect each analyte being measured by MS, and examples of how to achieve this have been published [4,7,15]. If matrix effects are found to negatively impact the sensitivity of the method required for steroid analysis, the chromatography may be altered, or the sample preparation can be improved in order to minimize these effects [15].

5.7 INTERFERENCES

As with other methodologies, MS assays should be evaluated for potential interferences, such as lipemia, icterus, and hemolysis. Additionally, isobaric and isomeric compounds must be evaluated to ensure that they do not contribute to false positive results for the analyte of interest [4,57]. A steroid hormone-specific example is testosterone and DHEA: they have the same molecular weight, same molecular formula, and only differ in the position of a hydroxyl group. Therefore, MS or MS/MS systems cannot discriminate between these compounds. The only way to measure them independently is to separate them via the chromatography column before they enter the MS [33] or to monitor different product ions. The same applies to: testosterone and epitestosterone; 17-hydroxyprogesterone and 11-deoxycorticosterone; 11-deoxycortisol and corticosterone; cortisol and 18-hydroxycorticosterone; and cortisone and aldosterone. Another factor to consider is that the steroid hormones are formed from a common cholesterol precursor molecule and so share common product ions in MS/MS analysis (e.g., 97 and 109). Therefore, it is important to check as many steroid hormones as possible on the developed method to ensure no interference exists [3]. Other potential interferences include over-the-counter medications and commonly administered or prescribed drugs.

For blood samples, the collection tube type should be investigated for interference in MS assays. The gel material in blood collection tubes has been shown to cause interference in testosterone LC-MS/MS assays and has to be chromatographically separated from the testosterone peak if blood collected in these tubes is going to be accepted as a sample type (Fig. 10.3) [33,58].

A way of monitoring for potential unknown interferences in a MS method is to employ ion ratios or confirmation ratios. In GC this can be accomplished using SIM and in LC using SRM. In these modes, only specific fragment or product ions from a specific precursor ion are monitored for each analyte, and the ratio of the relative abundance among each of these ions is stable and can be monitored. A common method is to determine the ion ratios in the calibrators of each patient batch or run and then set some acceptance criteria (e.g., ±20%) for each patient sample. If the ion ratios of a patient sample do not fall within these criteria, there may be a potential unknown interference, and the patient sample should be run using another method if available [3,4,38].

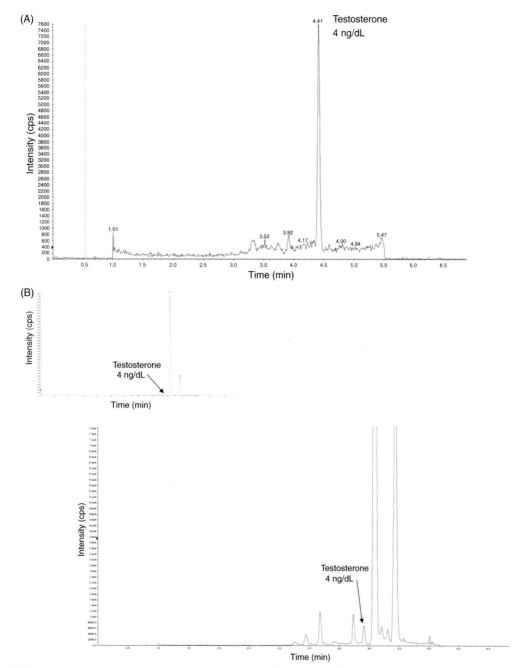

FIGURE 10.3 Comparison Between Collecting Serum in a Clot Activator Tube Versus a Serum Separator Tube for Total Testosterone Analysis

(A) The extracted ion chromatogram of a patient sample collected in a clot activator tube with a testosterone concentration of 4 ng/dL. (B) The extracted chromatogram of a sample collected from the same patient at the same time in a serum separator tube.

Reprinted from French D. Development and validation of a serum total testosterone liquid chromatography-tandem mass spectrometry (LC-MS/MS) assay calibrated to NIST SRM 971. Clin Chim Acta 2013; 415:109–117, [33] with permission from Elsevier.

6 POSTIMPLEMENTATION MONITORING

6.1 DATA REVIEW

Analyte and internal standard peaks from every chromatographic run should be reviewed to ensure that they are within the acceptance criteria determined during method validation (e.g., the retention time, peak shape, peak resolution, and background signal can be monitored). The relative retention time should not change by more than ± 2.5% between runs [38]. As mentioned earlier, since gel-containing blood collection tubes cause interference in testosterone LC-MS/MS methods, individual chromatograms should be reviewed to ensure that if the laboratory accepts these tubes, that the interference is separated from the testosterone peak [33,58].

Calibration curves from each run should be reviewed (e.g., the slope should be consistent between runs or batches and this can be monitored). The recovery of the calibrators when back-calculated from the calibration curve should be 100 ± 15% [45], and the correlation coefficient (r^2) should be ≥ 0.99 [4]. Acceptability criteria for QC samples should be determined on a lot-by-lot basis before putting them into use and QC rules are established by the laboratory. For steroid hormones that may have a wide range of concentrations in the general population, it is important to use QC samples at concentrations across this whole range (see Section 2.5). For steroid hormones that have low concentrations like total testosterone in pediatric and female patients, it is very important to use a QC sample with a concentration near the LLOQ to ensure that the assay is performing according to expectations at low concentrations [4,45].

6.2 PROFICIENCY TESTING

External quality assessment, or proficiency testing, is a required and useful part of postimplementation monitoring. For certain steroid hormones, a MS peer group may not be available due to a limited number of methods. Comparing results from a MS method to immunoassay results is not always informative due to the lack of specificity of these methods and the noncommutability of some proficiency testing materials. If available, it is important to utilize the accuracy-based proficiency programs for steroid hormones as these samples are processed minimally to ensure that the matrix is as close to human samples as possible. Additionally, target values for the analytes are assigned by reference measurement procedures to enable assessment of accuracy and traceability [3,4].

For some more esoteric or low-volume steroid hormone assays, there may not be a proficiency testing scheme available. In these cases, an alternative assessment procedure must be implemented, which could involve comparing results from the same patient samples with a different laboratory [59].

6.3 HARMONIZATION/STANDARDIZATION

Standardization of a laboratory test means that calibrator value assignments for a routine laboratory method are traceable to a reference method such that patient results from the routine method are comparable to the reference method. Harmonization of a laboratory test means that a reference method is not available, but numerical values for patient results are comparable. Harmonization can be achieved by use of reference materials or by calibrating assays to the same standard to eliminate differences in reported values (the consensus approach) [3]. Given that clinical practice guidelines may be based on test result values regardless of what methodology was used to determine the result, lack of harmonization or standardization may be very problematic [60]. As mentioned earlier (Section 5.1), reference measurement procedures exist for aldosterone, testosterone 17-hydroxyprogesterone, progesterone, estrone, estriol, and estradiol (www.bipm.org/jctlm/). For laboratories developing MS assays, one way of documenting traceability of the method to a reference measurement procedure is to participate in the CDC HoST program for testosterone and estradiol (http://www.cdc.gov/labstandards/hs.html). It is possible to prepare calibrators and have the CDC assign values to them using the reference measurement procedure. Alternatively, the CDC can provide laboratories with 40 samples that can be used to optimize the calibration of the LC-MS/MS assay, or for calibration verification of a developed assay. To participate in the full program, the CDC HoST program will send 40 samples to optimize calibration of the assay, and then once a quarter will send challenge samples. If the method shows an acceptable bias (determined based upon the biological variation of each analyte), it is certified by the CDC HoST program (http://www.cdc.gov/labstandards/hs.html) [49]. Another alternative is to purchase certified reference materials (CRMs) from NIST. CRMs are currently available for cortisol, progesterone, and testosterone. The CRMs are provided with certified concentrations in order for users to check the calibration of the developed MS assay (http://www.nist.gov/srm/).

7 PREANALYTICAL CONSIDERATIONS

7.1 SAMPLE TYPE

Steroid hormones are measured clinically in different matrices including serum, plasma, and urine. Testosterone, free testosterone, estradiol, free estradiol, estrone, estriol, DHEA, DHEAS, progesterone, 17-hydroxyprogesterone, 17-hydroxypregnenolone, 11-deoxycorticosterone, 11-deoxycortisol, and androstenedione are measured in serum or plasma. Aldosterone and cortisol are measured in serum, plasma, and urine.

It is important to document acceptable tube types for each analyte, the time the sample should be collected, and how the sample should be stored so that the analyte remains stable (see Sections 7.2, 7.5, and 7.6). There are also other preanalytical variables that also have to be taken into account for the steroid hormones (see Sections 7.3 and 7.4 later).

7.2 TIME OF SAMPLING

For some steroid hormones, the time of day of the sample collection is not important. However, for others it is extremely important in order to aid the clinician in determining if a pathologic process is occurring (Table 10.2).

7.3 OTHER PATIENT-DEPENDENT CONSIDERATIONS

When measuring the sex steroids, it is important to note the age and gender of the patient when the sample is drawn, as there are age-specific and gender-specific reference ranges (see Section 8.1). For measuring certain hormones in females, it is important to note the phase of the menstrual cycle at the current time (follicular, mid-cycle, or luteal). It should also be noted if the patient is postmenopausal, pregnant or not, and if so, the stage of the pregnancy. This is important for measurement of estradiol, estrone, progesterone, and 17-hydroxyprogesterone, as the reference ranges reflect these different phases or conditions.

Table 10.2 Correct Sample Collection Times for Selected Steroid Hormones

Analyte	Sample Type	Variation	Correct Sampling Time
Total testosterone [61–63]	Serum or plasma	Highest in the morning in males	First thing in the morning
Cortisol [61–64]	Serum or plasma	Shows diurnal variation; highest in the morning	Morning samples should be drawn between 0700 and 0900 h; afternoon samples should be drawn between 1500 and 1700 h
Dehydroepiandrosterone (DHEA) [61–63]	Serum or plasma	Shows diurnal variation; highest in the morning	Early morning sample preferred (0800 h)
Aldosterone [62]	Serum or plasma	Shows diurnal variation; highest in the morning.	Morning samples should be drawn between 0800 and 1000 h; afternoon samples should be drawn between 1600 and 1800 h
Androstenedione [63]	Serum or plasma	Shows diurnal variation; highest in the morning	Early morning sample preferred (0800 h)
11-Deoxycorticosterone [64]	Serum or plasma	Shows diurnal variation; highest in the morning	Early morning sample preferred (0800 h)
11-Deoxycortisol [63,64]	Serum or plasma	Shows diurnal variation; highest in the morning	Early morning sample preferred (0800 h)
Corticosterone [63,64]	Serum or plasma	Shows diurnal variation; highest in the morning	Early morning sample preferred (0800 h)
17-Hydroxyprogesterone [63]	Serum or plasma	Shows diurnal variation; highest in the morning	Early morning sample preferred (0800 h)

7.4 OTHER PREANALYTICAL CONSIDERATIONS

Before a patient has a sample for aldosterone drawn, they should be seated or lying down for at least 30 min as exercise can cause up to a 15-fold increase in plasma aldosterone concentrations [65]. It should also be documented whether the patient is supine or upright when the sample is drawn as the aldosterone concentration can be approximately 15% lower if the patient is supine compared to upright [66]. If a patient is having a sample for cortisol drawn, caffeine ingestion should be avoided for at least 24 h, because it increases cortisol to a level so pronounced that diurnal variation is no longer observed, nor is the increase in cortisol concentration as a stress response [67]. Smoking can cause approximately a twofold increase in both aldosterone and cortisol and should be avoided for as long as possible before samples for these tests are collected [68]. Aldosterone can also be increased by up to 10% in women who take oral contraceptives [69]. Cocaine use can cause up to a 50% increase in serum cortisol concentration [70], and if a patient is prescribed phenytoin, it can cause up to a 45% increase in serum estradiol concentration [71]. Further, the stress or anxiety that can be caused by a visit to a physician or phlebotomist can lead to higher serum concentrations of cortisol [72] and aldosterone [73].

7.5 SAMPLE TUBE TYPES

As mentioned in Section 5.7 and shown in Fig. 10.3, the choice of sample collection tube type can be important in steroid MS assays. For testosterone in particular, tubes containing separator gel cause an interference in the most commonly monitored SRM mass transition (289/97). This interference appears on the chromatogram as extra peaks, and if these extra peaks are not chromatographically separated from the real testosterone peak, the quantification of testosterone can be significantly impacted [24,33,58]. Use of tubes containing separator gel may also result in potentially interfering peaks for androstenedione and DHEA. Therefore, it is important that appropriate tube types are identified during the method validation process.

7.6 STABILITY AND STORAGE

The stability of each steroid hormone should be determined during the validation process, and tested conditions should match the requirements of the laboratory. For example, if the MS assay will not be run every day, is the analyte stable in the blood or urine collection tube? Is the sample stable in the refrigerator or should it be frozen before analysis? Is it stable during freeze/thaw cycles? If the MS should be down for a length of time, is the analyte stable in extracted samples? At what temperature and for how long? Limited stability studies have been published, but it is recommended that the laboratory determine the stability internally always using fresh matrix as the comparative sample type [11,21,32,74].

8 POSTANALYTICAL CONSIDERATIONS
8.1 REFERENCE RANGES

All steroid hormone assay results should have a reference range appended that should be laboratory and assay specific. Reference ranges should preferably be determined in the laboratory using the assay that will be used to measure patient samples, and there is guidance available [75]. Although reference ranges are helpful in guiding a clinician, patient results should always be interpreted in conjunction with the clinical picture of the patient.

A number of steroid hormones require more complex reference ranges than others, and these can be appreciated here; gender-specific reference ranges should be given for total and free testosterone, dihydrotestosterone, estradiol, estrone, 17-hydroxyprogesterone, progesterone, 17-hydroxyprogesterone, dehydroepiandrosterone sulfate, and androstenedione. Age-specific reference ranges and Tanner stage reference ranges, if available, should be given for total and free testosterone, dihydrotestosterone, estradiol, estrone, 17-hydroxyprogesterone, dehydroepiandrosterone, dehydroepiandrosterone sulfate, androstenedione, 17-hydroxypregnenolone, and aldosterone. Menstrual cycle stage-based and postmenopausal reference ranges should be given for estradiol, estrone, progesterone, 17-hydroxyprogesterone, and androstenedione. Finally, pregnancy trimester reference ranges should be given for progesterone and 17-hydroxyprogesterone.

8.2 UNITS AND REPORTING

The conventional units used to report steroid hormone concentrations vary, and reporting results in both conventional and Système International d'Unités (SI) units should be considered. Total testosterone, 17-hydroxyprogesterone, 17-hydroxypregnenolone, dehydroepiandrosterone, androstenedione, deoxycorticosterone, deoxycortisol, and aldosterone are typically measured in nanograms per deciliter (ng/dL; SI units: nmol/L). Estradiol, free estradiol, estrone, and free testosterone are measured in picograms per milliliter (pg/mL; SI units: pmol/L). Dehydroepiandrosterone sulfate and cortisol are measured in micrograms per deciliter (μg/dL; SI units: μmol/L) and progesterone is measured in nanograms per milliliter (ng/mL; SI units: nmol/L).

Since clinical interpretation of steroid hormone results often takes place at very low concentrations, it is critical for the laboratory to validate the LLOQ of the assay to ensure that the low concentrations are in fact clinically meaningful. For example, low concentrations will be found in estrogen measurement in postmenopausal women, testosterone measurement in women for the determination of polycystic ovary disease or testosterone measurement for diagnosis of precocious puberty in children.

DISCLAIMER

The findings and conclusions in this report are those of the author(s) and do not necessarily represent the views of the Centers for Disease Control and Prevention. Use of trade names and commercial sources is for identification only and does not constitute endorsement by the US Department of Health and Human Services, or the US Centers for Disease Control and Prevention.

REFERENCES

[1] Guo T, Taylor RL, Singh RJ, Soldin SJ. Simultaneous determination of 12 steroids by isotope dilution liquid chromatography-photospray ionization tandem mass spectrometry. Clin Chim Acta 2006;372(1–2):76–82.
[2] Kushnir MM, Blamires T, Rockwood AL, Roberts WL, Yue B, Erdogan E, Bunker AM, Meikle AW. Liquid chromatography-tandem mass spectrometry assay for androstenedione, dehydroepiandrosterone, and testosterone with pediatric and adult reference intervals. Clin Chem 2010;56(7):1138–47.
[3] CLSI. Mass spectrometry for androgen and estrogen measurements in serum; Approved Guideline. CLSI document C57-A. Wayne, PA: Clinical and Laboratory Standards Institute; 2015.
[4] CLSI. Liquid chromatography-mass spectrometry methods; Approved Guideline. CLSI document C62-A. Wayne, PA: Clinical and Laboratory Standards Institute; 2014.
[5] CLSI. Metreological Traceabililty and its Implementation: A Report. CLSI document EP32-R. Wayne, PA: Clinical and Laboratory Standards Institute; 2006.
[6] Vesper HW, Thienpont LM. Traceability in laboratory medicine. Clin Chem 2009;55(6):1067–75.
[7] Matuszewski BK, Constanzer ML, Chavez-Eng CM. Strategies for the assessment of matrix effect in quantitative bioanalytical methods based on HPLC-MS/MS. Anal Chem 2003;75(13):3019–30.
[8] CLSI. Evaluation of commutability of process samples; Approved Guideline. 3rd ed.. CLSI document EP14-A3. Wayne, PA: Clinical and Laboratory Standards Institute; 2014.
[9] Tai SS-C, Xu B, Welch MJ, Phinney KW. Development and evaluation of a candidate reference measurement procedure for the determination of testosterone in human serum using isotope dilution liquid chromatography/tandem mass spectrometry. Anal Bioanal Chem 2007;388:1087–94.

[10] Faix JD. Principles and measurement of free hormone measurements. Best Pract Res Clin Endocrinol Metab 2013;27(5):631–45.

[11] Ray JA, Kushnir MM, Bunker A, Rockwood AL, Meikle AW. Direct measurement of free estradiol in human serum by equilibrium dialysis-liquid chromatography-tandem mass spectrometry and reference intervals of free estradiol in women. Clin Chim Acta 2012;413(11–12):1008–14.

[12] Salameh WA, Redor-Goldman MM, Clarke NJ, Reitz RE, Caulfield MP. Validation of a total testosterone assay using high-turbulence liquid chromatography tandem mass spectrometry: total and free testosterone reference ranges. Steroids 2010;75:169–75.

[13] Viahos I, MacMahon W, Sgoutas D, Bowers W, Thompson J, Trawick W. An improved ultrafiltration method for determining free testosterone in serum. Clin Chem 1982;28(11):2286–91.

[14] Hauser B, Deschner T, Boesch C. Development of a liquid chromatography-tandem mass spectrometry method for the determination of 23 endogenous steroids in small quantities of primate urine. J Chromatogr B Analyt Technol Biomed Life Sci 2008;862(1–2):100–12.

[15] Annesley TM. Ion suppression in mass spectrometry. Clin Chem 2003;49(7):1041–4.

[16] Owen LJ, Kevil BG. Supported liquid extraction as an alternative to solid phase extraction for LC-MS/MS aldosterone analysis? Ann Clin Biochem 2013;50(5):489–91.

[17] Kushnir MM, Rockwood AL, Roberts WL, Yue B, Bergquist J, Meikle AW. Liquid chromatography tandem mass spectrometry for analysis of steroids in clinical laboratories. Clin Biochem 2011;44(1):77–88.

[18] Janzen N, Sander S, Terhardt M, Peter M, Sander J. Fast and direct quantitation of adrenal steroids by tandem mass spectrometry in serum and dried blood spots. J Chromatogr B Analyt Technol Biomed Life Sci 2008;861(1):117–22.

[19] Rauh M, Gröschl M, Rascher W, Dörr HG. Automated, fast and sensitive quantitation of 17 alpha-hydroxy-progesterone, androstenedione and testosterone by tandem mass spectrometry with on-line extraction. Steroids 2006;71(6):450–8.

[20] Kulle AE, Riepe FG, Melchior D, Hiort O, Holterhus PM. A novel ultrapressure liquid chromatography tandem mass spectrometry method for the simultaneous determination of androstenedione, testosterone, and dihydrotestosterone in pediatric blood samples: age- and sex-specific reference data. J Clin Endocrinol Metab 2010;95(5):2399–409.

[21] Gallagher LM, Owen LJ, Keevil BG. Simultaneous determination of androstenedione and testosterone in human serum by liquid chromatography-tandem mass spectrometry. Ann Clin Biochem 2007;44(Pt 1):48–56.

[22] Fanelli F, Belluomo I, Di Lallo VD, et al. Serum steroid profiling by isotopic dilution-liquid chromatography-mass spectrometry: comparison with current immunoassays and reference intervals in healthy adults. Steroids 2011;76(3):244–53.

[23] Chadwick CA, Owen LJ, Keevil BG. Development of a method for the measurement of dehydroepiandrosterone sulphate by liquid chromatography-tandem mass spectrometry. Ann Clin Biochem 2005;42(Pt 6): 468–74.

[24] Wang C, Shiraishi S, Leung A, et al. Validation of a testosterone and dihydrotestosterone liquid chromatography tandem mass spectrometry assay: interference and comparison with established methods. Steroids 2008;73(13):1345–52.

[25] Shiraishi S, Lee PW, Leung A, Goh VH, Swerdloff RS, Wang C. Simultaneous measurement of serum testosterone and dihydrotestosterone by liquid chromatography-tandem mass spectrometry. Clin Chem 2008;54(11):1855–63.

[26] Moal V, Mathieu E, Reynier P, Malthièry Y, Gallois Y. Low serum testosterone assayed by liquid chromatography-tandem mass spectrometry: comparison with five immunoassay techniques. Clin Chim Acta 2007;386(1–2):12–9.

[27] Singh RJ. Validation of a high throughput method for serum/plasma testosterone using liquid chromatography tandem mass spectrometry (LC-MS/MS). Steroids 2008;73(13):1339–44.

[28] Salameh WA, Redor-Goldman MM, Clarke NJ, Reitz RE, Caulfield MP. Validation of a total testosterone assay using high-turbulence liquid chromatography tandem mass spectrometry: total and free testosterone reference ranges. Steroids 2010;75(2):169–75.

[29] Vicente FB, Smith FA, Sierra R, Wang S. Measurement of serum testosterone using high-performance liquid chromatography/tandem mass spectrometry. Clin Chem Lab Med 2006;44(1):70–5.

[30] Cawood ML, Field HP, Ford CG, et al. Testosterone measurement by isotope-dilution liquid chromatography-tandem mass spectrometry: validation of a method for routine clinical practice. Clin Chem 2005;51(8): 1472–9.

[31] Rhea JM, French D, Molinaro RJ. Direct total and free testosterone measurement by liquid chromatography tandem mass spectrometry across two different platforms. Clin Biochem 2013;46(7–8):656–64.

[32] Kyriakopoulou L, Yazdanpanah M, Colantino DA, Chan MK, Daly CH, Adeli K. A sensitive and rapid mass spectrometric method for the simultaneous measurement of eight steroid hormones and CALIPER pediatric reference intervals. Clin Biochem 2013;46(7–8):642–51.

[33] French D. Development and validation of a serum total testosterone liquid chromatography-tandem mass spectrometry (LC-MS/MS) assay calibrated to NIST SRM 971. Clin Chim Acta 2013;415:109–17.

[34] Guo T, Gu J, Soldin OP, Singh RJ, Soldin SJ. Rapid measurement of estrogens and their metabolites in human serum by liquid chromatography-tandem mass spectrometry without derivatization. Clin Biochem 2008;41(9):736–41.

[35] Corona G, Elia C, Casetta B, et al. Liquid chromatography tandem mass spectrometry assay for fast and sensitive quantification of estrone-sulfate. Clin Chim Acta 2010;411(7–8):574–80.

[36] Giton F, Caron P, Bérubé R, Bélanger A, Barbier O, Fiet J. Plasma estrone sulfate assay in men: comparison of radioimmunoassay, mass spectrometry coupled to gas chromatography (GC-MS), and liquid chromatography-tandem mass spectrometry (LC-MS/MS). Clin Chim Acta 2010;411(17–18):1208–13.

[37] Guillarme D, Ruta J, Rudaz S, Veuthey JL. New trends in fast and high-resolution liquid chromatography: a critical comparison of existing approaches. Anal Bioanal Chem 2010;397(3):1069–82.

[38] CLSI. Mass spectrometry in the clinical laboratory: General principles and guidance; Approved Guideline. CLSI document C50-A. Wayne, PA: Clinical and Laboratory Standards Institute; 2007.

[39] Huhtaniemi IT, Tajar A, Lee DM, O'Neill TW, Finn JD, Bartfai G, Boonen S, Casaneuva FF, Giwercman A, Han TS, Kula K, Labrie F, Lean ME, Pendleton N, Punab M, Silman AJ, Vanderschueren D, Forti G, Wu FC. EMAS Group. Comparison of serum testosterone and estradiol measurements in 3174 European men using platform immunoassay and mass spectrometry; relevance for the diagnostics in ageing men. Eur J Endocrinol 2012;166(6):983–91.

[40] Harwood DT, Handelsman DJ. Development and validation of a sensitive liquid chromatography-tandem mass spectrometry assay to simultaneously measure androgens and estrogens in serum without derivatization. Clin Chim Acta 2009;409(1–2):78–84.

[41] Ray JA, Kushnir MM, Yost RA, Rockwood AL, Meikle WA. Performance enhancement in the measurement of 5 endogenous steroids by LC-MS/MS combined with differential ion mobility spectrometry. Clin Chim Acta 2015;438:330–6.

[42] Wu AHB, Gerona R, Armenian P, French D, Petrie M, Lynch KL. Role of liquid chromatography-high-resolution mass spectrometry (LC-HR/MS) in clinical toxicology. Clin Toxicol 2012;50(8):733–42.

[43] CLSI. Measurement procedure comparison and bias estimation using patient samples; Approved Guideline. 3rd ed. CLSI document EP09-A3. Wayne, PA: Clinical and Laboratory Standards Institute; 2013.

[44] CLSI. User verification of performance for precision and trueness; Approved Guideline. 2nd ed. CLSI document EP15-A2. Wayne, PA: Clinical and Laboratory Standards Institute; 2006.

[45] U.S. Department of Health and Human Services: FDA, CDER and CVM. Guidance for Industry: Bioanalytical Method Validation. Available from: http://www.fda.gov/downloads/Drugs/../Guidances/ucm070107

[46] Perich C, Minchinela J, Ricós C, Fernández-Calle P, Alvarez V, Doménech MV, Simón M, Biosca C, Boned B, García-Lario JV, Cava F, Fernández-Fernández P, Fraser CG. Biological variation database: structure and criteria used for generation and update. Clin Chem Lab Med 2015;53(2):299–305.

[47] CLSI. Evaluation of precision performance of quantitative measurement methods; Approved Guideline. 2nd ed. CLSI document EP05-A2. Wayne, PA: Clinical and Laboratory Standards Institute; 2004.

[48] Fraser CG. General strategies to set quality specifications for reliability performance characteristics. Scand J Clin Lab Invest 1999;59(7):487–90.

[49] Yun Y-M, Botelho JC, Chandler DW, Katayev A, Roberts WL, Stanczyk FZ, Vesper HW, Nakamoto JM, Garibaldi L, Clarke NJ, Fitzgerald RL. Performance criteria for testosterone measurements based on biological variation in adult males: recommendations from the partnership for the accurate testing of hormones. Clin Chem 2012;58(12):1703–10.

[50] Wang C, Catlin DH, Demers LM, Starcevic B, Swerdloff RS. Measurement of total serum testosterone in adult men: comparison of current laboratory methods versus liquid chromatography-tandem mass spectrometry. J Clin Endocrinol Metab 2004;89(2):534–43.

[51] Soldin SJ, Soldin OP. Steroid hormone analysis by tandem mass spectrometry. Clin Chem 2009;55(6): 1061–6.

[52] Bhasin S, Zhang A, Coviello A, Jasuja R, Ulloor J, Singh R, Vesper H, Vasan RS. The impact of assay quality and reference ranges on clinical decision making in the diagnosis of androgen disorders. Steroids 2008;73(13):1311–7.

[53] Santen RJ, Lee JS, Wang S, Demers LM, Mauras N, Wang H, Singh R. Potential role of ultra-sensitive estradiol assays in estimating the risk of breast cancer and fractures. Steroids 2008;73(13):1318–21.

[54] Van der Gugten JG, Dubland J, Liu HF, Wang A, Joseph C, Holmes DT. Determination of serum aldosterone by liquid chromatography and tandem mass spectrometry: a liquid-liquid extraction method for the AB-SCIEX API-5000 mass spectrometry system. J Clin Pathol 2012;65(5):457–62.

[55] CLSI. Evaluation of the linearity of quantitative measurement procedures: a statistical approach; Approved Guideline. 2nd ed. CLSI document EP06-A. Wayne, PA: Clinical and Laboratory Standards Institute; 2003.

[56] CLSI. Preliminary evaluation of quantitative clinical laboratory measurement procedures; Approved Guideline. 3rd ed. CLSI document EP10-A3. Wayne, PA: Clinical and Laboratory Standards Institute; 2006.

[57] CLSI. Interference testing in clinical chemistry; Approved Guideline. 2nd ed. CLSI document EP07-A2. Wayne, PA: Clinical and Laboratory Standards Institute; 2005.

[58] Shi RZ, van Rossum HH, Bowen RA. Serum testosterone quantitation by liquid chromatography-tandem mass spectrometry: interference from blood collection tubes. Clin Biochem 2012;45(18):1706–9.

[59] CLSI. Assessment of laboratory tests when proficiency testing is not available; Approved Guideline. 2nd ed. CLSI document GP29-A2. Wayne, PA: Clinical and Laboratory Standards Institute; 2008.

[60] Miller WG, Myers GL, Gantzer ML, Kahn SE, Schönbrunner ER, Thienpont LM, Bunk DM, Christenson RH, Eckfeldt JH, Lo SF, Nübling CM, Sturgeon CM. Roadmap for harmonization of clinical laboratory measurement procedures. Clin Chem 2011;57(8):1108–17.

[61] Weitzman ED. Circadian rhythms and episodic hormone secretion in man. Ann Rev Med 1976;27: 225–43.

[62] Portaluppi F, Bagni B, degli Uberti E, Montanari L, Cavallini R, Trasforini G, Marqutti A, Ferlini M, Zanella M, Parti M. Circadian rhythms of atrial natriuretic peptide, renin, aldosterone, cortisol, blood pressure and heart rate in normal and hypertensive subjects. J Hypertens 1990;8(1):85–95.

[63] Stolze BR, Gounden V, Gu J, Abel BS, Merke DP, Skarulis MC, Soldin SJ. Use of micro-liquid chromatography/tandem mass spectrometry method to assess diurnal effects on steroid hormones. Clin Chem 2015;61(3):556–8.

[64] Kage A, Fenner A, Weber B, Schöneshöfer M. Diurnal and ultradian variations of plasma concentrations of eleven adrenal steroids in human males. Klin Wochenschr 1982;60(13):659–66.

[65] Freund BJ, Shizuru EM, Hashiro GM, Claybaugh JR. Hormonal, electrolyte and renal responses to exercise are intensity dependent. J Appl Physiol 1985;70(2):900–6.

[66] Viti A, Lupo C, Lodi L, Bonifazi M, Martelli G. Hormonal changes after supine posture, immersion and swimming. Int J Sports Med 1989;10(6):401–5.

[67] Bennett JM, Rodrigues IM, Klein LC. Effects of caffeine and stress on biomarkers of cardiovascular disease in healthy men and women with a family history of hypertension. Stress Health 2013;29:401–9.

[68] Baer L, Radichevich I. Cigarette smoking in hypertensive patients. Blood pressure and endocrine responses. Am J Med 1985;78(4):564–8.

[69] Ribstein J, Halimi JM, du Cailar G, Mimran A. Renal characteristics and effect of angiotensin suppression in oral contraceptive users. Hypertension 1999;33(1):90–5.

[70] Heesch CM, Negus BH, Keffer JH, Snyder RW, Risser RC, Eichhorn EJ. Effects of cocaine on cortisol secretion in humans. Am J Med Sci 1995;310(2):61–4.

[71] Isojarvi JIT, Pakarinen AJ, Ylipalosaari PJ, Myllyla VV. Serum hormones in male epileptic patients receiving anticonvulsant medication. Arch Neurol 1990;47(6):670–6.

[72] Chrousos GP, Kino T. Glucocorticoid action networks and complex psychiatric and/or somatic disorders. Stress 2007;10(2):213–9.

[73] Kubzansky LD, Adler GK. Aldosterone: a forgotten mediator of the relationship between psychological stress and heart disease. Neurosci Biobehav Rev 2010;34(1):80–6.

[74] Stroud LR, Solomon C, Shenassa E, Papandonatos G, Niaura R, Lipsitt LP, LeWinn K, Buka SL. Long-term stability of maternal prenatal steroid hormones from the National Collaborative Perinatal Project: still valid after all these years. Psychoneuroendocrinology 2007;32:140–50.

[75] CLSI. Defining, establishing and verifying reference intervals in the clinical laboratory; Approved Guideline. 3rd ed. CLSI document EP28-A3C. Wayne, PA: Clinical and Laboratory Standards Institute; 2010.

MASS SPECTROMETRY IN THE CLINICAL MICROBIOLOGY LABORATORY

11

I.W. Martin

Dartmouth-Hitchcock Medical Center, One Medical Center Drive, Lebanon, NH, United States

1 INTRODUCTION

Timely, accurate pathogen detection and identification is a core task of the clinical microbiology laboratory. Over the past 5 years, matrix-assisted laser desorption ionization-time of flight mass spectrometry (MALDI-TOF MS) has gained prominence as a powerful tool for rapid bacterial, fungal, and mycobacterial identification. Advantages over prior phenotypic and biochemical identification methods include quicker time to results and low cost per test in the context of comparable or superior ability to identify organisms to the species level. However, MALDI-TOF MS does have its limitations, which include similarity in profiles between certain organisms, the inability thus far to generate definitive susceptibility data, and the inability to identify microorganisms directly from most specimen types, such as swabs, tissues, urines, and respiratory specimens.

In this chapter, we will review how MALDI-TOF MS works and discuss the two commercial MALDI-TOF MS systems that are available in the United States. Although the applications of MALDI-TOF MS in the clinical microbiology laboratory continue to evolve, we will touch on the current uses and challenges of MALDI-TOF MS specific to the domains of bacteriology, mycology, and mycobacteriology. Lastly we will discuss potential future applications of MALDI-TOF MS and other less common mass spectrometry techniques being explored for the diagnosis of clinical infections.

2 HOW MALDI-TOF MS WORKS

At the most basic level, MALDI-TOF MS works by matching the protein profile (spectrum) of an unknown isolate to the most similar profiles in a dedicated database (library) of known organism profiles. The specific proteins themselves are not identified; the overall pattern and abundance of proteins of different masses represents a unique fingerprint for each organism.

Before beginning MALDI-TOF MS analysis, one must ensure that the microorganism to be analyzed is a "pure" sample (not mixed with other microorganisms) of sufficient quantity (1.5×10^5 CFU, i.e., a fraction of one robust colony from an agar plate) [1]. Only pure samples generate representative spectra that can be reliably compared with database spectra. Additionally, the intracellular proteins of interest must be removed from their cells, rendering them available for analysis. The preparatory steps necessary to ensure protein availability depend on the type of microorganism in question. For most Gram-positive and Gram-negative bacteria as well as some yeasts, an isolated colony of intact microorganisms from solid media can be transferred directly (spotted) with a sterile implement without pretreatment because their cell walls will be lysed upon exposure to the solvent present in the MALDI matrix (usually acetonitrile and/or ethanol). This simple transfer from agar plate to target plate is the easiest in terms of laboratory workflow and can be performed with a simple sterile implement, such as a wooden stick, pipette tip, or loop. However, on-plate protein extraction with formic acid can optimize spectral identification and is necessary for some yeasts and spore-producing bacteria [2]. To simplify workflow in some laboratories as well as to optimize identification of Gram-positive organisms, on-plate extraction is performed on every bacterial isolate and reduces the number of unsuccessful identifications [3]. For mycobacterial isolates, molds, and dimorphic fungi, extensive additional pretreatment preparation is necessary as described later, in some cases for safety reasons and in all cases to enable successful identification.

Most current laboratory MALDI-TOF MS workflow starts with colony growth on solid agar media. However, bacterial growth in liquid media, such as blood culture bottles or mycobacterial liquid culture systems can also be used for identification once organisms are concentrated [4]. When growth in liquid media is used, manufacturer's guidelines should be followed and simultaneous subculturing to solid medium is important to ensure organism purity and enable susceptibility testing when appropriate. MALDI-TOF MS analysis cannot identify more than one microorganism growing simultaneously in a liquid medium.

Once spotted on the target plate, the sample of microorganism is overlaid with a matrix which is usually composed of a phenolic acid, such as α-cyano-4-hydroxycinnamic acid in purified water plus an organic solvent like acetonitrile and/or ethanol. As described earlier, the solvent lyses the target cells. Then as the matrix dries, a crystalline lattice is formed which serves to protect the large, nonvolatile proteins of interest from fragmentation by the laser, making MALDI-TOF MS a so-called "soft" method of ionization. A laser beam is then pulsed at the dried isolate/matrix mixture on the target plate, vaporizing (desorbing) and ionizing the protein molecules. An electrostatic field is then applied which accelerates the ions into the vacuum flight tube. The ions travel at differential speeds through the tube according to their mass to charge (m/z) ratio, with smaller molecules being the first to reach an ion detector at the opposite end (Fig. 11.1). The time-of-flight data of different ions hitting the ion detector is used to generate a mass spectrum with the m/z represented on the x-axis and the signal intensity (i.e., abundance of each ion) on the y-axis. Spectra are generally composed of 100–200 peaks, representing ribosomal proteins, nucleic acid-binding proteins, and heat-shock proteins [5].

Specialized software analyzes the data and a report is generated listing the best profile match (assuming successful identification) for each unknown isolate along with a value indicating the level of confidence in the identification. With increasing strength of the confidence value, either a family, genus, or species level identification can be made. The two major commercial systems perform their analyses slightly differently to match unknown profiles with those in the database and to generate confidence values. These differences will be discussed in Section 3. Control strains must be included in each run to calibrate the instrument. Generally, most common organisms will be correctly identified with a strong confidence value. If identification is not successful, the confidence value is usually low. This can be due to lack of spectra in the database or to technical error in target plating (inoculum too heavy or light).

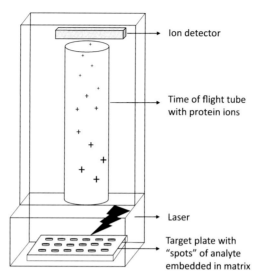

FIGURE 11.1 Schematic of How MALDI-TOF MS Works

A laser beam shoots a matrix-covered analyte on the target plate, ionizing it. The cloud of ions is then accelerated through the time of flight (TOF) tube with lighter protein ions traveling fastest and heavier ones slower. A detector at the opposing end of the TOF tube detects ions hitting its surface at different times. A spectrum is thereby generated showing relative quantities of proteins of different masses.

Misidentification with a strong confidence value is not common. However, even when the confidence value is high, it is important to correlate the MALDI-TOF MS identification with the colony morphology and Gram stain results.

Ultimately, performance of MALDI-TOF MS relies heavily on the quality and number of spectra in the reference library. Both of the commercially-available systems in the United States have produced several versions of their libraries, increasing the scope and depth with each iteration. When reviewing the MALDI-TOF literature to compare assay characteristics, it is important to take into account which library version was used and whether it was enhanced with user-generated spectra.

Another important consideration when implementing MALDI-TOF MS in the clinical laboratory is that identification is optimized when the extraction method used is the same as that used to generate the library spectra. Therefore, a laboratory's isolate preparation methods ideally should follow manufacturer's instructions, mirroring preparation methods used for isolates contained in the commercial library. However, through extensive validation, laboratories can develop in-house preparation methods that deviate from manufacturer's instructions. Additionally, the matrix used to cover an unknown isolate on the target plate must be the same as that which was used to create the reference library. Mass spectra from the same strain can differ when analyzed using a different matrix [6].

Several important variables affect performance of MALDI-TOF MS in the clinical microbiology laboratory: the method of protein extraction, the matrix, the version of library used, and whether it has been enhanced with laboratory-specific spectra, and differences in interpretive guidelines (i.e., variation in thresholds for acceptance of correct species identification).

3 COMMERCIAL SYSTEMS

Currently, two commercial assays are FDA-cleared for identification of cultured bacterial isolates: the MALDI Biotyper CA System (Bruker Daltonics, Inc., Billerica, MA) and the VITEK MS (bioMérieux, Durham, NC), both of which received their initial US regulatory clearance in 2013. A third major commercial system—the Andromas system (Paris, France)—is used in Europe and will not be discussed here.

Of the two available in the United States, the Bruker system is more widely used and has been described more extensively in the literature. The Bruker instrument fits on a benchtop and, when used at maximum capacity, can perform up to 200 identifications in an hour on 96-spot target plates which can be either reusable or disposable. The system's most recent FDA-clearance as of this writing includes 210 species and species groups covering 280 clinically relevant bacteria and yeast species. There is an additional research use only (RUO) library that can also be used to expand diagnostic breadth to 2,297 species including some mycobacteria and filamentous fungi. Specialized libraries for mycobacteria, filamentous fungi, and select agents (such as *Brucella* species, *Francisella tularensis* and *Burkholderia pseudomallei*) can also be purchased separately. The CDC recently offered a new library free to clinical microbiology and public health laboratories: the MicrobeNet MALDI-TOF reference library which encompasses 800 species of rare and unusual organisms. Regardless of which library is used, the Bruker analytic software uses spectral pattern matching to compare spectra and find the best match from the library. The accompanying confidence value assigned to identification results by Bruker software consists of a log score value ranging from 0.000 to 3.000, with values ≥2.000 representing a high-confidence identification at the species level and values from 1.700 to 1.999 representing a low-confidence identification at the genus level. A log score of <1.700 indicates no identification was possible.

The bioMérieux system was the first to receive FDA clearance and consists of a larger instrument which rests on the floor. Barcoded target plates are disposable and contain 48 spots. Up to four target plates can be loaded on the instrument in a single "run" allowing for high-throughput. As of this writing, 193 organisms are FDA-cleared for detection in the VITEK MS "KnowledgeBase" database with an additional 562 organisms available as RUO identifications which require verification by other laboratory methods. Of these organisms, 645 are bacteria and 110 are fungi. This database was built with a proprietary algorithm using a weighted bin matrix approach whereby multiple spectra of >2 isolates (average 10) for each reference species were analyzed. Each spectrum was divided into 1,300 predefined data "bins" from which bin weights were calculated based on the consensus presence or absence of peaks within the bins. Data from the spectrum of a test sample is similarly sorted into discrete data points in bins before being compared to the database. Bins are algorithmically weighted by importance to a particular species' identification. A confidence value is calculated to express the similarity between the unknown organism and every organism or organism group in the database. Then thresholds are applied to retain and report only the best matches. A single identification is reported with "good" confidence, represented by a probability percentage of 60–99.9%. Alternatively, a "low discrimination" confidence level is reported when 2–4 possible matches are identified. In this case, the sum of the probability percentages will equal 100%. Lastly, when no identification is achieved, greater than 4 possible matches will be displayed with the sum of probability percentages < 100%. An RUO library called the Spectral Archiving and Microbial Identification System (SARAMIS) database can be used which expands the total organism list to 1,388 and includes mycobacteria and filamentous fungi.

Both the Bruker and the bioMérieux systems allow for in-house expansion of the commercial database. Identification of many organisms can be optimized and confidence values increased by adding additional strains to the existing library [7–9]. Some laboratories have also observed that acceptance of identifications with modified cutoff values (lower than manufacturer's recommendations) in certain situations, such as the identification of filamentous fungi or mycobacteria results in increased successful identifications without significant loss of specificity [10,11].

The initial cost of either of these commercial instruments with their accompanying software and database(s) is substantial (approximately $200,000). Instruments from the FDA-approved commercial systems must be purchased outright since the lack of costly reagents involved in testing (target slides, matrix solution, formic acid, pipette tips, and disposable loops or toothpicks) renders reagent rentals economically unappealing for the manufacturers. This purchase represents a significant capital cost to an institution. Additionally, the annual maintenance contracts often range from $25,000 to $30,000. However, subsequent consumables and labor costs are inexpensive per sample (<$1) relative to those of traditional phenotypic tests. One large academic medical center's microbiology laboratory estimated a cost savings of just over $100,000 in reagent and labor costs in the first year of MALDI-TOF MS implementation [12]. Another large academic medical center found that their instrument costs were recouped in 3 years and subsequently resulted in significant cost savings to the institution [13].

In addition to cost savings, incorporation of MALDI-TOF MS also results in significantly faster time to identification when compared with traditional biochemical methods both from an individual isolate perspective [14] and when analyzed from the perspective of implementation into a laboratory's work-flow [15]. What's more, both systems have the future potential to be incorporated into total laboratory automation systems within the microbiology laboratory where automated streaking, incubation, and monitoring of cultures can further speed up organism growth and reduce time to identification. More importantly, significant positive effects have been observed on patient care after implementation of MALDI-TOF MS, including quicker time to effective/optimal antibiotic therapy and reduced length of hospitalization [16,17].

4 BACTERIOLOGY

In this section, we will highlight overall performance of the two FDA-approved commercial MALDI-TOF MS systems compared to automated systems and compared to each other. We will discuss the profound effect this technology has had on the identification of anaerobes and then consider two bacterial species whose distinction from other closely related species has been facilitated by MALDI-TOF MS. Lastly, we will briefly discuss the role of MALDI-TOF MS in identification of select agents.

In large studies, both commercial systems available in the United States have been shown to perform overall as well or better than automated biochemical systems for bacteria that are routinely encountered in the clinical microbiology laboratory with species level identification achieved for 85–97% of isolates [18–22]. Numerous studies evaluating specific classes of bacteria have also been performed and are summarized in a review by Clark and coworkers [23]. Gram-positive organisms, which possess a thick layer of peptidoglycan in their cell wall, benefit from a simple on-plate formic acid extraction prior to addition of matrix and mass spectrometry analysis [3]. MALDI-TOF MS can aid in correctly identifying myriad bacteria that are otherwise phenotypically difficult to identify and have in the past relied on 16S rRNA gene sequencing. A study by Lau and coworkers analyzed 67 "difficult-to-identify" isolates, such as *Facklamia hominis, Campylobacter* spp., and *Raoultella ornitholyica* and found that the Bruker system (MALDI Biotyper 3.0 with Reference Library v.3.1.0) correctly identified 45% of them to the species level and an additional 30% to the genus level [24]. Of the 13 isolates with a result of "no identification," 10 were species missing from the database.

Slight differences between the systems have been observed for certain organism groups in head-to-head comparison studies. In 2010, Cherkaoui and coworkers observed correct, high-confidence species level identifications in 94% of 720 isolates tested with the Bruker system (database version unspecified) compared with 89% of those tested with the Shimadzu MALDI-TOF system (an early version of MALDI-TOF MS that was bought by bioMérieux and became the VITEK MS; this study used the SARAMIS database, version unspecified) [25]. In a study of 200 glucose nonfermenting Gram-negative rods isolated from cystic fibrosis patients, the Bruker Biotyper (database v3.0) combined species/complex/genus level identification was 97% compared to 89.5% for the bioMérieux (SARAMIS database 3.62) [26]. However, the Bruker system required an extra extraction step more often (20 isolates vs. 2 isolates) to achieve this. In a study of 274 anaerobic isolates, the bioMérieux system (database v2.0) accurately identified 100% of organisms to the species level compared with the API20AN while the Bruker system (Biotyper DB Update v3.3) identified only 89.1% with high confidence [27]. Less than 1% of isolates were misidentified by either system. In a 2012 study, Martiny and coworkers found the Bruker (database v2.0) and the bioMérieux (SARAMIS database, version unspecified) systems to perform similarly (92.7% and 93.2% correct species identification, respectively) in comparison with phenotypic biochemical tests, API strips and the Vitek automated instrument in identifying 986 routinely encountered isolates [22]. As previously mentioned, when evaluating this literature, it is important to take note of the extraction method, database and acceptance threshold for confidence values used in comparisons. Failure to successfully identify an organism with either system can occur due to poor quality of the test sample's spectrum (usually due to poor spotting technique) or to a lack of strains in the commercial database [23].

Despite this generally good performance, there are instances where closely-related organisms are misidentified by MALDI-TOF MS. One well-described example of this is the difficulty discriminating between *Escherichia coli* and *Shigella* spp. which are so closely related genetically that they sometimes cannot even be differentiated with molecular tools, such as 16S rRNA sequencing [23]. If both *E. coli* and *Shigella* are possible pathogens in the specimen type in question (i.e., stool), one must confirm isolates identified as *E. coli* by MALDI-TOF MS with some other phenotypic test, such as indole production or by one of the automated commercial systems [15]. Another example of misidentification of closely-related organisms can be found in the viridans group streptococci, specifically the *Streptococcus mitis* group. Within this group, *S. pneumoniae, S. mitis, S. pseudopneumoniae,* and *S. oralis* are all colonizers of the human oral cavity. However, their virulence properties differ significantly: *S. pneumoniae* is a major human pathogen that can cause community-acquired pneumonia, meningitis, and otitis media whereas the others are less virulent and usually cause disease only if they find entry into the blood stream in immunocompromised patients. Despite the stark clinical difference between these isolates, they are genetically closely related and MALDI-TOF MS is not always successful at distinguishing one from another. Phenotypic tests, such as bile solubility and optochin disk susceptibility should therefore be performed prior to reporting identification [23]. However, two studies with the bioMérieux system (one with database v2.0 and one with database unspecified) reported excellent identification rates of *Streptococcus mitis* group species including differentiation of *S. pneumoniae* from its closely related but less pathogenic cousins [21,28].

The class of bacteria whose successful identification has been most profoundly impacted by MALDI-TOF MS is the anaerobic bacteria. Anaerobes are important causes of infections of the brain, lung, pelvis, and abdomen. However, they have traditionally been notoriously difficult to cultivate for several reasons. First, because oxygen is toxic to these organisms, they require special transport media and incubation in special anaerobic jars or chambers. Manipulation of cultures can therefore be cumbersome and time-consuming for technologists. Second, many anaerobic species grow more slowly than aerobic bacteria. Third, phenotypic tests sometimes lack discriminatory power between species and identification at the species level then requires 16S rRNA sequencing or gas liquid chromatography. Smaller laboratories without the capacity to perform such testing have historically opted not to perform anaerobic culture at all. However, species level discrimination of anaerobes is important because certain species of *Bacteroides* are associated with resistance to commonly used antibiotics [29]. Due to MALDI-TOF MS's ease of use, timeliness, and cost-effectiveness, it has replaced 16S rRNA gene sequencing and gas-liquid chromatography as the method of choice for anaerobe identification [30]. However, expanded user-enhanced databases have been found to optimize MALDI-TOF MS identification of anaerobes [8].

Another class of bacteria whose identification MALDI-TOF MS facilitates greatly is the coagulase-negative staphylococci (CoNS). These Gram-positive cocci are frequently encountered in the clinical microbiology laboratory. Many species are part of normal human skin microbiota and can thus represent laboratory contaminants. However, CoNS can also be important opportunistic pathogens and can cause infections associated with medical devices. The species in this group of bacteria are genetically similar and sometimes difficult to distinguish with phenotypic tests, often requiring gene sequencing. MALDI-TOF MS has been shown to be highly accurate in identifying these organisms and is a much simpler and more cost-effective method than gene sequencing.

In addition to facilitating improved and easy identification across the anaerobe and CoNS groups of bacteria, there are several examples where the clinical relevance of specific organisms has been brought to light by the implementation of MALDI-TOF MS. These include *Staphylococcus pseudintermedius* and *Streptococcus pseudoporcinus*. *S. pseudintermedius* is a coagulase-positive organism that is a common colonizer of dogs and a leading cause of canine pyoderma and otitis externa as well as a common urinary pathogen. While traditional phenotypic methods will misidentify this organism as the well-known pathogen *S. aureus*, MALDI-TOF MS discriminates the two readily, raising awareness of the clinical relevance of *S. pseudintermedius* in recent years as an organism capable of causing invasive infections in humans and associated with companion animal contact [31]. *S. pseudoporcinus*, on the other hand, has been identified in the female genital tract during routine screening for *Streptococcus agalactiae* during pregnancy. Its clinical significance has yet to be defined. In the days before MALDI-TOF MS, these isolates would have been identified phenotypically as *S. agalactiae* [32]. MALDI-TOF MS's ability to differentiate certain species with ease that were previously difficult or cumbersome to distinguish will likely enhance our understanding of other disease associations in the future.

Another realm of bacteriology where MALDI-TOF MS has the potential to prove itself useful is the identification of potential agents of bioterrorism. The US Government has designated several organisms as "select agents" that are a potential threat to public health, including *Bacillus anthracis*, *Brucella* spp., *Francisella tularensis*, *Yersinia pestis*, and *Burkholderia pseudomallei* [33]. Due to security concerns, these organisms are not included in the Bruker database. However, Bruker has a "Security-Relevant" library that can be purchased separately and has been found to perform well for

F. tularensis, *Brucella* spp., and *B. pseudomallei* [34]. When the regular Bruker Biotyper library was used to analyze isolates in the same study, *Francisella* and *Brucella* yielded no identification and *B. pseudomallei* was misidentified as *Burkholderia thailandensis*. The bioMérieux database contains a number of select agent spectra in its database, including *B. pseudomallei* but no study has analyzed its performance to date with these organisms. If a select agent is suspected based on colony morphology and Gram stain, laboratories should consult with their public health laboratory and should not run isolates on MALDI-TOF MS.

5 MYCOBACTERIOLOGY

Mycobacterium species are responsible for significant mortality and morbidity worldwide. The *M. tuberculosis* complex (MTB) is the most important, accounting for 1.5 million deaths in 2014 [35]. However, the over 150 described species of nontuberculous mycobacteria (NTM) have been increasingly recognized as a significant cause of infections in recent years as our ability improves to support patients through periods of profound immunosuppression associated with bone marrow and organ transplantation [36]. The laboratory identification of all of these organisms has traditionally relied on phenotypic and biochemical testing. However, these methods are extremely time and labor-intensive due to the slow growth rate of some species (up to several weeks) and the need for biosafety level 3 (BSL-3) practices and procedures when unknown isolates might be MTB. Any potentially aerosol-generating activity must be performed in a class II biosafety cabinet. Molecular methods—including DNA sequencing, [37] DNA probes, [38] and single-phase reverse hybridization [39]—have augmented diagnostic capabilities but are expensive and sometimes impractical in low-prevalence areas. Neither commercial MALDI-TOF MS system is currently FDA-cleared for the diagnosis of mycobacteria, but MALDI-TOF MS is being used increasingly as a faster, less expensive, and less labor-intensive method of mycobacterial identification [30].

To minimize exposure of laboratory personnel to MTB, mycobacterial isolates must undergo inactivation prior to analysis by MALDI-TOF MS. In addition and in contrast to most other bacteria which can be analyzed with whole cell preparations, mycobacterial clumps must be disrupted, cell walls broken down and cellular proteins extracted before MALDI-TOF MS analysis takes place. This extra preparation is due to the waxy hydrophobic nature of the cell wall which is composed of complex lipids [40]. Both commercially-defined and modified protocols have been described for the Bruker and bioMérieux platforms to accomplish inactivation, disruption, and tube extraction. These usually involve a combination of heat inactivation and mechanical disruption, such as vortexing the isolate in the presence of silica beads and ethanol. Early protocols were technically complex involving multiple centrifugation and wash steps that take over 2 h [41,42] and requiring materials not readily available in some laboratories [12]. However, recent studies have sought to simplify and shorten isolate preparation [43,44].

Performing MALDI-TOF MS directly from a liquid medium with mycobacterial growth shows promise as a way to shorten the time to organism identification, because organisms often grow more quickly in liquid media. However, challenges to using liquid medium growth include: (1) the proteins present in the medium itself that can affect the spectrum if not removed through wash steps; and (2) the necessity of timing MALDI-TOF MS analysis when sufficient growth is achieved. As with bacteria, when performing MALDI-TOF MS from liquid medium, a simultaneous subculture to solid media must be performed to ensure organism purity.

Most studies evaluating the identification of mycobacteria with MALDI-TOF MS have focused on the Bruker system [12,42,45] with fewer performed on the bioMérieux platform [46–48]. Bruker's most recent library, the Mycobacterial Library 3.0, became available in June, 2015 and includes 149 different mycobacterial species with 853 main spectrum profiles. As of 2013, the SARAMIS 4.12 RUO database had 1286 spectra from 37 species [43]. Mather and coworkers performed a head-to-head study between the Bruker (*Mycobacterium* database 2.0 as well as an in-house enhanced version of the same) and bioMérieux systems (SARAMIS 4.12 RUO database) using 198 clinical strains representing 18 species of *Mycobacterium* [43]. In this study, the unenhanced Bruker *Mycobacterium* v2.0 library correctly identified either 59.6% or 79.3% of strains depending on the extraction protocol used. However, by decreasing the confidence score threshold to ≥1.7, 93.3% of the isolates were correctly identified with only two misidentifications that were determined to be clinically insignificant. When the Bruker database was enhanced with in-house strains, it performed much better with 94.9% correct identification using manufacturer-recommended thresholds. The bioMérieux system correctly identified 94.4% of isolates with the SARAMIS 4.12 RUO database.

One limitation thus far of MALDI-TOF MS is the ability to identify MTB isolates only at the complex level [12]. Certain NTM species groups also present a challenge to MALDI-TOF MS for species differentiation and similarly require reporting at the complex level only. These include members of the *M. abscessus* complex and *M. avium* complex. However, MALDI-TOF MS is able to easily differentiate the "rapid growers" *M. abscessus* and *M. chelonae* [12,42,43]. Differentiation of these two species can enable more prompt initiation of appropriate antimicrobial therapy than traditional identification methods because of distinct species-specific susceptibility patterns. Two other closely related species pairs that MALDI-TOF MS differentiates with ease are *M. ulcerans/M. marinum* and *M. kansasii/M. gastri*.

Similarly to *Mycobacteria*, the aerobic actinomycetes require tube extraction for successful identification due to their cellular wall components. Buckwalter et al. analyzed 285 clinical isolates of mycobacteria and 197 aerobic actinomycetes (including 136 *Nocardia* species isolates) with the Bruker MALDI Biotyper 3.0 Microflex LT system (flexControl software v3.0 and library v3.3.1.0). They created an in-house-enhanced database and compared it with the basic Bruker library (Bruker BDAL 5627) and the *Mycobacteria* Library v2.0. Correct identification of *Nocardia* species was only 42% using the manufacturer's library but jumped to 90% when using the enhanced library. However, even with the enhanced library, the non-*Nocardia* aerobic actinomycetes strains yielded only 51% correct identification. These strains were enriched for *Streptomyces* species and poor performance was thought to be due to lack of spectra in the database. In this same study, a correct identification was obtained for 55% of mycobacterial isolates using the manufacturer's libraries. This rate went up to 88% when using the in-house supplemented version. Lowering the accepted score threshold to ≥1.7 resulted in an improvement in the number of correct identifications to 75% using manufacturer's libraries and 91% using the in-house enhanced library.

Some laboratories may choose to develop their own extraction methods or to validate lower thresholds for acceptable identification. However, extensive validation must be performed before either of these modifications is used on patient samples. Additionally, if a modified extraction procedure is used, a laboratory must ensure that complete organism inactivation occurs because MTB has a very low infectious dose.

6 MYCOLOGY

As medicine's ability to support patients through periods of profound immunosuppression associated with cancer treatments and organ or bone marrow transplantation has improved over the past few decades, the number of fungal infections has increased. As with mycobacteriology, traditional phenotypic and biochemical methods of identification take days to weeks, especially for mold isolates. MALDI-TOF MS therefore can offer significantly shortened times for identification, which can have a significant effect on patient care.

Reliable identification of the most commonly encountered yeasts has been documented for both FDA-approved platforms, although significantly more independent studies exist for the Bruker system than for the bioMérieux system [2]. For many common yeasts, direct on-plate extraction works just as well as it does for bacteria. However, like mycobacteria, some fungi require protein extraction steps due to their stronger cell walls. Successful identification of these yeasts relies heavily on the adequacy of library databases.

One advantage of MALDI-TOF MS over traditional biochemical identification of common yeasts is its ability to identify closely-related species which in the past have been difficult to distinguish. Species within groups, such as *Candida ortho-/meta-/parapsilosis*, [49] *Candida glabrata/bracarensis/nivariensis*, [50] *Candida albicans/dublinensis* [50] and the phenotypically similar *Candida palmioleophila*, *Candida famata*, and *Candida guillerimondii* [51] can be distinguished without difficulty by MALDI-TOF MS analysis. This ability to identify species at a more granular level enables more nuanced study of each species' clinical significance. Additionally, a handful of studies have shown that MALDI-TOF MS analysis can be useful in strain-typing analysis during outbreak settings. One such study involved a *C. parapsilosis* outbreak in a neonatal intensive care unit [52]. Additionally, MALDI-TOF MS can be used to discriminate between phenotypically different subtypes of the same species which could have clinical relevance [53].

In contrast to yeasts, molds represent a challenge for MALDI-TOF MS for several reasons and have not received FDA-clearance for identification by either MALDI-TOF MS system. First, the morphology of molds is more complex than bacteria, mycobacteria, or yeast and evolves over time. The presence or absence of reproductive conidial structures can alter the spectra [54]. Second, the stronger cell walls necessitate preparatory extraction steps. Third, the extent of agar invasion of a given mold isolate can affect the spectrum due to agar contamination of the analyte. Fourth, fungal taxonomy continues to evolve as molecular methods refine and redefine traditional phylogenetic relationships and the mycology community moves toward a more streamlined nomenclature [55]. Lastly, the presence of melanin pigment in some molds can inhibit ionization, although growth in liquid medium can reduce pigment production [56].

The Bruker system has a separate "Filamentous Fungi Library" which includes spectra obtained from filamentous fungi grown in standardized Sabouraud liquid culture. In the bioMérieux library, the consensus "superspectra" in theory help account for phenotypic heterogeneity introduced by various growth conditions, [2] although scant data has been published on the identification of filamentous fungi using this system. Regardless of which commercial library is used, in-house database enhancement is necessary to obtain acceptable results [57]. Unlike in many other realms of clinical microbiology, filamentous fungi remain a challenge for MALDI-TOF MS identification.

7 OTHER POSSIBLE MALDI-TOF MS APPLICATIONS

Much interest has been paid recently in exploring the potential of MALDI-TOF MS for performing antimicrobial susceptibility testing (AST) of clinical isolates. The gold standard for determining antimicrobial susceptibility has long been the growth of organism in the presence of antibiotic which takes an additional 24 h after initial growth of most bacteria, or longer for mycobacteria and some fungi. Using MALDI-TOF MS to help predict antimicrobial susceptibilities could help reduce time to result and enable narrowing of broad-spectrum antimicrobials when appropriate. Methods that have been explored using MALDI-TOF MS for AST include detecting degradation changes in the protein spectrum of an antibiotic after incubation with a test strain (i.e., hydrolysis by the presence of a beta-lactamase), detection of isotope-labeled amino acids incorporated into organism proteins, and identifying proteomic differences between susceptible and resistant strains [58]. However, these methods are still being defined and the ultimate usefulness to the clinical laboratory will be clarified in the years to come.

Another potential application of MALDI-TOF MS is microbial strain typing which identifies an isolate or group of isolates with phenotypic and/or genotypic traits that differ from those of other isolates of the same species. Strain typing enables important epidemiologic investigation of disease transmission and outbreaks. The gold standard for strain typing has long been multilocus sequence typing which is a laborious, multiday technique. Although no standardized, generalizable protocol exists for strain typing using MALDI-TOF MS, a statistical method called phyloproteomic principal component analysis hierarchical clustering can map the relatedness between protein spectra. Strain typing using MALDI-TOF MS has been described for many bacteria and yeast, including *Campylobacter jejuni*, *E. coli*, *Salmonella*, *Clostridium difficile*, *Listeria monocytogenes*, staphylococci, pneumococci, and *Candida auris* [59–61].

8 OTHER MS TECHNOLOGIES

In addition to these MALDI-TOF MS applications, there are other mass spectrometry technologies being explored for use in clinical microbiology. These include surface-enhanced laser desorption/ionization-time of flight (SELDI-TOF) MS, liquid chromatography-tandem mass spectrometry (LC-MS/MS, discussed in Chapter 12), and MS and polymerase chain reaction electrospray ionization mass spectrometry (PCR ESI-MS). Many of these technologies have greater quantitative capabilities than MALDI-TOF MS, but are currently used primarily in the research environment in pursuit of antibiotic resistance testing, virulence factor identification, novel antimicrobial therapy target identification or organism characterization.

Of all of these mass spectrometry technologies, PCR ESI-MS seems to be the closest to adoption for organism identification in the clinical microbiology laboratory and is available commercially as the IRIDICA System (Abbott Molecular Diagnostics, Des Plaines, IL). This system, which is CE marked but not FDA approved yet, can bypass culture and identify organisms present in whole blood or sterile fluids [62]. After chemical and mechanical lysis steps and nucleic acid extraction, polymerase chain reaction (PCR) is performed using broad-ranged primers that bind to conserved regions of bacterial, fungal, and viral genomes. The PCR products are dissolved in an organic aqueous phase carrier, passed through a heated capillary and exposed to a high voltage, generating an aerosol of ions which are analyzed by *m/z* ratio. The molecular weight of amplicons can then be used to calculate their unique nucleotide composition of A's, T's, C's, and G's and compared to a reference library. The IRIDICA library includes over 750 bacteria and *Candida* species as well as over 200 other fungi and over 130 viruses. PCR ESI-MS technology can detect quantities as low as 40 genome equivalents. Aside from the ability to interrogate directly from specimens and increased sensitivity to low-burden organisms, other advantages of this technology over MALDI-TOF MS include its ability to target viruses and noncultivatable/fastidious bacteria, its potential to identify genetic markers of antimicrobial resistance, and the theoretical ability to detect multiple pathogens present in a specimen (although this last point has yet to be seen in studies) [1]. The technology is limited by its high cost, its inability to provide comprehensive susceptibility testing, and the requirement of a preanalytic PCR step, which requires several hours and lends itself to batch processing, thus compromising some of the time-saving benefits of mass spectrometry.

9 SUMMARY AND CONCLUSIONS

Accurate and rapid diagnosis of infections in the clinical microbiology laboratory is of paramount importance to high-quality patient care. MALDI-TOF MS can accurately identify the vast majority of bacterial and yeast isolates commonly encountered in the clinical microbiology laboratory in just minutes per sample at relatively low cost. This new diagnostic tool has quickly changed the landscape of clinical microbiology and its full range of applications are still being explored. Optimal procedures for the identification of molds and mycobacteria by MALDI-TOF MS are still being defined. MALDI-TOF MS is limited by the need to analyze pure, cultured isolates and the inability to provide comprehensive definitive susceptibility testing. Despite these limitations, this method of organism identification has fundamentally changed the clinical microbiology landscape in the last decade. With possible applications of antimicrobial susceptibility testing and strain typing still being explored, the role of mass spectrometry in the clinical laboratory continues to evolve.

REFERENCES

[1] Buchan BW, Ledeboer NA. Emerging technologies for the clinical microbiology laboratory. Clin Microbiol Rev 2014;27(4):783–822.

[2] Bader O. MALDI-TOF-MS-based species identification and typing approaches in medical mycology. Proteomics 2013;13(5):788–99.

[3] McElvania Tekippe E, Shuey S, Winkler DW, Butler MA, Burnham CA. Optimizing identification of clinically relevant Gram-positive organisms by use of the Bruker Biotyper matrix-assisted laser desorption ionization-time of flight mass spectrometry system. J Clin Microbiol 2013;51(5):1421–7.

[4] Wieme AD, Spitaels F, Aerts M, De Bruyne K, Van Landschoot A, Vandamme P. Effects of growth medium on matrix-assisted laser desorption-ionization time of flight mass spectra: a case study of acetic acid bacteria. Appl Environ Microbiol 2014;80(4):1528–38.

[5] Dieckmann R, Helmuth R, Erhard M, Malorny B. Rapid classification and identification of salmonellae at the species and subspecies levels by whole-cell matrix-assisted laser desorption ionization-time of flight mass spectrometry. Appl Environ Microbiol 2008;74(24):7767–78.

[6] Carbonnelle E, Mesquita C, Bille E, et al. MALDI-TOF mass spectrometry tools for bacterial identification in clinical microbiology laboratory. Clin Biochem 2011;44(1):104–9.

[7] Christensen JJ, Dargis R, Hammer M, et al. Matrix-assisted laser desorption ionization-time of flight mass spectrometry analysis of Gram-positive, catalase-negative cocci not belonging to the Streptococcus or Enterococcus genus and benefits of database extension. J Clin Microbiol 2012;50(5):1787–91.

[8] Veloo AC, Welling GW, Degener JE. The identification of anaerobic bacteria using MALDI-TOF MS. Anaerobe 2011;17(4):211–2.

[9] Buckwalter SP, Olson SL, Connelly BJ, et al. Evaluation of matrix-assisted laser desorption ionization-time of flight mass spectrometry for identification of *Mycobacterium* species, *Nocardia* species, and other aerobic actinomycetes. J Clin Microbiol 2016;54(2):376–84.

[10] Goyer M, Lucchi G, Ducoroy P, Vagner O, Bonnin A, Dalle F. Optimization of the preanalytical steps of matrix-assisted laser desorption ionization-time of flight mass spectrometry identification provides a flexible and efficient tool for identification of clinical yeast isolates in medical laboratories. J Clin Microbiol 2012;50(9):3066–8.

[11] Steensels D, Verhaegen J, Lagrou K. Matrix-assisted laser desorption ionization-time of flight mass spectrometry for the identification of bacteria and yeasts in a clinical microbiological laboratory: a review. Acta Clin Belg 2011;66(4):267–73.

[12] Saleeb PG, Drake SK, Murray PR, Zelazny AM. Identification of mycobacteria in solid-culture media by matrix-assisted laser desorption ionization-time of flight mass spectrometry. J Clin Microbiol 2011;49(5):1790–4.

[13] Tran A, Alby K, Kerr A, Jones M, Gilligan PH. Cost savings realized by implementation of routine microbiological identification by matrix-assisted laser desorption ionization-time of flight mass spectrometry. J Clin Microbiol 2015;53(8):2473–9.

[14] Gaillot O, Blondiaux N, Loiez C, et al. Cost-effectiveness of switch to matrix-assisted laser desorption ionization-time of flight mass spectrometry for routine bacterial identification. J Clin Microbiol 2011;49(12):4412.

[15] Tan KE, Ellis BC, Lee R, Stamper PD, Zhang SX, Carroll KC. Prospective evaluation of a matrix-assisted laser desorption ionization-time of flight mass spectrometry system in a hospital clinical microbiology laboratory for identification of bacteria and yeasts: a bench-by-bench study for assessing the impact on time to identification and cost-effectiveness. J Clin Microbiol 2012;50(10):3301–8.

[16] Huang AM, Newton D, Kunapuli A, et al. Impact of rapid organism identification via matrix-assisted laser desorption/ionization time-of-flight combined with antimicrobial stewardship team intervention in adult patients with bacteremia and candidemia. Clin Infect Dis 2013;57(9):1237–45.

[17] Perez KK, Olsen RJ, Musick WL, et al. Integrating rapid pathogen identification and antimicrobial stewardship significantly decreases hospital costs. Arch Pathol Lab Med 2013;137(9):1247–54.

[18] Bessede E, Angla-Gre M, Delagarde Y, Sep Hieng S, Menard A, Megraud F. Matrix-assisted laser-desorption/ionization biotyper: experience in the routine of a University hospital. Clin Microbiol Infect 2011;17(4):533–8.

[19] Sogawa K, Watanabe M, Sato K, et al. Use of the MALDI BioTyper system with MALDI-TOF mass spectrometry for rapid identification of microorganisms. Anal Bioanal Chem 2011;400(7):1905–11.

[20] Neville SA, Lecordier A, Ziochos H, et al. Utility of matrix-assisted laser desorption ionization-time of flight mass spectrometry following introduction for routine laboratory bacterial identification. J Clin Microbiol 2011;49(8):2980–4.

[21] Dubois D, Grare M, Prere MF, Segonds C, Marty N, Oswald E. Performances of the Vitek MS matrix-assisted laser desorption ionization-time of flight mass spectrometry system for rapid identification of bacteria in routine clinical microbiology. J Clin Microbiol 2012;50(8):2568–76.

[22] Martiny D, Busson L, Wybo I, El Haj RA, Dediste A, Vandenberg O. Comparison of the Microflex LT and Vitek MS systems for routine identification of bacteria by matrix-assisted laser desorption ionization-time of flight mass spectrometry. J Clin Microbiol 2012;50(4):1313–25.

[23] Clark AE, Kaleta EJ, Arora A, Wolk DM. Matrix-assisted laser desorption ionization-time of flight mass spectrometry: a fundamental shift in the routine practice of clinical microbiology. Clin Microbiol Rev 2013;26(3):547–603.

[24] Lau SK, Tang BS, Teng JL, et al. Matrix-assisted laser desorption ionisation time-of-flight mass spectrometry for identification of clinically significant bacteria that are difficult to identify in clinical laboratories. J Clin Pathol 2014;67(4):361–6.

[25] Cherkaoui A, Hibbs J, Emonet S, et al. Comparison of two matrix-assisted laser desorption ionization-time of flight mass spectrometry methods with conventional phenotypic identification for routine identification of bacteria to the species level. J Clin Microbiol 2010;48(4):1169–75.

[26] Marko DC, Saffert RT, Cunningham SA, et al. Evaluation of the Bruker Biotyper and Vitek MS matrix-assisted laser desorption ionization-time of flight mass spectrometry systems for identification of nonfermenting gram-negative bacilli isolated from cultures from cystic fibrosis patients. J Clin Microbiol 2012;50(6):2034–9.

[27] Jamal WY, Shahin M, Rotimi VO. Comparison of two matrix-assisted laser desorption/ionization-time of flight (MALDI-TOF) mass spectrometry methods and API 20AN for identification of clinically relevant anaerobic bacteria. J Med Microbiol 2013;62(Pt. 4):540–4.

[28] Branda JA, Markham RP, Garner CD, Rychert JA, Ferraro MJ. Performance of the Vitek MS v2.0 system in distinguishing *Streptococcus* pneumoniae from *nonpneumococcal* species of the *Streptococcus* mitis group. J Clin Microbiol 2013;51(9):3079–82.

[29] Snydman DR, Jacobus NV, McDermott LA, et al. National survey on the susceptibility of Bacteroides fragilis group: report and analysis of trends in the United States from 1997 to 2004. Antimicrob Agents Chemother 2007;51(5):1649–55.

[30] Patel R. MALDI-TOF MS for the diagnosis of infectious diseases. Clin Chem 2015;61(1):100–11.

[31] Somayaji R, Priyantha MA, Rubin JE, Church D. Human infections due to *Staphylococcus* pseudintermedius, an emerging zoonosis of canine origin: report of 24 cases. Diagn Microbiol Infect Dis 2016;85(4):471–6.

[32] Suwantarat N, Grundy M, Rubin M, et al. Recognition of *Streptococcus* pseudoporcinus colonization in women as a consequence of using matrix-assisted laser desorption ionization-time of flight mass spectrometry for group B *Streptococcus* identification. J Clin Microbiol 2015;53(12):3926–30.

[33] Biosafety in Microbioloical and Biomedical Laboratories. 5th ed: US Department of Health and Human Services, Public Health Service, Centers for Disease Control and Prevention, National Institutes of Health; 2009.

[34] Cunningham SA, Patel R. Importance of using Bruker's security-relevant library for Biotyper identification of *Burkholderia* pseudomallei, *Brucella* species, and *Francisella* tularensis. J Clin Microbiol 2013;51(5):1639–40.

[35] 2015 Global Tuberculosis Report. Geneva, Switzerland: World Health Organization, 2015.

[36] Johnson MM, Odell JA. Nontuberculous mycobacterial pulmonary infections. J Thorac Dis 2014;6(3):210–20.

[37] Tortoli E. Standard operating procedure for optimal identification of mycobacteria using 16S rRNA gene sequences. Stand Genomic Sci 2010;3(2):145–52.

[38] Lumb R, Lanser JA, Lim IS. Rapid identification of mycobacteria by the Gen-Probe Accuprobe system. Pathology 1993;25(3):313–5.

[39] Richter E, Rusch-Gerdes S, Hillemann D. Evaluation of the GenoType Mycobacterium Assay for identification of *mycobacterial* species from cultures. J Clin Microbiol 2006;44(5):1769–75.

[40] Balazova T, Makovcova J, Sedo O, Slany M, Faldyna M, Zdrahal Z. The influence of culture conditions on the identification of *Mycobacterium* species by MALDI-TOF MS profiling. FEMS Microbiol Lett 2014;353(1):77–84.

[41] Inactivated Mycobacteria Bead Preparation Method (in Mbpm). In: Bruker Daltonics I, editor. Version 1.0 ed. Billerica, MA: Bruker Daltonics, Inc.; 2011.

[42] Balada-Llasat JM, Kamboj K, Pancholi P. Identification of mycobacteria from solid and liquid media by matrix-assisted laser desorption ionization-time of flight mass spectrometry in the clinical laboratory. J Clin Microbiol 2013;51(9):2875–9.

[43] Mather CA, Rivera SF, Butler-Wu SM. Comparison of the Bruker Biotyper and Vitek MS matrix-assisted laser desorption ionization-time of flight mass spectrometry systems for identification of mycobacteria using simplified protein extraction protocols. J Clin Microbiol 2014;52(1):130–8.

[44] Adams LL, Salee P, Dionne K, Carroll K, Parrish N. A novel protein extraction method for identification of mycobacteria using MALDI-ToF MS. J Microbiol Methods 2015;119:1–3.

[45] Buchan BW, Riebe KM, Timke M, Kostrzewa M, Ledeboer NA. Comparison of MALDI-TOF MS with HPLC and nucleic acid sequencing for the identification of *Mycobacterium* species in cultures using solid medium and broth. Am J Clin Pathol 2014;141(1):25–34.

[46] Chen TS, Chang MH, Kuo WW, et al. MALDI-TOF mass spectrometry analysis of small molecular weight compounds (under 10 KDa) as biomarkers of rat hearts undergoing arecoline challenge. Pharm Biol 2013; 51(4):488–91.

[47] Machen A, Kobayashi M, Connelly MR, Wang YF. Comparison of heat inactivation and cell disruption protocols for identification of mycobacteria from solid culture media by use of vitek matrix-assisted laser desorption ionization-time of flight mass spectrometry. J Clin Microbiol 2013;51(12):4226–9.

[48] Wilen CB, McMullen AR, Burnham CA. Comparison of sample preparation methods, instrumentation platforms, and contemporary commercial databases for identification of clinically relevant mycobacteria by matrix-assisted laser desorption ionization-time of flight mass spectrometry. J Clin Microbiol 2015;53(7):2308–15.

[49] Quiles-Melero I, Garcia-Rodriguez J, Gomez-Lopez A, Mingorance J. Evaluation of matrix-assisted laser desorption/ionisation time-of-flight (MALDI-TOF) mass spectrometry for identification of *Candida* parapsilosis, C. orthopsilosis and C. metapsilosis. Eur J Clin Microbiol Infect Dis 2012;31(1):67–71.

[50] Santos C, Lima N, Sampaio P, Pais C. Matrix-assisted laser desorption/ionization time-of-flight intact cell mass spectrometry to detect emerging pathogenic *Candida* species. Diagn Microbiol Infect Dis 2011;71(3):304–8.

[51] Castanheira M, Woosley LN, Diekema DJ, Jones RN, Pfaller MA. *Candida* guilliermondii and other species of *candida* misidentified as *Candida* famata: assessment by vitek 2, DNA sequencing analysis, and matrix-assisted laser desorption ionization-time of flight mass spectrometry in two global antifungal surveillance programs. J Clin Microbiol 2013;51(1):117–24.

[52] Pulcrano G, Roscetto E, Iula VD, Panellis D, Rossano F, Catania MR. MALDI-TOF mass spectrometry and microsatellite markers to evaluate *Candida* parapsilosis transmission in neonatal intensive care units. Eur J Clin Microbiol Infect Dis 2012;31(11):2919–28.

[53] Kubesova A, Salplachta J, Horka M, Ruzicka F, Slais K. *Candida* "psilosis"–electromigration techniques and MALDI-TOF mass spectrometry for phenotypical discrimination. Analyst 2012;137(8):1937–43.

[54] Alanio A, Beretti JL, Dauphin B, et al. Matrix-assisted laser desorption ionization time-of-flight mass spectrometry for fast and accurate identification of clinically relevant *Aspergillus* species. Clin Microbiol Infect 2011;17(5):750–5.

[55] Taylor JW. One fungus = one name: DNA and fungal nomenclature twenty years after PCR. IMA Fungus 2011;2(2):113–20.

[56] Buskirk AD, Hettick JM, Chipinda I, et al. Fungal pigments inhibit the matrix-assisted laser desorption/ionization time-of-flight mass spectrometry analysis of darkly pigmented fungi. Anal Biochem 2011;411(1):122–8.

[57] Lau AF, Drake SK, Calhoun LB, Henderson CM, Zelazny AM. Development of a clinically comprehensive database and a simple procedure for identification of molds from solid media by matrix-assisted laser desorption ionization-time of flight mass spectrometry. J Clin Microbiol 2013;51(3):828–34.

[58] Charretier Y, Schrenzel J. Mass spectrometry methods for predicting antibiotic resistance. Proteomics Clin Appl 2016. [Epub ahead of print].

[59] Cheng K, Chui H, Domish L, Hernandez D, Wang G. Recent development of mass spectrometry and proteomics applications in identification and typing of bacteria. Proteomics Clin Appl 2016;10(4):346–57.

[60] Girard V, Mailler S, Chetry M, et al. Identification and typing of the emerging pathogen Candida auris by matrix-assisted laser desorption ionisation time of flight mass spectrometry. Mycoses 2016;59(8):535–8.

[61] van Belkum A, Welker M, Erhard M, Chatellier S. Biomedical mass spectrometry in today's and tomorrow's clinical microbiology laboratories. J Clin Microbiol 2012;50(5):1513–7.

[62] Metzgar D, Frinder MW, Rothman RE, et al. The IRIDICA BAC BSI assay: rapid, sensitive and culture-independent identification of bacteria and *Candida* in blood. PLoS One 2016;11(7):e0158186.

HIGH RESOLUTION ACCURATE MASS (HRAM) MASS SPECTROMETRY

12

C.A. Crutchfield, W. Clarke

Johns Hopkins University School of Medicine, Baltimore, MD, United States

1 INTRODUCTION

Over the past decade, mass spectrometry has grown from a primary use as an analytical method relegated to research or industrial applications to a method routinely applied in clinical settings. The utilization of mass spectrometry in a clinical setting requires accurate and robust measurement of diverse sets of analytes from a variety of matrices. The adoption of mass spectrometry in the clinic has certainly been slow because in many cases, there are large-scale immunoassay analyzers that meet the clinical requirements and can provide thousands of results for several assays every day. In many cases, immunoassay-based platforms can be more sensitive than mass spectrometers, though in recent years this divide is closing and in some cases mass spectrometry has surpassed immunoassays in this respect [1]. However, there is a greater likelihood of analytical interference in immunoassay-based analyzers than mass spectrometers. This has caused some professional societies, such as the Endocrine Society, to mandate that in the near future steroid hormone levels be generated from a mass spectrometry based analyzer for consideration for publication in the Society Journal [2]. This mandate comes at a time when the mass spectrometer has started to see considerable adoption in several clinical fields: small molecule analysis (e.g., toxicology), microbial identification, and molecular diagnostics. However, these fields may have not realized the clinical utility of mass spectrometry if not for the advent of *high resolution* mass spectrometry (HRMS) [3,4].

2 HRMS FUNDAMENTALS

All clinical mass spectrometry, whether performed with high resolution or not, uses instrumentation that generates ions from the analyte(s) of interest. Depending on the application, contemporary ionization processes typically arise from either the liquid phase or the solid phase. Ionization of analytes from the liquid phase is typically accomplished via electrospray ionization (ESI), atmospheric pressure chemical ionization (APCI), or atmospheric pressure photoionization (APPI) in concert with liquid chromatography [5,6]. These ionizations sources have differential ionization efficiency for analytes depending on analyte polarity: ESI has the greatest efficiency at ionizing polar molecules and APPI being most adept at ionizing nonpolar molecules, though these effects are high analyte specific. Alternatively, ions can be generated from the solid phase via matrix-assisted laser desorption ionization

Mass Spectrometry for the Clinical Laboratory. http://dx.doi.org/10.1016/B978-0-12-800871-3.00012-2

(MALDI) [7]. Other techniques for generating ions exist, however ESI, APCI, APPI, and MALDI methods are the most prevalent in a clinical setting. While the specific physical mechanisms by which these different ionization sources differ, they all operate on the principle of applying energy to the liquid or solid medium carrying the analyte which results ion generation, typically by the addition or subtraction of a proton. However, the ionization process is complicated and depending on the analyte, ionization source, and source conditions analytes can pick up multiple charge states, can fragment within the source, and can have other molecular features associated with their ion (e.g., sodium addition to form a sodium "adduct").

Prior to discussing the applications of HRMS, it is important to emphasize how and why the capacity to measure the mass-to-charge ratio (*m/z*) of ions in high resolution is important analytically. First, the utility of a mass spectrometer is based on its ability to distinguish molecules based on the mass and the charge of the ions they generate in an ion source. The principle that distinguishes HRMS from conventional "unit" resolution mass spectrometry is rooted in atomic physics with regards to the distinct nuclear binding energy of each isotope of every element, the *mass defect* [8], which is the difference between the exact mass and nominal mass. With carbon-12 set at 12.00000 Da (where Da = Daltons, units designated 1/12 the mass of carbon-12), the exact mass of nitrogen-14 is 14.00307 Da. The mass defect for nitrogen-14 can then be calculated as 14.00307–14.00000 = 0.00307. The exact masses, mass defects, and relative abundance for the most pertinent isotopes of common organic elements are in Table 12.1. It is important to note that in mass spectrometry when referring to elemental isotopes that they are typically *stable* isotopes—hence when a method is developed that utilizes an isotopically-labeled analog of an analyte, the carbon atoms are typically replaced with C-13, not C-14; hydrogen atoms are typically replaced with deuterium, not tritium. However, as listed in Table 12.1, there is a natural abundance of heavy stable isotopes for a given element, and that abundance will determine the isotopic distribution. Mass spectra can be difficult to interpret; their interpretation is best facilitated when considering both the calculated mass defect of a peak as well as the presence of other peaks nearby. Interferences observed in method development may be part of the isotopic cluster of a compound

Table 12.1 Exact Masses of Common Organic Isotopes and Their Mass Defect

Isotope	Exact Mass (Da)	Mass Defect (Da)	Isotopic Composition
^{1}H	1.007825	0.007825	0.999885
^{2}H	2.014102	0.014102	0.000115
^{12}C	12.000000	0.000000	0.9893
^{13}C	13.003355	0.003355	0.0107
^{14}N	14.003074	0.003074	0.99636
^{15}N	15.000109	0.000109	0.00364
^{16}O	15.994915	−0.005085	0.99757
^{18}O	17.999161	−0.000839	0.00205
^{31}P	30.973762	−0.026238	1.0000
^{32}S	31.972071	−0.027929	0.92223

with a different *monoisotopic mass*—the mass of a molecule that consists of the most abundant isotopic species of each element (e.g., C-12, H-1, N-14, etc.). The consequence of each elemental isotope having a slightly different mass defect is that if the m/z of an analyte can be accurately measured, then the exact molecular formula can be determined computationally, provided the measurement was done with sufficiently high resolution. This is a distinct mechanism for analytical specificity from tandem mass spectrometry (MS/MS), where specificity is typically derived via an analyte-specific fragmentation event; in the clinical laboratory these MS/MS instruments are typically triple quadrupole mass spectrometers.

When discussing HRMS, it is important to consider how the resolution is determined. The peak width definition for resolution is:

$$r = \frac{m}{\Delta m} \tag{12.1}$$

where m is the mass of a single peak in a mass spectrum and Δm is the width of the peak at 50% of maximal peak height, or full width at half maximum (FWHM). The resolution is dependent on the mass at which it is measured therefore the mass should always be stated in the context of resolution. The exact dependence of resolution to mass is different for different types of mass analyzers. TOF analyzers have relatively constant resolution for m/z while orbitrap analyzers have relatively lower resolution at high m/z. A simulated spectrum is presented in Fig. 12.1 that demonstrates the mass resolution of two steroid hormones, testosterone and estriol, using high resolution instrumentation that would have the same nominal mass observed using standard instrumentation. This figure illustrates the mechanism HRMS uses to achieve high analytical specificity; specifically, the differences in the mass defects of carbon, hydrogen, and oxygen are leveraged to distinguish the two molecules.

In general, HRMS implies *accurate* mass measurements in addition to increased resolution. However, these parameters are not necessarily linked—different mass spectrometers address these issues, that of resolution and accuracy, differently. The definition of mass accuracy is usually given in parts per million (ppm):

$$ppm = \frac{1.0 \times 10^6 \, (\text{measured mass} - \text{theoretical mass})}{\text{theoretical mass}} \tag{12.2}$$

High mass accuracy is achieved by operating an instrument with high resolution as well as having the instrument sufficiently calibrated. An ideal accurate mass instrument would have zero systematic bias due to insufficient calibration or drift. In this circumstance the accuracy of the instrument would simply reflect the precision of the mass analyzer. In order to approach this ideal scenario, high resolution mass spectrometers require periodic calibration—ideally via both internal calibration and external calibration strategies. External calibration is the determination of parameters used to calculate mass external to the mass peak in a spectrum. Internal calibration uses mass peaks *within* a spectrum to mass correct. Internal calibration typically improves mass accuracy ~twofold relative to external calibration [3].

While the earliest high resolution mass analyzers for research were based on time-of-flight (TOF) or ion cyclotron resonance, clinical applications of HRMS have been primarily limited to TOF mass spectrometers and orbitrap mass spectrometers. These instruments operate on different technical principles to attain high mass resolution measurements. TOF instruments attain high mass resolution by precisely

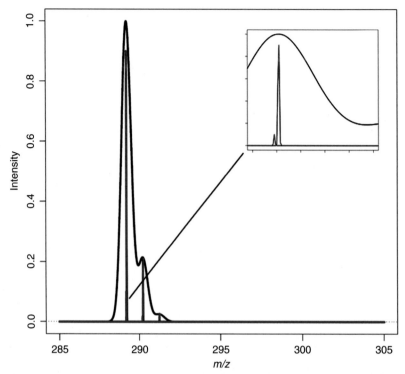

FIGURE 12.1 A Simulated Spectrum Showing the Mass Resolution of Testosterone (*Red*) and Estriol (*Blue*) With High Resolution That Could not be Resolved Using an Instrument With Low Resolution (Black)

The low resolution instrument combines signals from both components to make it appear as a single peak because both compounds have the same nominal protonated mass of 289 Da. The simulated high resolution traces have a resolution of 15,000 at 300 Da. Here the two analytes are clearly defined as different chemical components of the mixture.

Discoveries 2014; 2(2): e17.

measuring the time it takes for ions to traverse a flight tube and hit a detector. TOF instruments are extremely fast at scanning a large range of masses. They can also maintain higher resolution at higher m/z values ($> \sim 2000$ Da) relative to orbitrap mass spectrometers.

Orbitrap mass spectrometers generate high resolution mass measurements by detecting the axial frequency of an ion as it oscillates inside the trap. The orbitrap stores ions in a stable flight path (orbit around the inner spindle) by balancing their electrostatic attraction by their inertia coming from an RF only trap. The frequency of the axial motion around the inner electrode is related to the m/z of the ion. Orbitrap technology has only been recently developed, and was commercialized in 2005. It has gained substantial popularity for its ability to provide mass resolution much greater than TOF instruments, particularly at lower m/z ($\sim < 1,000$ Da). It should be noted that the capacity to generate very high resolution mass measurements comes at the expense of scan speed. However, with the widespread adoption of ultra-high performance liquid chromatography, analyte peak widths have been reduced to seconds—this forces a compromise between determining a high resolution mass measurement or comprehensive chromatographic peak analysis.

It is important to note that both conventional triple quadrupole MS instruments and hybrid HRMS instruments provide front-end ion optics to operate in MS/MS (product ion mode) mode. Even operating at the highest resolution settings, HRMS alone does not differentiate isobaric (equal mass) analytes. When isobaric analytes are present in a sample, additional analytical strategies are required to differentiate these compounds; primarily chromatographic separation or MS/MS fragmentation are used. In fact, these strategies are often employed even in the absence of isobaric interference in order to further improve analytical specificity. There are other acquisition modes in which a mass spectrometer can be operated in addition to product ion mode that include neutral loss scan and precursor ion scan, among others [9]. These acquisition modes are not routinely used in the clinic, but are worth consideration, as they have been useful for some class specific analyses, such as phospholipids [10]. Full scan analysis commonly denotes using HRMS to analyze the signal intensity from all ions generated in the ion source. This acquisition mode is much more frequently used with HRMS compared to standard mass spectrometric analyses because it has much greater analytical selectivity compared to full scan acquisition using a lower resolution instrument.

3 CLINICAL AND FORENSIC TOXICOLOGY

The toxicology laboratory is one area that has adopted LC-HRMS as a routine analytical methodology. Analytical methods that were historically widely used, such as GC-MS are labor intensive, and rely on analytes compatible with GC analysis (primarily analytes that are volatile or can be derivatized to become volatile). Where LC-MS/MS methodology requires careful methodological design to generate analyte specific fragmentation products for monitoring, LC-HRMS has helped alleviate the pitfalls of previous methods. As an LC based methodology, analytes typically do not require chemical derivatization for their analysis. In addition, with high resolution accurate mass, the instrument can provide exquisite analytical specificity for the presence or absence of toxicological substances present in a biofluid. Most analytical workflows depend on a platform-specific accurate mass database of known substances. After analysis of a specimen, the platform software will match mass peaks from the sample with values from the database and provide a list of putative matched compounds. More specific strategies for toxicological substance detection use more a traditional "targeted" approach, whereby retention time windows are identified for specific analytes. Using stable-labeled internal standards as an internal QC metric, the presence of analytes can be qualitatively determined using an intensity cutoff. HRMS applications to toxicology have primarily been developed using either orbitrap or TOF based technology; as previously discussed, the choice between technologies is highly dependent on the specific toxicology application [11].

One appealing feature of HRMS is the ease of assay development compared to conventional immunoassay approaches. Immunoassay method development can take considerable time due to the process of developing an antibody with high affinity for the analyte of interest. However, HRMS-based methods only require the drug (or its metabolite) to be obtainable for use in method development. HRMS enables rapid implementation of drug testing in clinical settings that involve patients consuming novel or uncommon toxic substances. The past decade has seen large growth in the abuse of designer drugs, particularly synthetic cannabinoids [12] and stimulant cathinones, "bath salts" [13]. In 2012, Wu et al., [14] remarked on the lack of available commercial immunoassays for these newer drugs.

In addition, specific immunoassays are remarkably difficult to design for these drugs because there are a large number of analogs. Kronstrand et al., recently remarked on the superiority of LC-Q-TOF screening relative to immunoassay for comprehensive analysis of synthetic cannabinoids in urine [15]. They demonstrated that although an immunoassay test had reactivity with MAM-2201 and JWH-122 metabolites, it did not react with UR-144 metabolites. UR-144 metabolites were found in specimens analyzed and demonstrate the risk of false negatives when relying on immunoassay methods. In 2013, Concheiro et al., developed the most comprehensive to-date quantitative screen of cathinones, measuring 28 synthetic cathinones in urine [16]. Notably, the authors used an orbitrap with a quadrupole frontend that can operate in MS/MS mode. Additionally, Andersson et al., were able to use this technology for metabolic profiling of synthetic cannabinoids after hepatocyte and liver microsome incubations, in order to elucidate pertinent metabolic pathways and compounds for monitoring [17].

Another significant benefit of HRMS based detection in toxicology testing is the ability to perform retrospective analysis when full-scan data are obtained. With full-scan acquisition, the data acquired during LC-MS acquisition are archived after reporting the presence or absence of known substances, and can be further queried to generate extracted ion chromatograms for analytes of interest as long as the exact mass is known. In 2013, Rosano et al., describe a postmortem drug screening method using UPLC-MS^E-TOF that evaluated the presence of over *950* toxicologically relevant drugs and metabolites [18]. In their study, they claim a 99% detection rate using UPLC-MS^E-TOF, which is a unique feature of Waters Q-TOFs (Waters, Milford, MA). For comparison, in the same study a low resolution mass spectrometer only provided 80% detection rate. It is important to note that this particular study did not have positive controls for ~10% of the compounds screened. These particular drugs without control material were only determined via a database search of the MS^E data acquisition. MS^E is an example of data *independent* acquisition (DIA). DIA strategies acquire product ion information of ions independent of intensity; different manufacturers use different DIA strategies. For this particular platform, MS^E ramps collision energy and fragments *all* ions generated by the source, generating a convoluted product ion spectrum that results from all initial ions. Software algorithms then process these data and attempt to deconvolute the spectra into individual features.

An alternate DIA strategy includes SWATH, which is featured on AB Sciex Triple-TOFs (AB Sciex, Foster City, CA), which generates convoluted spectra at variable (typically ~25 Da) isolation bins across a variable (typically ~600 Da) window. These data are empirically less convoluted than the MS^E data because the ions inserted into the collision cell consist of a few ion species. These DIA strategies have been developed in response to data dependent acquisition strategies (DDA), which operate on the basis of fragmenting ions on the priority of ion intensity. A significant drawback with DDA is that low intensity ions, though they may be clinically significant, may not be fragmented and generated into a spectral feature. Database-driven toxicology is problematic; even with extraordinarily good spectral match scores, the possibility of a false positive is significant. One place where database-driven toxicology has a future is in hypothesis generation for drug confirmation. If a compound is preliminarily identified by fragmentation pattern matching the database, the analyst can perform a spike-recovery experiment using the reference compound. In the future, in combination with a standard mixture of isotopically labeled internal standards, a standard UPLC method may produce a sufficiently interlab reproducible retention times to approximate a retention-index-like parameter to combine both MS/MS and retention time to better predict analyte identity using a single database.

Marzinke and coworkers [19] present an application of HRMS screening of antiretroviral (ARV) drugs for evaluating enrollment eligibility in a HIV clinical trial. The study examines a significant challenge in some HIV clinical trials—in these trials, investigators commonly attempt to recruit newly diagnosed, drug naïve patients; however, patients may be untruthful with regards to their knowledge of their HIV status as an attempt to obtain care by enrolling in the trial. For this study, the assay was a qualitative screen analyzing for the presence of 15 ARV drugs in patient plasma using liquid chromatography coupled to an orbitrap mass spectrometer. This study found that of 155 newly diagnosed HIV positive patients, 45.8% had at least 1 ARV drug detected during their enrollment period, indicating that almost half on the enrollees were not drug naïve. This importance of this study is that it provides a framework for establishing analytical prerequisites in clinical trials of chronic conditions where rejection criteria include previous exposure to a drug.

In addition, a French study by Roche et al., demonstrates the use of HRMS to build a spectral library for screening of different classes of drugs [20]. This method used a simple protein precipitation of samples, followed by analysis in positive and negative modes, both with and without fragmentation. Using this approach with reference standards, they were able to build a library containing 616 compounds that has been accredited by COFRAC (i.e., the French Accreditation Committee) according to the ISO 15189 standard. This library-based approach has also been published by Concheiro et al., where they were able to use a library generated by HRAM mass spectrometry to screen for 40 novel psychoactive stimulants in urine [21].

4 MICROBIOLOGY

Recently, HRMS methods, principally in the form of matrix assisted laser desorption ionization time-of-flight mass spectrometry (MALDI-TOF), has shifted the microbial screening paradigm from antiquated phenotypic algorithms to a more cost-effective, rapid, and accurate mode of microbial species identification [22]. Classical methods for this testing include staining, culturing, biochemical testing, and susceptibility testing, where some analyses could last weeks. However, the MALDI-TOF workflow has unified testing strategies across several classes of microbes. Tan et al., evaluated the cost and time savings from adopting MALDI-TOF and found that, on average, identification of a pathogen was produced 1.45 days earlier and at roughly half the cost, with the majority of those savings realized in reagent costs [23]. A separate investigation by Huang et al., found that institutional and society cost savings are even further amplified when an antimicrobial stewardship team is implemented using MALDI-TOF data. They found that this new paradigm can drop the length of intensive care unit stays from 14.9 to 8.3 days [24].

For MALDI analysis, a colony is placed on a MALDI target directly, after extraction with acid, or after bead lysis (depending on the cell wall strength of the microbial class). The preparation is then finished with an overlay of chemical matrix, typically 4-α-cyano-4-hydroxycinnamic acid (CCA). Next, the target is irradiated by a laser, which mediates the generation of protonated gas ions of polypeptides and small proteins—the majority of which are ribosomal or DNA binding in origin. The spectra generated are then computationally scored against reference spectra in a database supplied with the instrument, and a report is generated providing the confidence of microbial species identification. In some cases, the MALDI-TOF analysis fails to provide an accurate identification—when that occurs, the microbe can be reflexed to a more sensitive method of analysis and detection. Interestingly, an alternative to the MALDI-TOF analysis of proteins from microbiological specimens has to use ESI-TOF analysis of nucleic acids (which will be discussed in further detail in Section 5).

There are two major platforms for MALDI-TOF microbial identification analyzers: the Bruker Biotyper (Bruker Daltonik GmbH, Bremen, Germany) and the bioMérieux Vitek MS (bioMérieux, Marcy l'Etoile, France). In a direct comparison for assessment of bacteria in real patient samples, both devices performed well with regard to accurate identification of microbes, with genus agreement $>\sim 90\%$ and species agreement $>\sim 70\%$ [25]. These instruments are not limited to characterization of bacteria, but can be used in assessment of fungal infections [26], and many groups continue to search for new ways to apply this technology in a clinical setting. Both instruments have their own databases. These databases can be periodically updated to include additional organisms and improve overall accuracy. The most compelling aspect of these instruments for the clinical laboratory is how relatively little training is required to operate the instrument. Because the interpretation of the microbes is automated via the computer, the most major concern is appropriate sampling and preparation of the specimens.

5 MOLECULAR DIAGNOSTICS

Mass spectrometers as detectors for nucleic acids have been overshadowed by massively parallel next generation sequencers (NGS), which conventionally have light-based or electrochemical based detectors. However, it is important to note that nucleic acid mass spectrometry does have its own broad, clinically relevant applications: microbial identification [27], microbial susceptibility testing, viral identification, SNP genotyping, methylation analysis, gene expression analysis, copy number variation, and comparative sequencing [28,29]. All of these applications rely on polymerase chain reaction (PCR) amplification of targets sequences, or other offline nucleic acid processing prior to detection. It is worthwhile to emphasize that NGS sequencers can only analyze nucleic acid sequence, while mass spectrometry as a technology enables the analysis of broad classes of clinically relevant analytes including, but not limited to, nucleic acids. This analytical flexibility may eventually lead to an automated analyzer with mass spectrometry as the detector, and would provide routine testing capabilities in the clinical applications of toxicology, endocrinology, microbiology, and molecular diagnostics, augmenting traditional immunoassay and nucleic acid testing.

Nucleic acid mass spectrometry has been predominantly commercialized via the Sequenom MassARRAY [30] and Ibis T5000 [31] platforms. These two platforms differ substantially in how the analysis of nucleic acid molecules is performed. The Sequenom platform uses MALDI-TOF, and the Ibis platform uses ESI-TOF. In addition, the sample preparation for the two methods differs slightly in that the Sequenom platform operates on the principle of primer extension and the Ibis platform operates on the principle of "triangulation identification for the genetic evaluation of risk (TIGER)" [31]. Regardless of the methodologic differences between the two platforms, the analytical measurements the instruments make are the same, that is, measuring the m/z of an amplified nucleic acid product. However, the Sequenom platform is measuring the mass addition of a primer extension product and the Ibis platform is measuring the m/z of several amplicon products. The Sequenom platform result gives true "sequencing" data, that is, the difference in m/z corresponds to the base added during primer extension. Conversely, the Ibis platform never provides sequence information; rather, it relies on the accuracy of the m/z measurement to determine amplicon base composition. The amplicon base composition is used in a calculation to determine the origin of the DNA, typically a microbe or a virus.

The majority of the infectious disease field has focused on the capacity of mass spectrometry to impact microbial identification. However, viral infections are extraordinarily important from both a clinical and epidemiological standpoint. The standard MALDI-TOF based platforms for microbial identification require colony isolation before identification. Some infectious disease agents are more appropriately identified using direct sampling, for example, virus or difficult to culture microbes. This workflow has been commercialized as PLEX-ID using the Ibis platform. PLEX-ID takes advantage of PCR to amplify relatively low concentrations of nucleic acid to detectable concentrations by mass spectrometry. Another benefit to this approach is the selectivity afforded by the PCR reaction, which limits the potential amplicon products that will be analyzed by the mass spectrometer. With carefully designed sequence primers, subtyping has become possible. Tang et al., recently reported on the accuracy of the influenza A and B assay, PLEX-ID flu [32]. In this study, they found that the overall accuracy when evaluating the presence of either influenza A or B and if found to be influenza A subtyped as H1N1-p, H1N1-s, or H3N2 compared to a RT-PCR based assay Prodesse ProFLU+ for Influenza A or B typing, and Prodesse ProFAST+ for A subtyping to be 97.1–100%.

The Sequenom platform provides a flexible platform for high throughput genotype screening. After primer extension and target preparation, the nucleotide analysis is very rapid due to the speed of MALDI ionization. Moreover, as the specimen preparation utilizes PCR, the assay is not limited by analytical sensitivity, which sometimes is a challenge for MALDI-based methods. The device has seen application in screening for SNPs, such as in a study done by Sinotte et al., evaluating mutations in vitamin D binding protein and their consequence in plasma concentrations of 25-hydroxy vitamin D in premenopausal women [33]. Godfrey et al., use the Sequenom platform for assessing methylation status of five candidate genes and associated them with risk for adiposity [34]—this work has contributed to the epigenetic component of risk for metabolic disease and obesity. Additionally, Holzinger and coworkers have demonstrated the utility of the Sequenom platform for evaluating CYP2B6 polymorphisms associated with efavirenz pharmacokinetics [35].

Outside of commercial applications and platforms, others are exploring the use of HRMS for genetic analyses. One such example by Hemeryck and coworkers explores the use of UHPLC-HRMS/MS (with quadrupole-orbitrap instrumentation) for characterization of diet-related DNA adducts [36]. The authors of this study used relative retention time, accurate mass of positive ions, and characteristic fragments to construct a database of DNA adducts used for characterization of in vitro and in vivo samples. The study assessed full scan, SIM-MS, and SIM-MS/MS data acquisition modes. The benefit of the 2 SIM data acquisition modes is that they can be used for quantitative analysis, and not simply qualitative detection of the adducts.

6 FUTURE HRMS CLINICAL APPLICATIONS AND DEVELOPMENTS

The role of mass spectrometry in the clinic is growing rapidly. While the past decade has seen many new clinical applications developed for routine use, other technological developments are emerging that can change the way patients have a diagnosis determined or have an operation performed. One of the biggest developments is direct tissue sampling for mass spectrometric analysis. One of the first applications of direct tissue sampling has been posited as the beginning of a new era for histology—mass spectrometry imaging (MSI). MSI attempts to molecularly characterize a tissue with resolution in the range of microns by precisely ionizing an area of a tissue and recording the relative intensities of the ions generated [37–40]. MSI has been conventionally performed using MALDI, however other ionization mechanisms have been applied as well [41]. Like microbial identification, MSI relies solely on the quality of the mass spectrum to make chemical assignments. As a result, utilization of HRMS for the MS acquisition in a MSI application can improve analytical specificity for this application. The primary challenges of MSI are the cost (traditional staining strategies are routine in the clinical setting), the time for analysis (an image may take several hours for its acquisition), relative lack of spatial resolution (laser spot sizes can be as large as several 100 µm in diameter), and ability to generate reliably quantitative information on analyte concentration. Nonetheless, clinical applications have been developed that could supplant or complement traditional approaches when the clinical utility of multimarker analysis has been demonstrated. For example, Rauser et al., [42] demonstrate the analysis of HER2 receptor status in breast cancer tissue using MALDI MSI. In the future MSI data may be used to direct treatment and resection of a tumor, —for example, Li and Hummon have demonstrated the application of MSI to analyze the relative distribution of proteins in a carcinoma spheroid in an attempt to better characterize the effect of differential hypoxia on tumor expression [43].

Another exciting tissue application where HRMS could prove useful is rapid evaporative ionization mass spectrometry (REIMS). REIMS takes advantage of an existing byproduct of electrosurgical devices, "smoke," to whether the tissue origin of the smoke is cancerous [44]. This is made possible by the molecular ion formation in the smoke discharge when tissue is excised using an electrocautery device. Mass spectra are acquired throughout the surgery, and then a multivariate classification model is used to determine similarity of intraoperative spectra with spectra from a known database. Balog and corworkers [44] used a moderate resolution mass spectrometer for their initial in vivo component of the report using an LCQ Deca XP Plus. However, their ex vivo work utilized an Orbitrap Discovery mass spectrometer with > 10,000 FWHM. Their classification rates were very good, with misclassification rates depending on tissue type between 0% and 7.7%. The only analytical separation during the analysis is via the time and space dimensions of the usage of the electrocautery device and the mass differences observed in the mass spectra. Without a physical separation the analytical determination relies solely on the quality of the mass spectra. The classification rate of this method might further improve with a higher resolution mass spectrometer driving the analysis.

Another potential clinical application of mass spectrometry—although it is too early to determine whether it has clinical utility—is metabolic flux analysis [45,46]; a biological application uniquely suited for HRMS. This type of analysis investigates differential utilization of biochemical pathways, typically in a cell line. When applied to cancer, for example, when using an isotopically labeled glucose or glutamine, aberrant increased production of lactate (Warburg effect) from glucose or production of fatty acids from glutamine can be seen. HRMS facilely enables the differentiation of the isotopomer species of these analytes. The research is still very early in development, so it is difficult to ascertain the clinical implications of subtyping cancers based on differential metabolic flux analysis [47,48]. However, for example, there may be therapeutic interventions that apply strictly to cancers that have reorganized their glutamine metabolism versus those that have not based on drugging different targets [49]. These interventions could be implemented as a personalized approach to cancer therapeutics. The benefit to this type of analysis relative to a standard targeted molecular test is that the result could have an established reference range based on flux—it could be independent of an individual cancer genotype. This is in contest with next generation sequencing assays that rely on genomic databases to establish if a molecular variant is associated with malignancy. This type of analysis could finally close the diagnostic loop. PET scanning often relies on the consumption of 18fluorodeoxyglucose (18F-FDG) to accumulate at the site(s) of cancer; however, the metabolic fate of the increased glucose influx is never determined. HRMS may eventually determine this metabolic fate in order to direct clinical decision-making.

7 CONCLUSIONS

HRMS has started to find its way into the clinical laboratory. Not only has it been established as a viable alternative to traditional triple quadrupole mass spectrometer based quantification, but has enabled application developments that would have been impossible to develop without it, such as microbial identification or molecular assays. The widespread adoption of HRMS and mass spectrometry in general will certainly fuel more technological developments. With additional developments in quantification strategies, the high resolution mass spectrometer is positioning itself into a space where it could be a new cornerstone in all areas clinical decision-making: a device that at the point of care simultaneously determine all toxicological and hormonal disorders presenting in a patient, in the microbiology laboratory will determine the identification and drug susceptibility of infections, and in the surgical suite help elucidate tumor margins during resection. The versatility of HRMS, especially with the availability of hybrid HRMS instruments capable of more conventional quantitative applications, allow these instruments to be used for both highly specific qualitative and quantitative assays. These applications and more will make for a very exciting next decade of clinical mass spectrometry.

REFERENCES

[1] Wu AHB, French D. Implementation of liquid chromatography/mass spectrometry into the clinical laboratory. Clin Chim Acta 2013;420:4–10.

[2] Handelsman DJ, Wartofsky L. Requirement for mass spectrometry sex steroid assays in the journal of clinical endocrinology and metabolism. J Clin Endocrinol Metab 2013;98(10):3971–3.

[3] Marshall AG, Hendrickson CL. High-resolution mass spectrometers. Annu Rev Anal Chem 2008;1(1):579–99.

[4] Xian F, Hendrickson CL, Marshall AG. High resolution mass spectrometry. Anal Chem 2012;84(2):708–19.

[5] Covey TR, Thomson BA, Schneider BB. Atmospheric pressure ion sources. Mass Spectrom Rev 2009;28(6):870–97.

[6] Ho C, Lam C, Chan M, Cheung R, Law L, Lit L, et al. Electrospray ionisation mass spectrometry: principles and clinical applications. Clin Biochem Rev 2003;24(1):3–12.

[7] Zenobi R, Knochenmuss R. Ion formation in MALDI mass spectrometry. Mass Spectrom Rev 1998;17(5):337–66.

[8] Sleno L. The use of mass defect in modern mass spectrometry. J Mass Spectrom 2012;47(2):226–36.

[9] De Hoffmann E. Tandem mass spectrometry: a primer. J Mass Spectrom 1996;31(2):129–37.

[10] Ekroos K, Chernushevich IV, Simons K, Shevchenko A. Quantitative profiling of phospholipids by multiple precursor ion scanning on a hybrid quadrupole time-of-flight mass spectrometer. Anal Chem 2002;74(5):941–9.

[11] Eichhorn P, Pérez S, Barceló D. Time-of-flight mass spectrometry versus orbitrap-based mass spectrometry for the screening and identification of drugs and metabolites: Is there a winner? In: Amadeo R, Fernandez-Alba, editor. Comprehensive analytical chemistry [Internet]. Elsevier; 2012. p. 217–272, Chapter 5. Available from: http://www.sciencedirect.com/science/article/pii/B9780444538109000092

[12] Schneir AB, Cullen J, Ly BT. "Spice" girls: synthetic cannabinoid intoxication. J Emerg Med 2011;40(3):296–9.

[13] Centers for Disease Control and Prevention (CDC). Emergency department visits after use of a drug sold as "bath salts"--Michigan, November 13, 2010-March 31, 2011. MMWR Morb Mortal Wkly Rep 2011;60(19):624–7.

[14] Wu AH, Gerona R, Armenian P, French D, Petrie M, Lynch KL. Role of liquid chromatography–high-resolution mass spectrometry (LC-HR/MS) in clinical toxicology. Clin Toxicol 2012;50(8):733–42.

[15] Kronstrand R, Brinkhagen L, Birath-Karlsson C, Roman M, Josefsson M. LC-QTOF-MS as a superior strategy to immunoassay for the comprehensive analysis of synthetic cannabinoids in urine. Anal Bioanal Chem 2014;.

[16] Anizan S, Concheiro M. Simultaneous quantification of 28 synthetic cathinones and metabolites in urine by liquid chromatography-high resolution mass spectrometry. Anal Bioanal Chem 2013;405(29):9437–948.

[17] Andersson M, Diao X, Wohlfarth A, Scheidweiler KB, Huestis MA. Metabolic profiling of new synthetic cannabinoids AMB and 5F-AMB by human hepatocyte and liver microsome incubations and high-resolution mass spectrometry. Rapid Commun Mass Spectrom 2016;30(8):1067–78.

[18] Rosano TG, Wood M, Ihenetu K, Swift TA. Drug screening in medical examiner casework by high-resolution mass spectrometry (UPLC–MSE-TOF). J Anal Toxicol 2013;37(8):580–93.

[19] Marzinke MA, Clarke W, Wang L, Cummings V, Liu T-Y, Piwowar-Manning E, et al. Nondisclosure of HIV status in a clinical trial setting: antiretroviral drug screening can help distinguish between newly diagnosed and previously diagnosed HIV infection. Clin Infect Dis 2014;58(1):117–20.

[20] Roche L, Pinguet J, Herviou P, Libert F, Chenaf C, Eschalier A, Authier, Richard D. Fully automated semi-quantitative toxicological screening in three biological matrices using turbulent flow chromatography/high resolution mass spectrometry. Clin Chim Acta 2016;455:46–54.

[21] Conchiero M, Castaneto M, Kronstrand R, Huestis MA. Simultaneous determination of 40 novel psychoactive stimulants in urine by liquid chromatography-high resolution mass spectrometry and library matching. J Chromatogr A 2015;1397:32–42.

[22] Clark AE, Kaleta EJ, Arora A, Wolk DM. Matrix-assisted laser desorption ionization-time of flight mass spectrometry: a fundamental shift in the routine practice of clinical microbiology. Clin Microbiol Rev 2013;26(3):547–603.

[23] Tan KE, Ellis BC, Lee R, Stamper PD, Zhang SX, Carroll KC. Prospective evaluation of a matrix-assisted laser desorption ionization-time of flight mass spectrometry system in a hospital clinical microbiology laboratory for identification of bacteria and yeasts: a bench-by-bench study for assessing the impact on time to identification and cost-effectiveness. J Clin Microbiol 2012;50(10):3301–8.

[24] Huang AM, Newton D, Kunapuli A, Gandhi TN, Washer LL, Isip J, et al. Impact of rapid organism identification via matrix-assisted laser desorption/ionization time-of-flight combined with antimicrobial stewardship team intervention in adult patients with bacteremia and candidemia. Clin Infect Dis 2013;57:1237–45.

[25] Marko DC, Saffert RT, Cunningham SA, Hyman J, Walsh J, Arbefeville S, et al. Evaluation of the Bruker Biotyper and Vitek MS matrix-assisted laser desorption ionization–time of flight mass spectrometry systems for identification of nonfermenting gram-negative bacilli isolated from cultures from cystic fibrosis patients. J Clin Microbiol 2012;50(6):2034–9.

[26] McMullen AR, Wallace MA, Pincus DH, Wilkey K, Burnham CA. Evaluation of the VITEK MS MALDI-TOF MS system for identification of clinically relevant filamentous fungi. J Clin Microbiol 2016;54(8):2068–73.

[27] Dingle TC, Butler-Wu SM. MALDI-TOF mass spectrometry for microorganism identification. Clin Lab Med 2013;33(3):589–609.

[28] Pusch W, Kostrzewa M. Application of MALDI-TOF mass spectrometry in screening and diagnostic research. Curr Pharm Des 2005;11(20):2577–91.

[29] Gao X, Tan B-H, Sugrue RJ, Tang K. MALDI mass spectrometry for nucleic acid analysis. Top Curr Chem 2013;331:55–77.

[30] Gabriel S, Ziaugra L, Tabbaa D. SNP genotyping using the sequenom massARRAY iPLEX platform. Current protocols in human genetics [Internet]. John Wiley & Sons, Inc.; 2001. Available from: http://onlinelibrary.wiley.com/doi/10.1002/0471142905.hg0212s60/abstract

[31] Ecker DJ, Sampath R, Massire C, Blyn LB, Hall TA, Eshoo MW, et al. Ibis T5000: a universal biosensor approach for microbiology. Nat Rev Microbiol 2008;6(7):553–8.

[32] Tang Y-W, Lowery KS, Valsamakis A, Schaefer VC, Chappell JD, White-Abell J, et al. Clinical accuracy of a PLEX-ID flu device for simultaneous detection and identification of influenza viruses A and B. J Clin Microbiol 2013;51(1):40–5.

[33] Sinotte M, Diorio C, Berube S, Pollak M, Brisson J. Genetic polymorphisms of the vitamin D binding protein and plasma concentrations of 25-hydroxyvitamin D in premenopausal women. Am J Clin Nutr 2009;89(2):634–40.

[34] Godfrey KM, Sheppard A, Gluckman PD, Lillycrop KA, Burdge GC, McLean C, et al. Epigenetic gene promoter methylation at birth is associated with child's later adiposity. Diabetes 2011;60(5):1528–34.

[35] Holzinger ER, Grady B, Ritchie MD, Ribaudo HJ, Acosta EP, Morse GD, et al. Genome-wide association study of plasma efavirenz pharmacokinetics in AIDS clinical trials group protocols implicates several CYP2B6 variants. Pharmacogenet Genomics 2012;22(12):858–67.

[36] Hemeryck LY, Decloedt AI, Vanden Bussche J, Geboes KP, Vanhaecke L. High resolution mass spectrometry based profiling of diet-related deoxyribonucleic acid adducts. Anal Chim Acta 2015;892:123–31.

[37] Caprioli RM, Farmer TB, Gile J. Molecular imaging of biological samples: localization of peptides and proteins using MALDI-TOF MS. Anal Chem 1997;69(23):4751–60.

[38] Ellis SR, Bruinen AL, Heeren RMA. A critical evaluation of the current state-of-the-art in quantitative imaging mass spectrometry. Anal Bioanal Chem 2014;406(5):1275–89.

[39] McDonnell LA, Heeren RMA. Imaging mass spectrometry. Mass Spectrom Rev 2007;26(4):606–43.

[40] Caprioli RM. Imaging mass spectrometry: molecular microscopy for enabling a new age of discovery. Proteomics 2014;14(7–8):807–9.

[41] Wu C, Dill AL, Eberlin LS, Cooks RG, Ifa DR. Mass spectrometry imaging under ambient conditions. Mass Spectrom Rev 2013;32(3):218–43.

[42] Rauser S, Marquardt C, Balluff B, Deininger S-O, Albers C, Belau E, et al. Classification of HER2 receptor status in breast cancer tissues by MALDI imaging mass spectrometry. J Proteome Res 2010;9(4):1854–63.

[43] Li H, Hummon AB. Imaging mass spectrometry of three-dimensional cell culture systems. Anal Chem 2011;83(22):8794–801.

[44] Balog J, Sasi-Szabó L, Kinross J, Lewis MR, Muirhead LJ, Veselkov K, et al. Intraoperative tissue identification using rapid evaporative ionization mass spectrometry. Sci Transl Med 2013;5(194):194ra93.

[45] Wiechert W. C-13 metabolic flux analysis. Metab Eng 2001;3(3):195–206.

[46] Munger J, Bennett BD, Parikh A, Feng X-J, McArdle J, Rabitz HA, et al. Systems-level metabolic flux profiling identifies fatty acid synthesis as a target for antiviral therapy. Nat Biotechnol 2008;26(10):1179–86.

[47] Hiller K, Metallo CM. Profiling metabolic networks to study cancer metabolism. Curr Opin Biotechnol 2013;24(1):60–8.

[48] Fan TWM, Lane AN, Higashi RM. The promise of metabolomics in cancer molecular therapeutics. Curr Opin Mol Ther 2004;6(6):584–92.

[49] Wise DR, Thompson CB. Glutamine addiction: a new therapeutic target in cancer. Trends Biochem Sci 2010;35(8):427–33.

EVOLVING PLATFORMS FOR CLINICAL MASS SPECTROMETRY

13

J.Y. Yang*, D.A. Herold*,**

*Department of Pathology, University of California San Diego, La Jolla, CA, United States;
**VAMC-San Diego, San Diego, CA, United States

1 INTRODUCTION

The push to increase value for the patient drives the evolution of clinical mass spectrometry (MS) platforms through improvements in accuracy, micro sampling or less invasive sampling, and yielding more information within decreased turnaround time. As laboratorians and mass spectrometrists, we aim to minimize error through standardization and automation to reproducibly achieve accurate results. With the current speed of data acquisition and stability of mass spectrometers, we optimize signal to noise with creative strategies to introduce analytes of interest into the instrument. These strategies may involve enrichment or clean-up, non-invasive sampling, automation, ambient ionization technologies, and combinations of existing technologies and approaches. The dream would be to walk-up to an instrument with a noninvasive sample collection and get reliable real time analysis.

The modular nature of mass spectrometers allows for instruments with alternative fragmentation techniques, front end separation strategies, ambient ionization approaches, and real time analysis in addition to less invasive sampling techniques to overlap and emerge in the clinical laboratory.

This chapter reviews existing MS instrumentation, how their applications are evolving in the context of clinical applications, high sensitivity or high resolution mass analyzers and combinations of technologies that are making their way in, and additional aspects that will affect future clinical mass spectrometry, excluding software and computational tools. However, at the time of publication of this book, with the anticipated implementation of the FDA Laboratory Developed Tests (LDT) guidance, the future evolution of clinical mass spectrometry platforms remains to be determined.

2 TRIPLE QUADRUPOLE

The workhorse in the clinical laboratory is the triple quadrupole (triple quad) MS/MS platform, also referred to as tandem MS, MS2, or the all encompassing MS.

2.1 SINGLE/TRIPLE QUAD GC—MS

Gas chromatography (GC) mass spectrometry, considered the gold standard for urine drug screens, has been validated for hundreds of drugs detected from blood [1]. Breath testing, an application of GC, is making a comeback in the clinical laboratory. GC-MS has been used to analyze breath samples since the 1990s [2–4] and more currently has been used to identify volatile organic compound biomarkers [5].

2.2 TRIPLE QUAD LC-MS

Coupled with liquid chromatography on the front end, the selectivity and sensitivity achieved from multiple reaction monitoring (MRM) with triple quad LC-MS, and detection of fragment ions with stable isotope dilution approaches for calculation of analyte to internal standard ratios, allows for quantification of multiple analytes within minutes. Clinical Biochemical Genetics assays for targeted metabolomics from extracted dried blood spots [6], therapeutic drug monitoring (TDM) [7], and other small molecule analyses in serum and urine utilize triple quad-MS. Mass spectrometer vendors and diagnostic companies are developing reagent kits for specific assays, such as TDM of immunosuppressant drugs. Triple quad LC-MS/MS is also used for targeted peptide and protein quantitation [8], which is beginning to be clinically implemented for identification of isoforms, such as vitamin D binding globulin (VDBG) [9], and can be extended to identify protein variants, uncovering biological differences based on genetics from plasma or serum without genetic sequencing.

3 QUADRUPOLE-TIME OF FLIGHT (TOF) AND TOF

Time of flight (TOF) and quadrupole-TOF (qTOF; QqTOF) MS platforms, hereafter referred to collectively as TOF, are surfacing in toxicology for broad spectrum drug screens. One advantage of using a TOF drug screen is that one does not need to preemptively know which specific drug is present [10–12], dissimilar to the triple quad, which requires a priori knowledge of specific fragments from known compounds at determined retention time windows. Additionally, TOF platforms allow the simultaneous screening of multiple drug classes within one LC run. TOF drug screens have been validated for 60–100 of compounds from various drug classes [13]. The comparison of acquired LC-MS/MS data from specimens containing unknown drugs to a library of spectra of standard compounds requires the same method and platform used to query the sample and to generate the libraries. This also allows the utilization of chromatographic retention times as additional criterion. The confidence in molecular identification relies on the match of precursor, fragment, or both ions along with retention time to data in the library and removes the bias and cross reactivity that is present with immunoassays. However, this approach is limited to drugs with available standards and will miss illicit analogs. A true broad screen would resemble a metabolomics approach with data independent acquisition. TOF MS broad spectrum drug screens may one day replace immunoassay screens.

4 ION TRAP AND ORBITRAP

Similar to the TOF broad spectrum drug screen, the ion trap (IT) mass analyzer is being used for identification of small molecules by matching spectra to a library of standards. Trap instruments increase selectivity and sensitivity by trapping ions of interest for a specified accumulation time, before allowing ions to hit the detector. Bruker has an IT set up with accompanying software called ToxTyper that ranks library matches by confidence [14].

Orbitrap, like qTOF, is categorized as HRMS. Orbitrap MS is being applied to endocrinology, TDM of immunosuppressants, anticancer drugs, and antifungals, as well as proteomics [15,16]. One example of proteomics is electron transfer dissociation (ETD) top down protein sequencing of immunoglobulins [17,18]. ETD-orbitrap MS is also useful in studying posttranslational modifications of proteins, such as glycosylation patterns in cancer biomarker discovery [19].

Orbitrap MS has been compared to GC-MS to look at derivatized and underivatized estrogen steroid hormones, and has been evaluated against triple quad MS for use in TDM [20–23].

5 MALDI-TOF

After receiving FDA clearance in the US for microbial identification in 2013, the utility of MALDI-TOF MS, expanded to identification of yeasts, fungi, and antibiotic resistance [24–30]. The potential for the use of MALDI-TOF platforms in other sections of the clinical lab has yet to be exploited [31]. MALDI-TOF mass spectrometers possess a large dynamic range and between all configurations, TOF can analyze most analytes of interest. One limitation of MALDI is considered to be matrix interference, matrix referring to the small organic acid used to assist the laser desorption and ionization process. However, with the precision and accuracy of current MALDI-TOF mass spectrometers, this is less of a limitation for the analysis of larger biomolecules. For MALDI-TOF, the sample preparation can be simple and sample requirement minimal. Sample deposition on the target plate is automatable and data acquisition can be batched to yield rapid results (80 spectra, each an average of 25,000 laser shots, in less than 30 min with a laser operating at 1 kHz).

MALDI-TOF, an approach that harbors a stigma of being solely qualitative, has been shown to be quantitative. In conjunction with immunoaffinity capture technology, such as stable isotope standards and capture by anti-peptide antibodies (SISCAPA Assay Technologies), MALDI-TOF was used to quantify proteins [32], such as cardiac marker BNP [33], for biomarker evaluation for quantitative proteomics [33–36], to differentiate different subtypes of influenza A virus within one hour of sample preparation [37], for absolute quantitation of CRP [38], and to quantitate a protein C inhibitor peptide with CV 2.2% over a 100-fold dynamic range [39]. Percent Hemoglobin A1c (HbA1c) a diagnostic and therapeutic marker for diabetes, is a relative quantitation with a strong linear correlation with total glycated beta chain. MALDI-TOF MS of diluted whole blood detects total glycated beta chain of hemoglobin, which linearly correlates with HbA1c determined by cation exchange HPLC [40].

Other applications of MALDI-TOF include detection of human herpesvirus [41], oligonucleotide analysis to discover genomic single nucleotide polymorphisms [42], and imaging MS, which is discussed later in this chapter.

6 HYBRID INSTRUMENTS
6.1 QUADRUPOLE-ORBITRAP MS

The versatility of a quadrupole-orbitrap mass spectrometer covers qualitative and quantitative, targeted and untargeted approaches, exemplified by one example determining presence of 40 psychoactive drugs and metabolites in urine [43].

6.2 CYTOF (ICP-MS)

Mass cytometry couples flow cytometry inline with inductively coupled plasma (ICP) MS, extending single-cell analysis into the realm of proteomics and phosphoproteomics [44]. ICP-MS has been utilized in the clinical space since the 1990s for trace elements in body fluids and tissues [45–48], and is making its way into clinical research [49]. Flow cytometry counts and sorts particles based on size and complexity—side scatter and forward scatter of light—as is done with peripheral blood mononuclear cells (PBMCs) on counts upward of 10^5 cells or particles, giving this approach inherent statistical power. Particles get shuttled one at a time in sheath fluid flow past a light source. Larger and more complex (highly nucleated) particles will increase the forward and side scatter of the light beam that orthogonally shines through the flow. Particles, or cells, can then be sorted based on light scatter properties. Multiplexing with markers, such as fluorescently labeled cell surface markers, enables further separation. Through gating, subsets and subsets of subsets of cells can be sorted, collected, and studied subsequently, in this context with ICP-MS.

Mass cytometry for single-cell analysis was commercialized by DVS Sciences as the CyTOF, which was acquired by Fluidigm in 2014. Cells are stained with an antibody tagged with a purified isotope of a lanthanide metal, which are then resolved and quantified via ICP (qTOF) MS at a rate of >1000 cells per second. Mass cytometry can also be used to generate fluorescent cell barcodes for cellular phenotypes [50].

Forward and side scatter are not available in first generation mass cytometer. The second generation mass cytometer CyTOF2 can resolve more than 120 metal probes, more than 36 proteins simultaneously in a single tube. The most current generation mass cytometer, the Helios (CyTOF3) has 135 detection channels, which in conjunction with the resolving power of the TOF MS, enables monitoring of more than 40 markers per cell.

CyTOF has applications in immunology, hematology, and oncology research [51,52]. Studies have focused on immunophenotyping, intracellular cytokines profiling, and characterizing phosphorylation signaling pathways [53]. Mass-tag cell bar-coding has been used on induced pluripotent stem cell reprogramming [54] and in multiple myeloma [55,56].

The major limitations of this platform include reliance on antibody specificity, sample type and preparation—purifying isotopes and attaching reporter isotopes to antibodies (which is a commercially available service through Fluidigm), vigilance in avoiding heavy metal contamination by avoiding chelators, clean time, oxidation of reporter tags, cost, and expertise required for manual analysis.

7 PCR-MS

PCR-MALDI-TOF has been used to detect genetic polymorphisms in the transmission of a rapidly mutating pathogen, hepatitis C virus [57–59].

PCR-qTOF MS has been applied to the identification of microbes and the quantitation of virus [60]. Abbott Technologies' IRIDICA platform, the second generation of Ibis Biosciences' Plex-ID, relies on MS detection of amplicons directly from whole blood, bronchial lavage, and other bodily fluids for microbial identification. This technology targets critically ill patients by reducing the time of bacterial, fungal, and viral identification from days and weeks to hours and including select antibiotic resistance markers. There are different workflows for the types of identification; bacterial workflow differs from the fungal and the viral. The throughput is low, six samples total per 8 h, and a control sample should be included with each batch. Also, there are some challenges with complete identification in mixed bacterial infections. A similar approach is used to rapidly identify microbes from blood cultures [61]. And a type-specific quantitative detection of human papillomavirus (HPV) by a PCR-MS based method was developed and evaluated for cervical cancer screening [62].

PCR-MS based approaches are not trivial to implement. PCR requires appropriate primers for amplification of oligonucleotides of unique masses, DNA extraction, vigilance in technique to reduce and avoid DNA contamination, optimization, and thorough evaluation and analysis to link the oligonucleotides to the pathology of interest. However, once these parameters are set up, PCR-MS can provide reliable and accurate results.

8 FRONT END—SAMPLE PREPARATION AND COLLECTION/AMBIENT IONIZATION/REAL TIME ANALYSIS

The evolution of clinical mass spectrometry is a balance between utilizing technologies that exist and creating technologies that fit clinical needs. A large portion of the turnaround time from collection to result is sample preparation and/or cleans up with minutes-long LC methods. Hence, with easier, faster, or standardized sample collection and preparation with removal of LC, the time to result would be reduced. This can be accomplished through efforts in non- or less-invasive sampling techniques, ambient ionization approaches, and real-time analysis [63].

8.1 SAMPLE PREPARATION/SAMPLE COLLECTION

The push to gather more information from small sample volumes has spawned creative exploration of different sample types, collection methods or devices for sample collection, and methods of preparation. In addition to previously mentioned immunocapture and SISCAPA strategies to pull out analytes of interest from small sample volumes, there are specialized sample collection devices. A few examples include blood spot collection devices that wick the plasma away from the red blood cells, devices that can collect a fixed, small volume of sample on the order of 10–20 μL, and breathe collection devices that have a filtered mouthpiece to capture aerosols. Non- or less-invasive sample types, such as breath, saliva, tears, or surface swabs or direct solvent extraction may play a larger role in the near future.

8.2 VOLATILE SAMPLING AND REAL TIME ANALYSIS

Analysis of breath samples is used for drug detection and has been used in forensic applications with flowing-afterglow (FA) and selected ion flow tube (SIFT) MS for real-time breath printing and metabolomics [64,65]. Breath analysis has implications in the clinical lab [66,67]. Chemical signatures of volatile organic compounds (VOCs) on breath from lung infections, acetone (a reflection of blood sugar levels), acetonitrile (smoking), and ethanol have been evaluated [68]. Real time breath analysis can be used to monitor the health status of CF patients through exhaled volatile biomarkers of *Pseudomonas aeruginosa* infection [69] or indicators of acidity of the airway mucosa [70]. These breath studies have been conducted in real time with SIFT MS and proton transfer reaction (PTR) MS [71,72]. These MS approaches are faster than GC and also allow collection of breath into bags or onto traps/filters for later analysis. SIFT-MS and PTR-MS are both chemical ionization (CI) methods that yield little to no fragmentation. And quantitation is accomplished via internal ion ratios, similar to analysis of MRM data from triple quad MS platforms.

Similar to breath testing, a group developed a man-portable membrane inlet mass spectrometer (MIMS) to generate "odorprints" by detecting odors generated from humans, breath and sweat, in a confined space [73]. However, generation of "odorprints" is not time efficient due to the large sampling space of a room. MIMS was also evaluated for narcotics, explosives, and chemical warfare agents on different membranes for use in homeland security measures. MIMS can be used for water, air, and some solvents for real-time analysis of VOCs and some small molecules [74].

9 AMBIENT AND DIRECT IONIZATION APPROACHES

9.1 DESI

Desorption electrospray ionization (DESI), developed in Graham Cook's laboratory and commercially available by Prosolia, Inc., relies on solvent extraction directly on the sample surface for localized information. DESI, which has been shown to ionize large proteins and complexes [75], is primary used on tissues in imaging mass spectrometry.

Iterations of DESI hold more promise in clinical applications. DESI combined with solid-phase microextraction was used to screen and quantitate drugs in urine [76]. Single droplet microextraction with DESI MS has been shown for trace analysis of methamphetamine in aqueous solution [77] and can be applied to other small molecules. Transmission mode (TM) DESI is more conducive to throughput, by reducing optimization of solvent delivery and sample introduction angles through the use of a mesh material as a sample substrate, and placing the mesh with 1 μL of sample in-line with the ESI source and mass spectrometer inlet [78,79]. However, TM-DESI has mainly been used for proof of principle studies of standards in solution, not for biofluid samples with complex biometrics.

9.2 PAPER SPRAY

LC-MS/MS analysis of dried blood spots from heel sticks have been used for biochemical genetics since the 90s [80,81]. Punch disks of the dried blood spots, used in an attempt to normalize the amount of sample, are extracted in 96 well plates then introduced into a triple quad mass spectrometer. The next step to minimize sample handling would be to combine the dried blood spots with paper spray.

Paper spray (PS) MS by Prosolia, Inc. is a combination of paper chromatography and electrospray ionization, and has been used to analyze immunosuppressant's in whole blood collected on triangular filter paper [82–84]. In PS MS the dried sample spot undergoes slight separation based on the wicking of the solvent through the filter paper. A voltage is applied to the filter paper and the pointed tip of the triangle, closest to the mass spectrometer inlet, is the location where sample enters the mass spectrometer. PS MS has also been used to detect microorganisms [85], tobacco derived nicotine alkaloids [86], and other molecules from biofluids, blood, urine, and saliva in both biofluid and dried spots [87–90].

9.3 TOUCH SPRAY

Another iteration of placing analyte in the stream of the electrospray path is touch spray (TS) [85,91]. Touch spray takes advantage of existing swab sampling for collection of surface molecules. Unlike the pointed filter paper in paper spray, the swab is physically placed in the spray path, partially obstructing the spray, analytes caught in the stream get ionized, and travel into the mass spectrometer for detection. This has been used to capture the phospholipid and glycerophospholipid profile of *Streptococcus pyogenes* responsible for strep throat infections. However, in order to be conductive, the swab used in the studies has a metal handle, which would constitute a significant change in cost and clinical practice.

Similar to DESI, Zhang et al. developed internal extractive electrospray ionization (iEESI) [92], in which charged solvent extraction through a sample (1–100 mm^3) travels to the mass spectrometer. The authors claim sampling occurs deeper than merely the surface.

Direct analysis in real time (DART) MS, developed in 2005, is an ambient ionization approach that does not require sample preparation. Ionization occurs directly on the sample, which can be liquid or solid as in pills or hair. DART MS has been used to detect or quantify drugs in toxicology and forensics, steroid hormones in clinical trials, fragrances, food contaminants, and pesticides with masses up to 800 Da [93]. A study used DART to analyze changes in the mouse skin metabolome when exposed to UV B [94] and one can deduce a parallel study in humans.

Laser ablation electrospray ionization (LAESI) (Protea Biosciences) uses a laser to generate a plume that gets caught in the path of electrospray and can detect masses up to 66 kDa. LAESI, another chromatography free technique, is most useful with samples that contain water and has been primarily used in toxicology. LAESI has also been used to detect drugs from human hair [95].

DART, DESI, TS, and PS can be paired with the mass analyzer that provides the mass resolution and sensitivity required for the study at hand. For example, there is DESI-LIT and DESI-TOF. The ideal dynamic range to cover biomolecules, from small molecules (10s Da) and metabolites to lipids (100s Da) to proteins (100,000s Da) needs to span ~5 orders of magnitude.

10 IMAGING MS

Imaging mass spectrometry (IMS, not to be confused with ion mobility mass spectrometry) or mass spectrometry imaging (MSI) produces a chemical snapshot, revealing the spatial distribution of molecules of one point in time. IMS is label-free and does not require the numerous steps in histological staining for tissue analysis. However, sample preparation of tissues requires cryosectioning, which requires practiced hands. In theory, any mass analyzer is capable of IMS as long as the front end is conducive to scanning over an area and the data can be processed and displayed in a two-dimensional manner.

10.1 MALDI-TOF IMAGING

Since its introduction in 1997 [96], MALDI-TOF IMS has been applied to direct lipid profiling in clear cell renal cell carcinoma [97], human colorectal cancers [98], gastric and breast cancer tissues, and tissue microarrays for different cancers [99]. MALDI-TOF mass spectrometers are powerful and can image at resolution approaching single cell resolution [100]. In addition to MALDI-TOF, desorption electrospray ionization (DESI)-qTOF, DESI-IT, and TOF-SIMS [101] are IMS platforms used in research.

11 DESI IMAGING

DESI, like MALDI-TOF, has been used for label-free, stain-free tissue imaging, with the advantage of ambient ionization. Detection of molecules via DESI depends on the extent of solvent permeation, sample extraction. Spatial resolution for the image is dictated by the size of the solvent droplet or the splash zone and ionization is affected by the spray angle. The types of molecules detected are dependent on the solvents used. DESI of biopsy tissue sections allows a deeper analysis, complementing traditional H&E staining for histopathology [102,103] distinguish two meningioma subtypes based on lipid profiles (negative mode MS), correlates well with histopathological analyses [104,105]. Most DESI studies have been conducted in negative ion mode, favoring the ionization and detection of lipids to profile tumor heterogeneity [106–109]. The turnaround time of DESI imaging is faster than histology, but slower than the iKnife.

The advantages of imaging mass spectrometry include the ability to profile in situ cell heterogeneity label-free, to differentiate between healthy tissue and cancer. One major challenge with imaging mass spectrometry platforms still is translating the chemical or molecular information to biological information. Parallel processing with H&E and other immunohistochemical staining and close collaboration with histologists/pathologists are required at least initially, to corroborate the imaging data generated from mass spectrometry. Co-registering the immunohistochemical (IHC) images helps connect the two.

Tissue samples for imaging mass spectrometry require either surgery or biopsy, which is invasive, and the majority of IMS results are retrospective. The turnaround time is days to weeks, and data analysis requires a database or library of spectra. Data files are large, so computing power, data management, and programs that support large files are concerns.

Complementary to histology, laser micro dissection and mass spectrometry was used to evaluate a renal biopsy for monoclonal IgG deposition disease [110]. And less invasive means for cellular collection, needle biopsy or fine needle aspirate smears can be used with DESI-MS imaging for complementary molecular information toward diagnosis [111,112].

One technique that demonstrates direct clinical application and is already being used in operating rooms is rapid evaporative ionization mass spectrometry (REIMS) [113]. The intelligent knife (iKnife) takes advantage of the laser used in surgery, was developed by Zoltan Takats's group, and is commercially available through Waters Corporation. The iKnife provides real time feedback to surgeons, enabling on-the-fly decisions about margin control. The boundary between tumor and healthy tissues is detected via comparison of real time laser ablation mass spectra during surgery to a database of spectra profiles of healthy and diseased tissues [114,115]. Surgeons with the iKnife rely on mass spectra to distinguish the boundaries of tumor versus healthy tissues and/or different cancer grades. REIMS also has utility in microbial identification [116].

12 ION MOBILITY MASS SPECTROMETRY

Ion mobility mass spectrometry, separation of molecules in the gas phase, has applications in breath analysis [117], "skin-sniffing" or surface analysis, and lipoprotein particle size and concentration analysis [118].

13 PROTEOMICS

Targeted proteomics has made its debut in the clinical lab and can be accomplished by immunoaffinity capture and triple quad LC-MS or MALDI-TOF MS as a combination of SID and MRM. Targeted proteomics requires rigorous front end analysis and a level of expertise to ensure that the monitored peptide/s reflect the amount of protein present. The immunoaffinity capture requires the generation of specific antibodies to the biomarker, or analyte of interest. This approach is dubbed mass spectrometric immunoassay (MSIA) [119–121]. Initially a biomarker/protein that has clinical significance or differential expression—quantitative results will guide therapy, treatment, or clinical decision making—is selected. This protein or peptide must contain a unique peptide/be unique, with little to no modification sites for accurate quantification. Second, detectability (coverage) of the candidate biomarker must be evaluated via mass spectrometry. Third, candidate peptide/s and fragment peptides are selected to serve as a proxy for the entire protein. And a synthetic labeled peptide must be synthesized to serve as an internal standard for quantitation. Fourth, the protocol and selection of peptide/biomarker must be evaluated in context of patient samples and clinical utility. With this approach, protein isoforms/variants can also be monitored [122–126].

With the increasing use of biologics, stem cell therapy, protein based therapies; this approach will become more and more prominent in clinical mass spectrometry.

The main advantages of this approach are avoiding the interference from endogenous immunoglobulins in immunoasssays, and sensitivity, or being able to quantitate down to picomolar quantities. The main challenge or bottleneck lies in selecting a peptide to serve as a protein surrogate, or biomarker discovery and validation. Another possible disadvantage is the immunoaffinity enrichment step in the development process if the protein has different forms or if protein degradation products containing the epitope are present [127].

Targeted quantitative proteomics may directly affect patient care in the near future, as a predictor of patient response to therapy. Assay development for clinical proteomics is not trivial, in addition to the limitations/challenges for mass spectrometry.

Additional exciting advances in mass spectrometry may have future clinical impact. Electrochemical microfluidic chips to detect short-lived drug metabolites [128], slug-flow microextraction [129] deal with generating MS data from small volumes of sample. Liquid electrospray laser desorption/ionization (liquid EDLI) allows native protein ions directly from aqueous solutions and biofluids (ref). However, most of the emerging technologies in clinical mass spectrometry are qualitative. The demand for quantitation has been addressed by Liu et al. by coating the inside of glass capillaries with internal standard prior to use for trace analysis in complex biological mixtures [130]. And the field is quickly moving toward quantitation [131].

14 FUTURE IMPLICATIONS

There is a demand in clinical mass spectrometry toward quantitation of multiple analytes from small volumes of sample, drop of blood or tear drop, swabbing or skin sensors, or noninvasive breath collection. Monitoring multiple analytes falls in the realm of omics studies, and with metadata can be put in the category of personalized medicine.

The push for untargeted omics approaches, such as the Human Longevity, Inc. (HLI) of JCVI, yields massive amounts of data, shifting the bottleneck to data processing, and management. However, with the complexity of gene expression profiles, temporal changes, splice variants, posttranslational modifications of proteins, and the interplay of metabolites with the microbiome, monitoring all genes, proteins and metabolites over time is not a trivial task. Despite the ability to acquire data, we yet don't know the full meaning of all this information. The search for biomarkers elevated in disease continues.

Advantages of the evolving platforms in clinical mass spectrometry include decreased cost of healthcare overall, the capacity for easy, high throughput assays, accurate results, and improved turnaround time.

Technologically, these future implications push for an increase in detection sensitivity and reliable and reproducible ionization from any biological matrix. Ambient sampling/ionization reduces the reliance on liquid chromatography for separation.

This will generate a lot of data, which also comes with the responsibility of educating the public appropriately.

REFERENCES

[1] Nair H, Woo F, Hoofnagle AN, Baird GS. Clinical validation of a highly sensitive GC-MS platform for routine urine drug screening and real-time reporting of up to 212 drugs. J Toxicol 2013;2013:329407.

[2] Phillips M, Sabas M, Greenberg J. Increased pentane and carbon-disulfide in the breath of patients with schizophrenia. J Clin Pathol 1993;46(9):861–4.

[3] Phillips M, Greenberg J, Sabas M. Alveolar gradient of pentane in normal human breath. Free Radic Res 1994;420(5):333–7.

[4] Phillips M, Herrera J, Zain M, et al. Variation in volatile organic compounds in the breath of normal humans. J Chromatogr 1999;729(1–2):75–88.

[5] Kumar S, Huang J, Abbassi-Ghadi N, et al. Mass spectrometric analysis of exhaled breath for the identification of volatile organic compound biomarkers in esophageal and gastric adenocarcinoma. Ann Surg 2015;262(6):981–90.

[6] Goulle J-P, Le Roux P, Castanet M, Mahieu L, Guyet-Job S, Guerbet M. Metallic profile of whole blood and plasma in a series of 99 healthy children. J Anal Toxicol 2015;39(9):707–13.

[7] Syed M, Srinivas NR. A comprehensive review of the published assays for the quantitation of the immunosuppressant drug mycophenolic acid and its glucuronidated metabolites in biological fluids. Biomed Chromatogr 2016;30(5):721–48.

[8] Janecki DJ, Bemis KG, Tegeler TJ, et al. A multiple reaction monitoring method for absolute quantification of the human liver alcohol dehydrogenase ADH1C1 isoenzyme. Anal Biochem 2007;369(1):18–26.

[9] Henderson CM, Lutsey PL, Misialek JR, et al. Measurement by a novel LC-MS/MS methodology reveals similar serum concentrations of vitamin D-binding protein in blacks and whites. Clin Chem 2016;62(1):179–87.

[10] Thoren KL, Colby JM, Shugarts SB, Wu AHB, Lynch KL. Comparison of information-dependent acquisition on a tandem quadrupole TOF vs a triple quadrupole linear ion trap mass spectrometer for broad-spectrum drug screening. Clin Chem 2016;62(1):170–8.

[11] Chindarkar NS, Wakefield MR, Stone JA, Fitzgerald RL. Liquid chromatography high-resolution TOF analysis: investigation of MSE for broad-spectrum drug screening. Clin Chem 2014;60(8):1115–25.

[12] Chindarkar NS, Park H-D, Stone JA, Fitzgerald RL. Comparison of different time of flight-mass spectrometry modes for small molecule quantitative analysis. J Anal Toxicol 2015;39(9):675–85.

[13] Guale F, Shahreza S, Walterscheid JP, et al. Validation of LCTOF-MS screening for drugs, metabolites, and collateral compounds in forensic toxicology specimens. J Anal Toxicol 2013;37(1):17–24.

[14] Kempf J, Traber J, Auwaerter V, Huppertz LM. Psychotropics caught in a trap—Adopting a screening approach to specific needs. Forensic Sci Int 2014;243:84–9.

[15] McShane AJ, Bunch DR, Wang S. Therapeutic drug monitoring of immunosuppressants by liquid chromatography-mass spectrometry. Clin Chim Acta 2016;454:1–5.

[16] Anderson LC, Karch KR, Ugrin SA, et al. Analyses of histone proteoforms using front-end electron transfer dissociation-enabled orbitrap instruments. Mol Cell Proteomics 2016;15(3):975–88.

[17] Tsybin YO, Fornelli L, Stoermer C, et al. Structural analysis of intact monoclonal antibodies by electron transfer dissociation mass spectrometry. Anal Chem 2011;83(23):8919–27.

[18] Zhang L, English AM, Bai DL, et al. Analysis of monoclonal antibody sequence and post-translational modifications by time-controlled proteolysis and tandem mass spectrometry. Mol Cell Proteomics 2016;15(4):1479–88.

[19] Liu H, Zhang N, Wan D, Cui M, Liu Z, Liu S. Mass spectrometry-based analysis of glycoproteins and its clinical applications in cancer biomarker discovery. Clin Proteomics 2014;11:14.

[20] Franke AA, Custer LJ, Morimoto Y, Nordt FJ, Maskarinec G. Analysis of urinary estrogens, their oxidized metabolites, and other endogenous steroids by benchtop orbitrap LCMS versus traditional quadrupole GCMS. Anal Bioanal Chem 2011;401(4):1319–30.

[21] Maskarinec G, Morimoto Y, Heak S, et al. Urinary estrogen metabolites in two soy trials with premenopausal women. Eur J Clin Nutr 2012;66(9):1044–9.

[22] Maskarinec G, Beckford F, Morimoto Y, Franke AA, Stanczyk FZ. Association of estrogen measurements in serum and urine of premenopausal women. Biomarkers Med 2015;9(5):417–24.

[23] Li X, Franke AA. Improved profiling of estrogen metabolites by orbitrap LC/MS. Steroids 2015;99:84–90.

[24] Levesque S, Dufresne PJ, Soualhine H, et al. A side by side comparison of Bruker Biotyper and VITEK MS: utility of MALDI-TOF MS technology for microorganism identification in a public health reference laboratory. PLoS One 2015;10.(12):e0144878.

[25] Ferreira L, Sanchez-Juanes F, Vega S, et al. Identification of fungal clinical isolates by matrix-assisted laser desorption ionization-time-of-flight mass spectrometry. Revista Espanola De Quimioterapia 2013;26(3):193–7.

[26] Hrabak J, Chudackova E, Walkova R. Matrix-assisted laser desorption ionization-time of flight (MALDI-TOF) mass spectrometry for detection of antibiotic resistance mechanisms: from research to routine diagnosis. Clin Microbiol Rev 2013;26(1):103–14.

[27] Foschi C, Compri M, Smirnova V, et al. Ease-of-use protocol for the rapid detection of third-generation cephalosporin resistance in Enterobacteriaceae isolated from blood cultures using matrix-assisted laser desorption ionization-time-of-flight mass spectrometry. J Hosp Infect 2016;93(2):206–10.

[28] Lau SKP, Lam CSK, Ngan AHY, et al. Matrix-assisted laser desorption ionization time-of-flight mass spectrometry for rapid identification of mold and yeast cultures of *Penicillium marneffei*. BMC Microbiol 2016;16:36.

[29] Panda A, Ghosh AK, Mirdha BR, et al. MALDI-TOF mass spectrometry for rapid identification of clinical fungal isolates based on ribosomal protein biomarkers. J Microbiol Methods 2015;109:93–105.

[30] Yunoki T, Matsumura Y, Nakano S, et al. Genetic, phenotypic and matrix-assisted laser desorption ionization time-of-flight mass spectrometry-based identification of anaerobic bacteria and determination of their antimicrobial susceptibility at a university hospital in Japan. J Infec Chemother 2016;22(5):303–7.

[31] Duncan MW, Nedelkov D, Walsh R, Hattan SJ. Applications of MALDI mass spectrometry in clinical chemistry. Clin Chem 2016;62(1):134–43.

[32] Kiernan UA, Phillips DA, Trenchevska O, Nedelkov D. Quantitative mass spectrometry evaluation of human retinol binding protein 4 and related variants. PLoS One 2011;6(3.):e17282.

[33] Kiernan UA, Nedelkov D, Nelson RW. Multiplexed mass spectrometric immunoassay in biomarker research: a novel approach to the determination of a myocardial infarct. J Proteome Res 2006;5(11):2928–34.

[34] Hoofnagle AN, Becker JO, Wener MH, Heinecke JW. Quantification of serum tumor markers: a general analytical approach based on anti-peptide immunoaffinity purification and isotope dilution tandem mass spectrometry. Clin Chem 2008;54(6):A206–7.

[35] Hoofnagle AN, Becker JO, Wener MH, Heinecke JW. Quantification of thyroglobulin, a low-abundance serum protein, by immunoaffinity peptide enrichment and tandem mass spectrometry. Clin Chem 2008;54(11):1796–804.

[36] Hoofnagle AN, Roth MY. Improving the measurement of serum thyroglobulin with mass spectrometry. J Clin Endocrinol Metab 2013;98(4):1343–52.

[37] Chou T-C, Hsu W, Wang C-H, Chen Y-J, Fang J-M. Rapid and specific influenza virus detection by functionalized magnetic nanoparticles and mass spectrometry. J Nanobiotechnol 2011;9:52.

[38] Kiernan UA, Addobbati R, Nedelkov D, Nelson RW. Quantitative multiplexed C-reactive protein mass spectrometric immunoassay. J Proteome Res 2006;5(7):1682–7.

[39] Anderson NL, Razavi M, Pearson TW, Kruppa G, Paape R, Suckau D. Precision of heavy-light peptide ratios measured by MALDI-TOF mass spectrometry. J Proteome Res 2012;11(3):1868–78.

[40] Hattan SJ, Parker KC, Vestal ML, Yang JY, Herold DA, Duncan MW. Analysis and quantitation of glycated hemoglobin by matrix assisted laser desorption/ionization time of flight mass spectrometry. J Am Soc Mass Spectr 2016;27(3):532–41.

[41] Cobo F. Application of MALDI-TOF mass spectrometry in clinical virology: a review. Open Virol J 2013;7:84–90.

[42] Stanssens P, Zabeau M, Meersseman G, et al. High-throughput MALDI-TOF discovery of genomic sequence polymorphisms. Genome Res 2004;14(1):126–33.

[43] Concheiro M, Castaneto M, Kronstrand R, Huestis MA. Simultaneous determination of 40 novel psychoactive stimulants in urine by liquid chromatography-high resolution mass spectrometry and library matching. J Chromatogr 2015;1397:32–42.

[44] Proserpio V, Loennberg T. Single-cell technologies are revolutionizing the approach to rare cells. Immunol Cell Biol 2016;94(3):225–9.

[45] Kerger BD, Gerads R, Gurleyuk H, Urban A, Paustenbach DJ. Total cobalt determination in human blood and synovial fluid using inductively coupled plasma-mass spectrometry: method validation and evaluation of performance variables affecting metal hip implant patient samples. Toxicol Environ Chem 2015;97(9):1145–63.

[46] Klencsar B, Bolea-Fernandez E, Florez MR, et al. Determination of the total drug-related chlorine and bromine contents in human blood plasma using high performance liquid chromatography-tandem ICP-mass spectrometry (HPLC-ICP-MS/MS). J Pharma Biomed Anal 2016;124:112–9.

[47] Trzcinka-Ochocka M, Brodzka R, Janasik B. Useful and fast method for blood lead and cadmium determination using ICP-MS and GF-AAS; validation parameters. J Clin Lab Anal 2016;30(2):130–9.

[48] Ulanova TS, Gileva OV, Stenno EV, Veikhman GA, Nedochitova AV. Determination of strontium content in whole blood and urine by ICP-MS. Biomeditsinskaya Khimiya 2015;61(5):613–6.

[49] Nassar AF, Wisnewski AV, Raddassi K. Mass cytometry moving forward in support of clinical research: advantages and considerations. Bioanalysis 2016;8(4):255–7.

[50] Bendall SC, Nolan GP, Roederer M, Chattopadhyay PK. A deep profiler's guide to cytometry. Trends Immunol 2012;33(7):323–32.

[51] Bendall SC, Nolan GP. From single cells to deep phenotypes in cancer. Nature Biotechnol 2012;30(7):639–47.

[52] Bendall SC, Simonds EF, Qiu P, et al. Single-cell mass cytometry of differential immune and drug responses across a human hematopoietic continuum. Science 2011;332(6030):687–96.

[53] Bisneto JV, Ansell SM. Multiparametric analysis of intra-tumoral T-cells in Hodgkin's lymphoma using mass cytometry (CyTOF). Blood 2015;126(23):1438.

[54] Zunder ER, Finck R, Behbehani GK, et al. Palladium-based mass tag cell barcoding with a doublet-filtering scheme and single-cell deconvolution algorithm. Nat Protoc 2015;10(2):316–33.

[55] Ghosh A, Carreau N, Moscatello A, et al. Flow cytometry based detection of MRD in bone marrow of patients with multiple myeloma: a comparison between fluorescent-based cytometry versus cytof. Blood 2015;126(23.):4195.

[56] Hansmann L, Blum L, Ju C-H, Liedtke M, Robinson WH, Davis MM. mass cytometry analysis shows that a novel memory phenotype B cell is expanded in multiple myeloma. Cancer Immunol Res 2015;3(6):650–60.

[57] Ganova-Raeva LM, Dimitrova ZE, Campo DS, et al. Detection of hepatitis C virus transmission by use of DNA mass spectrometry. J Infect Dis 2013;207(6):999–1006.

[58] Ganova-Raeva LM, Khudyakov YE. Application of mass spectrometry to molecular diagnostics of viral infections. Expert Rev Mol Diagn 2013;13(4):377–88.

[59] Kaleta EJ, Clark AE, Cherkaoui A, et al. Comparative analysis of PCR-electrospray ionization/mass spectrometry (MS) and MALDI-TOF/MS for the identification of bacteria and yeast from positive blood culture bottles. Clin Chem 2011;57(7):1057–67.

[60] Jordana-Lluch E, Gimenez M, Dolores Quesada M, et al. Evaluation of the broad-range PCR/ESI-MS technology in blood specimens for the molecular diagnosis of bloodstream infections. PloS One 2015;10(10):e0140865.

[61] Kaleta EJ, Clark AE, Johnson DR, et al. Use of PCR coupled with electrospray ionization mass spectrometry for rapid identification of bacterial and yeast bloodstream pathogens from blood culture bottles. J Clin Microbiol 2011;49(1):345–53.

[62] Patel DA, Shih Y-J, Newton DW, et al. Development and evaluation of a PCR and mass spectroscopy (PCR-MS)-based method for quantitative, type-specific detection of human papillomavirus. J Virol Methods 2009;160(1–2):78–84.

[63] Ferreira CR, Yanne KE, Jarmusch AK, Pirro V, Ouyang Z, Cooks RG. Ambient ionization mass spectrometry for point-of-care diagnostics and other clinical measurements. Clin Chem 2016;62(1):99–110.

[64] Smith D, Spanel P. Status of selected ion flow tube MS: accomplishments and challenges in breath analysis and other areas. Bioanalysis 2016;8(11):1183–201.

[65] Martinez-Lozano Sinues P, Landoni E, Miceli R, et al. Secondary electrospray ionization-mass spectrometry and a novel statistical bioinformatic approach identifies a cancer-related profile in exhaled breath of breast cancer patients: a pilot study. J Breath Res 2015;9(3):031001.

[66] Rattray NJW, Hamrang Z, Trivedi DK, Goodacre R, Fowler SJ. Taking your breath away: metabolomics breathes life in to personalized medicine. Trends Biotechnol 2014;32(10):538–48.

[67] Delfino RJ, Gong H, Linn WS, Hu Y, Pellizzari ED. Respiratory symptoms and peak expiratory flow in children with asthma in relation to volatile organic compounds in exhaled breath and ambient air. J Expo Anal Environ Epidemiol 2003;13(5):348–63.

[68] Wang GZ, Maranelli G, Perbellini L, Raineri E, Brugnone F. Blood acetone concentration in normal people and in exposed workers 16-H after the end of the workshift. Int Arch Occup Environ Health 1994;65(5):285–9.

[69] Scott-Thomas AJ, Syhre M, Pattemore PK, et al. 2-Aminoacetophenone as a potential breath biomarker for Pseudomonas aeruginosa in the cystic fibrosis lung. BMC Pulm Med 2010;10:56.

[70] Smith D, Sovova K, Dryahina K, Dousova T, Drevinek P, Spanel P. Breath concentration of acetic acid vapour is elevated in patients with cystic fibrosis. J Breath Res 2016;10(2):021002.

[71] Spesyvyi A, Smith D, Spanel P. Selected ion flow-drift tube mass spectrometry: quantification of volatile compounds in air and breath. Anal Chem 2015;87(24):12151–60.

[72] Smith D, Spanel P. Pitfalls in the analysis of volatile breath biomarkers: suggested solutions and SIFT-MS quantification of single metabolites. J Breath Res 2015;9(2):022001.

[73] Giannoukos S, Brkic B, Taylor S, France N. Membrane inlet mass spectrometry for homeland security and forensic applications. J Am Soc Mass Spectrom 2015;26(2):231–9.

[74] Davey NG, Bell RJ, Krogh ET, Gill CG. A membrane introduction mass spectrometer utilizing ion-molecule reactions for the on-line speciation and quantitation of volatile organic molecules. Rapid Commun Mass Spectrom 2015;29(23):2187–94.

[75] Ferguson CN, Benchaar SA, Miao Z, Loo JA, Chen H. Direct ionization of large proteins and protein complexes by desorption electrospray ionization-mass spectrometry. Anal Chem 2011;83(17):6468–73.

[76] Kennedy JH, Aurand C, Shirey R, Laughlin BC, Wiseman JM. Coupling desorption electrospray ionization with solid-phase microextraction for screening and quantitative analysis of drugs in urine. Anal Chem 2010;82(17):7502–8.

[77] Sun X, Yuan Z, de B, Harrington P, et al. Coupling of single droplet micro-extraction with desorption electrospray ionization-mass spectrometry. Int J Mass Spectrom 2011;301(1–3):102–8.

[78] Chipuk JE, Brodbelt JS. Transmission mode desorption electrospray ionization. J Am Soc Mass Spectrom 2008;19(11):1612–20.

[79] Chipuk JE, Gelb MH, Brodbelt JS. Rapid and selective screening for sulfhydryl analytes in plasma and urine using surface-enhanced transmission mode desorption electrospray ionization mass spectrometry. Anal Chem 2010;82(10):4130–9.

[80] Clark PT, Rice JD Jr. The use of filter paper PKU test specimen cards in the automated determination of blood phenylalanine concentration. Tech Bull Regist Med Technol 1966;36(9):224–7.

[81] Wagner M, Tonoli D, Varesio E, Hopfgartner G. The use of mass spectrometry to analyze dried blood spots. Mass Spectrom Rev 2016;35(3):361–438.

[82] Shi R-Z, El Gierari ETM, Faix JD, Manicke NE. Rapid measurement of cyclosporine and sirolimus in whole blood by paper spray-tandem mass spectrometry. Clin Chem 2016;62(1):295–7.

[83] Shi R-Z, El Gierari ETM, Manicke NE, Faix JD. Rapid measurement of tacrolimus in whole blood by paper spray-tandem mass spectrometry (PS-MS/MS). Clin Chim Acta 2015;441:99–104.

[84] Wiseman JM, Kennedy J, Manicke NE. Quantitation of tacrolimus in dried blood spots using paper spray mass spectrometry. LC GC N Am 2014;32(12):69.

[85] Hamid AM, Jarmusch AK, Kerian KS, Cooks RG. Rapid identification of micro-organisms by touch spray and paper spray ambient ionization. Abstr Pap Am Chem Soc 2013;246.

[86] Wang H, Ren Y, McLuckey MN, et al. Direct quantitative analysis of nicotine alkaloids from biofluid samples using paper spray mass spectrometry. Anal Chem 2013;85(23):11540–4.

[87] Damon DE, Davis KM, Moreira CR, Capone P, Cruttenden R, Badu-Tawiah AK. Direct biofluid analysis using hydrophobic paper spray mass spectrometry. Anal Chem 2016;88(3):1878–84.

[88] Ren Y, Chiang S, Zhang W, Wang X, Lin Z, Ouyang Z. Paper-capillary spray for direct mass spectrometry analysis of biofluid samples. Anal Bioanal Chem 2016;408(5):1385–90.

[89] Wang H, Manicke NE, Yang Q, et al. Direct analysis of biological tissue by paper spray mass spectrometry. Anal Chem 2011;83(4):1197–201.

[90] Vega C, Spence C, Zhang C, Bills BJ, Manicke NE. Ionization suppression and recovery in direct biofluid analysis using paper spray mass spectrometry. J Am Soc Mass Spectrom 2016;27(4):726–34.

[91] Jarmusch AK, Pirro V, Kerian KS, Cooks RG. Detection of strep throat causing bacterium directly from medical swabs by touch spray-mass spectrometry. Analyst 2014;139(19):4785–9.

[92] Zhang H, Lu H, Huang H, et al. Quantification of 1-hydroxypyrene in undiluted human urine samples using magnetic solid-phase extraction coupled with internal extractive electrospray ionization mass spectrometry. Anal Chim Acta 2016;926:72–8.

[93] Gross J. Direct analysis in real time—a critical review on DART-MS. Anal Bioanal Chem 2014;406(1):63–80.

[94] Park HM, Kim HJ, Jang YP, Kim SY. Direct analysis in real time mass spectrometry (DART-MS) analysis of skin metabolome changes in the ultraviolet B-induced mice. Biomol Ther 2013;21(6):470–5.

[95] Deimler RE, Razunguzwa TT, Reschke BR, Walsh CM, Powell MJ, Jackson GP. Direct analysis of drugs in forensic applications using laser ablation electrospray ionization-tandem mass spectrometry (LAESI-MS/MS). Anal Methods 2014;6(13):4810–7.

[96] Caprioli RM, Farmer TB, Zhang H, Stoeckli M. Molecular imaging of biological samples by MALDI MS. Abstr Pap Am Chem Soc 1997;214(1–2). 113-ANYL.

[97] Jones EE, Powers TW, Neely BA, et al. MALDI imaging mass spectrometry profiling of proteins and lipids in clear cell renal cell carcinoma. Proteomics 2014;14(7–8):924–35.

[98] Kurabe N, Hayasaka T, Ogawa M, et al. Accumulated phosphatidylcholine (16:0/16:1) in human colorectal cancer; possible involvement of LPCAT4. Cancer Sci 2013;104(10):1295–302.

[99] Balluff B, Frese CK, Maier SK, et al. De novo discovery of phenotypic intratumour heterogeneity using imaging mass spectrometry. J Pathol 2015;235(1):3–13.

[100] Zavalin A, Yang J, Hayden K, Vestal M, Caprioli RM. Tissue protein imaging at 1 m laser spot diameter for high spatial resolution and high imaging speed using transmission geometry MALDI TOF MS. Anal Bioanal Chem 2015;407(8):2337–42.

[101] Jungnickel H, Laux P, Luch A. Time-of-Flight secondary ion mass spectrometry (ToF-SIMS): a new tool for the analysis of toxicological effects on single cell level. Toxics 2016;4(1):5.

[102] Eberlin LS. DESI-MS imaging of lipids and metabolites from biological samples. Methods Mol Biol 2014;1198:299–311.

[103] Eberlin LS, Liu X, Ferreira CR, Santagata S, Agar NYR, Cooks RG. Desorption electrospray ionization then MALDI mass spectrometry imaging of lipid and protein distributions in single tissue sections. Anal Chem 2011;83(22):8366–71.

[104] Calligaris D, Caragacianu D, Liu X, et al. Application of desorption electrospray ionization mass spectrometry imaging in breast cancer margin analysis. Proc Natl Acad Sci USA 2014;111(42):15184–9.

[105] Calligaris D, Norton I, Feldman DR, et al. Mass spectrometry imaging as a tool for surgical decision-making. J Mass Spectrom 2013;48(11):1178–87.

[106] Chiou AS, Eberlin LS, Planell-Mendez I, et al. Two dimensional imaging of basal cell carcinoma using desorption electrospray ionization mass spectrometry (DESI-MS). J Invest Dermatol 2015;135. S36.

[107] Eberlin LS, Tibshirani RJ, Zhang J, et al. Molecular assessment of surgical-resection margins of gastric cancer by mass-spectrometric imaging. Proc Natl Acad Sci USA 2014;111(7):2436–41.

[108] Jarmusch AK, Pirro V, Baird Z, Hattab EM, Cohen-Gadol AA, Cooks RG. Lipid and metabolite profiles of human brain tumors by desorption electrospray ionization-MS. Proc Natl Acad Sci USA 2016;113(6):1486–91.

[109] Santagata S, Eberlin LS, Norton I, et al. Intraoperative mass spectrometry mapping of an onco-metabolite to guide brain tumor surgery. Proc Natl Acad Sci USA 2014;111(30):11121–6.

[110] Royal V, Quint P, Leblanc M, et al. IgD heavy-chain deposition disease: detection by laser microdissection and mass spectrometry. J Am Soc Nephrol 2015;26(4):784–90.

[111] Jarmusch AK, Kerian KS, Pirro V, et al. Characteristic lipid profiles of canine non-Hodgkin's lymphoma from surgical biopsy tissue sections and fine needle aspirate smears by desorption electrospray ionization—mass spectrometry. Analyst 2015;140(18):6321–9.

[112] Amann JM, Chaurand P, Gonzalez A, et al. Selective profiling of proteins in lung cancer cells from fine-needle aspirates by matrix-assisted laser desorption ionization time-of-flight mass spectrometry. Clin Cancer Res 2006;12(17):5142–50.

[113] Schafer KC, Denes J, Albrecht K, et al. In vivo, in situ tissue analysis using rapid evaporative ionization mass spectrometry. Angew Chem Int Ed Engl 2009;48(44):8240–2.

[114] Balog J, Kumar S, Alexander J, et al. In vivo endoscopic tissue identification by rapid evaporative ionization mass spectrometry (REIMS). Angew Chem Int Ed 2015;54(38):11059–62.

[115] Balog J, Sasi-Szabo L, Kinross J, et al. Intraoperative tissue identification using rapid evaporative ionization mass spectrometry. Sci Transl Med 2013;5(194):194ra93.

[116] Strittmatter N, Rebec M, Jones EA, et al. Characterization and identification of clinically relevant microorganisms using rapid evaporative ionization mass spectrometry. Anal Chem 2014;86(13):6555–62.

[117] Schneider T, Hauschild AC, Baumbach JI, Baumbach J. An integrative clinical database and diagnostics platform for biomarker identification and analysis in ion mobility spectra of human exhaled air. J Integr Bioinform 2013;10(2):218.

[118] Caulfield MP, Li S, Lee G, et al. Direct determination of lipoprotein particle sizes and concentrations by ion mobility analysis. Clin Chem 2008;54(8):1307–16.

[119] Krastins B, Prakash A, Sarracino DA, et al. Rapid development of sensitive, high-throughput, quantitative and highly selective mass spectrometric targeted immunoassays for clinically important proteins in human plasma and serum. Clin Biochem 2013;46(6):399–410.

[120] Nedelkov D, Trenchevska O, Pupinoska A, et al. Mass spectrometric immunoassays for quantitative determination of protein biomarker variants. Cancer Res 2011;71.

[121] Zhao L, Whiteaker JR, Voytovich UJ, Ivey RG, Paulovich AG. Antibody-coupled magnetic beads can be reused in immuno mrm assays to reduce cost and extend antibody supply. J Proteome Res 2015;14(10):4425–31.

[122] Niederkofler EE, Kiernan UA, O'Rear J, et al. Detection of endogenous B-type natriuretic peptide at very low concentrations in patients with heart failure. Circ Heart Fail 2008;1(4):258–64.

[123] Oran PE, Trenchevska O, Nedelkov D, et al. Parallel workflow for high-throughput (> 1,000 samples/day) quantitative analysis of human insulin-like growth factor 1 using mass spectrometric immunoassay. PLos One 2014;9(3):e92801.

[124] Peterman S, Niederkofler EE, Phillips DA, et al. An automated, high-throughput method for targeted quantification of intact insulin and its therapeutic analogs in human serum or plasma coupling mass spectrometric immunoassay with high resolution and accurate mass detection (MSIA-HR/AM). Proteomics 2014;14(12):1445–56.

[125] Trenchevska O, Schaab MR, Nelson RW, Nedelkov D. Development of multiplex mass spectrometric immunoassay for detection and quantification of apolipoproteins C-I, C-II, C-III and their proteoforms. Methods 2015;81:86–92.

[126] Xu Q, Zhu M, Yang T, Xu F, Liu Y, Chen Y. Quantitative assessment of human serum transferrin receptor in breast cancer patients pre- and post-chemotherapy using peptide immunoaffinity enrichment coupled with targeted proteomics. Clin Chim Acta 2015;448:118–23.

[127] Kushnir MM, Rockwood AL, Roberts WL, Abraham D, Hoofnagle AN, Meikle AW. Measurement of thyroglobulin by liquid chromatography-tandem mass spectrometry in serum and plasma in the presence of antithyroglobulin autoantibodies. Clin Chem 2013;59(6):982–90.

[128] Floris A, Staal S, Lenk S, et al. A prefilled, ready-to-use electrophoresis based lab-on-a-chip device for monitoring lithium in blood. Lab Chip 2010;10(14):1799–806.

[129] Ren Y, McLuckey MN, Liu J, Ouyang Z. Direct mass spectrometry analysis of biofluid samples using slugflow microextraction nano-electrospray ionization. Angew Chem Int Ed 2014;53(51):14124–7.

[130] Liu J, Cooks RG, Ouyang Z. Enabling quantitative analysis in ambient ionization mass spectrometry: internal standard coated capillary samplers. Anal Chem 2013;85(12):5632–6.

[131] Shiea J, Yuan C-H, Huang M-Z, et al. Detection of native protein ions in aqueous solution under ambient conditions by electrospray laser desorption/ionization mass spectrometry. Anal Chem 2008;80(13):4845–52.

Index

277